The Mellin Transformation and
Fuchsian Type Partial Differential Equations

Mathematics and Its Applications (*East European Series*)

Volume 56

The Mellin Transformation and Fuchsian Type Partial Differential Equations

by

Zofia Szmydt
Department of Mathematics,
University of Warsaw,
Warsaw, Poland

and

Bogdan Ziemian
Institute of Mathematics,
Polish Academy of Sciences,
Warsaw, Poland

SPRINGER-SCIENCE+BUSINESS MEDIA, B.V.

Library of Congress Cataloging-in-Publication Data

Szmydt, Zofia.
The Mellin transformation and Fuchsian type partial differential equations / by Zofia Szmydt and Bogdan Ziemian.
p. cm. -- (Mathematics and its applications (Kluwer Academic Publishers). East European Series ; v. 56)
Includes index.
ISBN 978-94-010-5069-2 ISBN 978-94-011-2424-9 (eBook)
DOI 10.1007/978-94-011-2424-9
1. Differential equations, Partial. 2. Mellin transform. I. Ziemian, Bogdan. II. Title. III. Title: Fuchsian type partial differential equations. IV. Series.
QA377.S969 1992
515'.353--dc20 92-4675

ISBN 978-94-010-5069-2

To the memory of our great friend and adviser
Professor Andrzej Pliś

SERIES EDITOR'S PREFACE

'Et moi, ..., si j'avait su comment en revenir, je n'y serais point allé.'

Jules Verne

The series is divergent; therefore we may be able to do something with it.

O. Heaviside

One service mathematics has rendered the human race. It has put common sense back where it belongs, on the topmost shelf next to the dusty canister labelled 'discarded nonsense'.

Eric T. Bell

Mathematics is a tool for thought. A highly necessary tool in a world where both feedback and nonlinearities abound. Similarly, all kinds of parts of mathematics serve as tools for other parts and for other sciences.

Applying a simple rewriting rule to the quote on the right above one finds such statements as: 'One service topology has rendered mathematical physics ...'; 'One service logic has rendered computer science ...'; 'One service category theory has rendered mathematics ...'. All arguably true. And all statements obtainable this way form part of the raison d'être of this series.

This series, *Mathematics and Its Applications*, started in 1977. Now that over one hundred volumes have appeared it seems opportune to reexamine its scope. At the time I wrote

> "Growing specialization and diversification have brought a host of monographs and textbooks on increasingly specialized topics. However, the 'tree' of knowledge of mathematics and related fields does not grow only by putting forth new branches. It also happens, quite often in fact, that branches which were thought to be completely disparate are suddenly seen to be related. Further, the kind and level of sophistication of mathematics applied in various sciences has changed drastically in recent years: measure theory is used (non-trivially) in regional and theoretical economics; algebraic geometry interacts with physics; the Minkowsky lemma, coding theory and the structure of water meet one another in packing and covering theory; quantum fields, crystal defects and mathematical programming profit from homotopy theory; Lie algebras are relevant to filtering; and prediction and electrical engineering can use Stein spaces. And in addition to this there are such new emerging subdisciplines as 'experimental mathematics', 'CFD', 'completely integrable systems', 'chaos, synergetics and large-scale order', which are almost impossible to fit into the existing classification schemes. They draw upon widely different sections of mathematics."

By and large, all this still applies today. It is still true that at first sight mathematics seems rather fragmented and that to find, see, and exploit the deeper underlying interrelations more effort is needed and so are books that can help mathematicians and scientists do so. Accordingly MIA will continue to try to make such books available.

If anything, the description I gave in 1977 is now an understatement. To the examples of interaction areas one should add string theory where Riemann surfaces, algebraic geometry, modular functions, knots, quantum field theory, Kac-Moody algebras, monstrous moonshine (and more) all come together. And to the examples of things which can be usefully applied let me add the topic 'finite geometry'; a combination of words which sounds like it might not even exist, let alone be applicable. And yet it is being applied: to statistics via designs, to radar/sonar detection arrays (via finite projective planes), and to bus connections of VLSI chips (via difference sets). There seems to be no part of (so-called pure) mathematics that is not in immediate danger of being applied. And, accordingly, the applied mathematician needs to be aware of much more. Besides analysis and numerics, the traditional workhorses, he may need all kinds of combinatorics, algebra, probability, and so on.

In addition, the applied scientist needs to cope increasingly with the nonlinear world and the extra mathematical sophistication that this requires. For that is where the rewards are. Linear models are honest and a bit sad and depressing: proportional efforts and results. It is in the nonlinear world that infinitesimal inputs may result in macroscopic outputs (or vice versa). To appreciate what I am hinting at: if electronics were linear we would have no fun with transistors and computers; we would have no TV; in fact you would not be reading these lines.

There is also no safety in ignoring such outlandish things as nonstandard analysis, superspace and anticommuting integration, *p*-adic and ultrametric space. All three have applications in both electrical engineering and physics. Once, complex numbers were equally outlandish, but they frequently proved the shortest path between 'real' results. Similarly, the first two topics named have already provided a number of 'wormhole' paths. There is no telling where all this is leading - fortunately.

Thus the original scope of the series, which for various (sound) reasons now comprises five subseries: white (Japan), yellow (China), red (USSR), blue (Eastern Europe), and green (everything else), still applies. It has been enlarged a bit to include books treating of the tools from one subdiscipline which are used in others. Thus the series still aims at books dealing with:

- a central concept which plays an important role in several different mathematical and/or scientific specialization areas;
- new applications of the results and ideas from one area of scientific endeavour into another;
- influences which the results, problems and concepts of one field of enquiry have, and have had, on the development of another.

The present volume in the series is a book about two things, maybe two and a half. The two are: the theory of the Mellin transform, a very useful integral transform that, till now, has had no systematic treatment (in more than one dimension), and Fuchsian type singular differential equations, the subject of Chapter III which includes the authors' own important results. The half is an appendix on Ecalle's resurgent functions, a most significant topic in my view, which can do with a few extra clear expositions here and there.

This is deep and up-to-date mathematics at the cutting edge of research, but, thanks to the authors, still accessible to all those with a standard background. That, as one of my teachers once remarked, is a sign of good research mathematics; within a few years of when they were obtained, the results should be explainable to graduate students. All this gives me something like 3½ reasons to welcome this volume in this series, and I do so with pleasure.

The shortest path between two truths in the real domain passes through the complex domain.
J. Hadamard

La physique ne nous donne pas seulement l'occasion de résoudre des problèmes ... elle nous fait pressentir la solution.
H. Poincaré

Never lend books, for no one ever returns them; the only books I have in my library are books that other folk have lent me.
Anatole France

The function of an expert is not to be more right than other people, but to be wrong for more sophisticated reasons.
David Butler

Bussum, 10 February 1992

Michiel Hazewinkel

CONTENTS

PREFACE

The purpose of this book is to provide a systematic introduction which paves the way to the results of Chapter III on Fuchsian type singular differential equations. This chapter consists of 3 sections. The first of these, Section 13, is devoted to ordinary Fuchsian equations. It contains known results as well as their extensions recently obtained by the authors. Elliptic Fuchsian partial differential equations are treated in Sections 14 and 15. Section 14 presents the authors' results ([Sz-Zie1], [Sz-Zie2], [Sz-Zie3]) concerning the local existence and regularity of solutions in suitable weighted spaces. However, it is only Section 15 that presents a complete geometric regularity result for such solutions, obtained in the recent papers [Zie4], [Zie5], [Zie6] of the second author.

The main tool used in Chapter III is the Mellin transformation, and in Section 15 also the generalized Cauchy and Hilbert transformations originating from [Zie3] and presented in Section 12.

Ordinary Fuchsian differential operators, their generalizations and "parameter versions" have been investigated by many authors in different contexts (see [El], [Rem–S], [T], [B–G], [Le–P], [M1], [W], [Mel–Men] to quote but a few). It should be underlined that the operators which we study in dimensions higher than 2 are not of the Baouendi–Goulaouic type [B–G]. They arise naturally, for example, as Laplace–Beltrami operators on manifolds with corners, and fall within the scope of the global theories developed in [Mel]. Systems of such operators (holonomic systems) were considered in [K–K], see also [Schl].

Due to numerous applications of the Mellin transformation (see e.g. [I], [Le–P], [La–M], [Mel–Men], [Mel], [Z], [C–L], [P–B–M], [Sn], [Wi], [O]) and the lack of text-book treatment of the Mellin transformation in several dimensions, we give its systematic presentation in Chapter II. In contrast to the classical texts on the Mellin and Laplace transformations (see for instance [Sm], [Wi], [C–L], [La–M], [Z]) we are interested in the *local* properties of the Mellin transforms, i.e. those properties of the Mellin transform of functions f which are preserved under multiplication of f by cut-off functions (of various types). We treat only certain aspects. Other important topics, including the relation with the theory of resurgent functions of J. Ecalle, are outlined in the Appendix.

The reading of the book requires no knowledge which exceeds the first 3 years of university studies of mathematics. Among the theorems which may not be included in standard university courses and whose proofs are not given in the text are the "division theorem" and the theorem on the Fourier transform of a convolution. We have confined ourselves to stating such theorems and indicating the pertinent references. Basic theorems and definitions of the theory of distributions and of the Fourier transformation used in the book are collected in Chapter I with proofs partially transformed into exercises with hints. The proofs omitted can be found in most textbooks on distributions (e.g. [Sz1]).

Sections 12 and 15 are more difficult due to the application of more advanced tools of complex analysis.

The book originated from a seminar conducted by the authors in the years 1983–1990 at Warsaw University. The authors wish to thank Dr. G. Łysik for his active participation in the seminar and valuable help in the preparation of the book.

The text was typeset in TEX at the Institute of Mathematics of the Polish Academy of Sciences in Warsaw.

Chapter I

INTRODUCTION

§1. TERMINOLOGY AND NOTATION

We employ the usual notation of set theory. The union, the intersection and the difference of sets A and B are denoted by $A \cup B$, $A \cap B$, $A \setminus B$, respectively. $\emptyset$ is the empty set. We write $a \in A$ if a is an element of A, otherwise we write $a \notin A$. The notation $A \subset B$ (or $B \supset A$) means that A is a subset of B.

The set of all elements of a set A satisfying condition $\mathcal{R}(\cdot)$ is denoted by $\{a \in A\colon \mathcal{R}(a)\}$, or more concisely $\{a\colon \mathcal{R}(a)\}$ or $\{\mathcal{R}(a)\}$ if the set A is fixed and no confusion is likely to arise.

The *product* of sets A and B is the set $A \times B$ consisting of all ordered pairs (a, b) with $a \in A$, $b \in B$.

The notation

$$f\colon A \to B \qquad \text{or} \qquad A \ni a \mapsto f(a) \in B$$

means that f is a function of the set A into the set B. A *functional* is a number-valued function. The supremum of a real-valued function on a set A is denoted by $\sup u$ or $\sup\limits_A u$, or $\sup\limits_{a \in A} u(a)$. Similarly the infimum of u on A is written as $\inf u$ or $\inf\limits_A u$, or $\inf\limits_{a \in A} u(a)$. Analogous notation is employed for $\max u$ and $\min u$ on A.

If $f\colon A \to B$ is any mapping and if $\widetilde{A}$ is a subset of A, then the *restriction* of f from A to $\widetilde{A}$, denoted by $f|_{\widetilde{A}}$, is the mapping: $\widetilde{A} \ni a \mapsto f(a) \in B$.

If $f\colon A \to B$, $g\colon B \to C$, then the *superposition* of f and g, i.e. the mapping $A \ni a \mapsto g(f(a)) \in C$, is denoted by $g \circ f$.

Suppose that T is a function defined in the product of two sets A, B and has values in a set C:

$$A \times B \ni (a, b) \mapsto T(a, b) \in C.$$

Then for any fixed values $\tilde{b} \in B$, $\tilde{a} \in A$ there are well defined mappings

$$A \ni a \mapsto T(a, \tilde{b}) \in C, \qquad B \ni b \mapsto T(\tilde{a}, b) \in C.$$

We denote these mappings by the symbols $T(\,\cdot\,, \tilde{b})$, $T(\tilde{a}, \,\cdot\,)$, respectively.

$\mathbf{R}$ denotes the set of real numbers. We denote by $\mathbf{R} \cup \{\infty\}$ the right-sided compactification of $\mathbf{R}$. The *n-dimensional Euclidean space* — denoted by $\mathbf{R}^n$ — is the product of n copies of the set $\mathbf{R}$. We also consider $(\mathbf{R} \cup \{\infty\})^n$, i.e., the set of all points $x = (x_1, \ldots, x_n)$, where $x_1, \ldots, x_n$ are real numbers or ∞. By e^j we denote the vector

$$(1) \qquad e^j = (e_1^j, \ldots, e_n^j) \qquad \text{with } e_i^j = 0 \text{ for } j \neq i, \quad e_j^j = 1.$$

The norm $\|x\|$ of a vector $x = (x_1, \ldots, x_n) \in \mathbf{R}^n$ is given by

$$\|x\|^2 = x_1^2 + \cdots + x_n^2.$$

We also denote $\langle x \rangle = 1 + |x_1| + \cdots + |x_n|$. If $x = (x_1, \ldots, x_n) \in \mathbf{R}^n$, $y = (y_1, \ldots, y_n) \in \mathbf{R}^n$, we write

$$xy = x_1 y_1 + \cdots + x_n y_n$$

and call this number the *scalar product* of x and y.

The smallest closed set containing a given set $A \subset \mathbf{R}^n$ is called the *closure* of A and is denoted by $\overline{A}$. The *boundary* of A is the set $\partial A = \overline{A} \cap \overline{\mathbf{R}^n \setminus A}$. The biggest open set contained in A is called the *interior* of A and denoted by $\operatorname{Int} A$.

We call a set $\Omega \subset \mathbf{R}^n$ *connected*, if Ω admits no decomposition of the form $\Omega = A \cup B$ with $\overline{A} \cap B = \emptyset = A \cap \overline{B}$.

Any set of the form

$$\{(x_1, \ldots, x_n) \colon a_i < x_i < b_i \text{ for } i = 1, \ldots, n\},$$

where $a_1, \ldots, a_n$, $b_1, \ldots, b_n$ are given real numbers or $\pm\infty$ with $a_i < b_i$ for $i = 1, \ldots, n$ is called an *open polyinterval* denoted by (a, b). Any set of the form

$$\{(x_1, \ldots, x_n) \colon -\infty < a_i \leq x_i \leq b_i < +\infty \text{ for } i = 1, \ldots, n\}$$

is called a *closed polyinterval* denoted by $[a, b]$. In an analogous way, we define polyintervals $(a, b]$ and $[a, b)$.

$\mathbf{N}$ denotes the set of positive integers (natural numbers), $\mathbf{N}_0$ is the set of non-negative integers, and $\mathbf{N}_0^n$ is the set of all multi-indices $\alpha = (\alpha_1, \ldots, \alpha_n)$, $\alpha_i \in \mathbf{N}_0$ for $i = 1, \ldots, n$. We write:

$$|\alpha| = \alpha_1 + \cdots + \alpha_n, \qquad \alpha! = \alpha_1! \ldots \alpha_n!.$$

Let $\alpha, \beta \in \mathbf{N}_0^n$, $\alpha = (\alpha_1, \dots, a_n)$, $(\beta_1, \dots, \beta_n)$. The Newton symbol

$$\binom{n}{k} = \frac{n!}{k!(n-k)!} \qquad \text{for } n, k \in \mathbf{N}_0, \quad n \geq k$$

extends to multi-indices as follows:

$$\binom{\alpha}{\beta} = \binom{\alpha_1}{\beta_1} \cdot \ldots \cdot \binom{\alpha_n}{\beta_n} \qquad \text{for } \alpha, \beta \in \mathbf{N}_0^n, \ \alpha_j \geq \beta_j \ \text{for } j = 1, \dots, n.$$

By $\mathbf{Z}$ we denote the set of integers, $\mathbf{C}$ denotes the set of complex numbers $z = a + ib$, $a, b \in \mathbf{R}$. We write $\bar{z} = a - ib$ (the *conjugate* of z), $a = \operatorname{Re} z$ (the *real part*), $b = \operatorname{Im} z$ (the *imaginary part*), $|z| = \sqrt{a^2 + b^2}$ (the *modulus* of z). If $z \in \mathbf{C}^n$ we write $\operatorname{Re} z = (\operatorname{Re} z_1, \dots, \operatorname{Re} z_n)$ and similarly $\operatorname{Im} z = (\operatorname{Im} z_1, \dots, \operatorname{Im} z_n)$. We also define

$$\langle z \rangle = 1 + |z_1| + \cdots + |z_n|$$

which agrees with the convention for $x \in \mathbf{R}^n$.

For convenience, we recall the basic notation and facts of the theory of the Lebesgue integral. By $L^1(A)$ we denote the space of integrable functions on a measurable set $A \subset \mathbf{R}^n$ and by $L^2(A)$ the space of square integrable functions. The respective norms are denoted by $\|\cdot\|_{L^1}$ and $\|\cdot\|_{L^2}$. The symbol $L^\infty(A)$ stands for the space of essentially bounded measurable functions on A.

The Lebesgue dominated convergence theorem. *Let A be a measurable set in $\mathbf{R}^n$, let $\{f_\nu\}$ be a sequence of integrable complex-valued functions defined a.e. on A and convergent a.e. to a function f. Suppose that there exists an integrable function $g \geq 0$ which is a common majorant for all f_ν's: $|f_\nu| \leq g$ $(\nu = 1, 2, \dots)$. Then f is integrable and we have*

$$\int_A f\, dx = \lim_{\nu \to \infty} \int_A f_\nu \, dx.$$

The Fubini–Tonelli theorem. *Let $A \subset \mathbf{R}^m$, $B \subset \mathbf{R}^n$, $m, n \in \mathbf{N}$, be measurable sets and let f: $A \times B \to \mathbf{C}$ be a measurable function. Suppose that either* (i) *f is integrable, or* (ii) *f is positive-valued. Then the following three quantities exist and are equal:*

$$\int_{A \times B} f(x, y)\, d(x, y) = \int_A \left(\int_B f(x, y)\, dy \right) dx = \int_B \left(\int_A f(x, y)\, dx \right) dy.$$

(In case (ii) *their common value may be infinite).*

Throughout the book we use the following vector notation: if $a, b \in \mathbf{R}^n$, $a = (a_1, \ldots, a_n)$, $b = (b_1, \ldots, b_n)$ then $a < b$ ($a \leq b$, resp.) denotes $a_j < b_j$ ($a_j \leq b_j$, resp.) for $j = 1, \ldots, n$. We denote $\mathbf{R}^n_+ = \{x \in \mathbf{R}^n \colon 0 < x\}$, $\mathbf{R}_- = \{x \in \mathbf{R} \colon x < 0\}$, $I = (0, t] = \{x \in \mathbf{R}^n_+ \colon x \leq t\}$ where $t \in \mathbf{R}^n_+$. $B(x, r) = \{y \in \mathbf{R}^n \colon \|y - x\| < r\}$, $r > 0$.We write shortly $B(r)$ instead of $B(0, r)$ and put $B = B(1)$, $B_+ = B \cap \mathbf{R}^n_+$ $S^{n-1} = \{x \in \mathbf{R}^n \colon \|x\| = 1\}$, $S^{n-1}_+ = S^{n-1} \cap \mathbf{R}^n_+$.

If $r \in \mathbf{R}$ we write $\mathbf{r} = (r_1, \ldots, r_n) \in \mathbf{R}^n$ where $r_1 = r_2 = \cdots = r_n = r$. For instance $\mathbf{1} = (1, \ldots, 1) \in \mathbf{R}^n$ and $\mathbf{e} = (e, \ldots, e) \in \mathbf{R}^n$ for the constant e.

For $x \in \mathbf{R}^n_+$ and $z = (z_1, \ldots, z_n) \in \mathbf{C}^n$ we write

$$x^z = x_1^{z_1} \cdot \ldots \cdot x_n^{z_n}.$$

Thus for $y = (y_1, \ldots, y_n) \in \mathbf{R}^n$, $\mathbf{e}^{-y} = e^{-y_1} \cdot \ldots \cdot e^{-y_n}$ where $\mathbf{R} \ni r \mapsto e^r \in \mathbf{R}_+$ is the exponential function. We also denote $e^{-y} = (e^{-y_1}, \ldots, e^{-y_n})$ for $y \in \mathbf{R}^n$ and similarly, if $x \in \mathbf{R}^n_+$, $\ln x = (\ln x_1, \ldots, \ln x_n)$ for the logarithmic function $\ln r$ inverse to e^r. In particular for $x \in \mathbf{R}^n_+$ and $\alpha \in \mathbf{N}^n_0$: $(\ln x)^\alpha = (\ln x_1)^{\alpha_1} \cdot \ldots \cdot (\ln x_n)^{\alpha_n}$. Vector notation is also used for differentiations. Namely we write

$$\frac{\partial}{\partial x} = \left(\frac{\partial}{\partial x_1}, \ldots, \frac{\partial}{\partial x_n}\right), \qquad x\frac{\partial}{\partial x} = \left(x_1\frac{\partial}{\partial x_1}, \ldots, x_n\frac{\partial}{\partial x_n}\right)$$

and if $\nu \in \mathbf{N}^n_0$ then

$$\left(\frac{\partial}{\partial x}\right)^\nu = \frac{\partial^{\nu_1}}{\partial x_1^{\nu_1}} \cdots \frac{\partial^{\nu_n}}{\partial x_n^{\nu_n}}, \qquad \left(x\frac{\partial}{\partial x}\right)^\nu = \left(x_1\frac{\partial}{\partial x_1}\right)^{\nu_1} \cdots \left(x_n\frac{\partial}{\partial x_n}\right)^{\nu_n}.$$

For points $a \in \mathbf{R}^n$ we write $a = (a_1, a')$ where $a_1 \in \mathbf{R}$, $a' \in \mathbf{R}^{n-1}$, similarly for $\zeta \in \mathbf{C}^n$, $\zeta = (\zeta_1, \zeta')$, $\zeta_1 \in \mathbf{C}$, $\zeta' \in \mathbf{C}^{n-1}$, we also consider sets $W \subset \mathbf{C}^n$ of the form $W = W^1 \times W'$ where $W^1 \subset \mathbf{C}$, $W' \subset \mathbf{C}^{n-1}$. For a set $W \subset \mathbf{C}^n$ and a vector $a \in \mathbf{R}^n$ we write $W + a = \{z \in \mathbf{C}^n \colon z - a \in W\}$.

Let U be an open subset of $\mathbf{R}^n$. Let f be a real-valued or complex-valued function defined in U and let m be an integer, $m \geq 0$. We say that f is *of class* C^m *in* U (in short, f is a *$C^m(U)$-function*) iff all the derivatives $\left(\frac{\partial}{\partial x}\right)^\alpha f$, $|\alpha| \leq m$, exist and are continuous functions in U; f is *of class* $C^\infty(U)$ (*smooth function*) iff it is of class $C^m(U)$ for all $m \in \mathbf{N}_0$. By $C_0^\infty(U)$ we denote the set of all functions in $C^\infty(U)$ vanishing outside a compact set.

Among all C^∞-functions on U we distinguish the important class $\mathcal{A}(U)$ of analytic functions: a (real- or complex-valued) function f is *analytic* in U iff for every point $\tilde{x} \in U$ there exists $r > 0$, so that f can be expanded into a *power series*:

$$f(x) = \sum_{\alpha \in \mathbf{N}^n_0} a_\alpha (x - \tilde{x})^\alpha \qquad \text{for } \|x - \tilde{x}\| < r.$$

Similarly, if $\Omega \subset \mathbf{C}^n$ we denote by $\mathcal{O}(\Omega)$ the set of holomorphic functions on Ω i.e. the functions F such that for every point $\tilde{z} \in \Omega$, F can be expanded into a power series

$$F(z) = \sum_{\alpha \in \mathbf{N}_0^n} a_\alpha (z - \tilde{z})^\alpha$$

convergent for $\|z - \tilde{z}\| < r$ for some $r > 0$ (depending on $\tilde{z}$).

§2. ELEMENTARY FACTS ON COMPLEX TOPOLOGICAL VECTOR SPACES

In this section, we recall some basic notions and theorems, referring the reader to [Sz1] for the proofs of the theorems quoted.

1. Multinormed Complex Vector Spaces and Their Duals

Let P be a complex vector space. A functional q on P is called a *seminorm* provided the following two conditions hold:

1) $\quad q(\xi + \eta) \leq q(\xi) + q(\eta) \qquad \text{for} \quad \xi, \eta \in P,$

2) $\quad q(\lambda \xi) = |\lambda|\, q(\xi) \qquad \text{for} \quad \xi \in P, \lambda \in \mathbf{C}.$

A seminorm q is a *norm* if and only if $q(\xi) = 0$ for a $\xi \in P$ implies $\xi = 0$.

We equip P with a convergence topology defined by a sequence of seminorms $\{q_k\}_{k=0}^{\infty}$. Since P is a linear space it is sufficient to define the convergence to zero.

Definition 1. Let P be a complex vector space and $\{q_k\}_{k=0}^{\infty}$ a sequence of seminorms on P. We say that a sequence $\{\xi_s\}_{s=1}^{\infty}$ of elements of P converges to zero if

(1) $$\lim_{s \to \infty} q_k(\xi_s) = 0 \qquad (k = 0, 1, \ldots).$$

The space P with the convergence defined above is denoted by $\mathcal{P}_{\{q_k\}}$ or by $\mathcal{P}$ for short. It is called a *multinormed space.*

It is easy to note that a necessary and sufficient condition in order that every convergent sequence had a unique limit, is that for every $\xi \in P$, $\xi \neq 0$ there exist $k \in \mathbf{N}_0$ such that $q_k(\xi) \neq 0$. In particular, this is the case if at least one of the seminorms q_k is a norm. In the following, we consider only the sequences of seminorms possessing the above property.

It can be proved (see [H] §§ 3 and 4) that $\mathcal{P}$ is a Hausdorff topological space.

The *sequences of seminorms* $\{q_k\}$ considered in the sequel will be *increasing* in the sense that

$$q_k(\xi) \le q_{k+1}(\xi) \qquad \text{for} \quad \xi \in P \quad (k = 0, 1, 2, \ldots).$$

This can always be achieved by replacing an arbitrary sequence of seminorms $\{q_k\}$ by the sequence $\{r_k\}$:

$$r_k(\xi) = \sum_{j=0}^{k} q_j(\xi) \qquad \text{or} \qquad r_k(\xi) = \max\big(q_0(\xi), \ldots, q_k(\xi)\big),$$

without affecting the topology of $\mathcal{P}$.

The theorem below characterizes linear continuous functionals on $\mathcal{P}$ with values in $\mathbf{C}$.

Theorem 1. *A linear functional f on $\mathcal{P}_{\{q_k\}}$ is continuous if and only if there exists a constant $C < \infty$ and $k \in \mathbf{N}_0$ such that*

$$|f(\xi)| \le C q_k(\xi) \qquad \text{for} \quad \xi \in \mathcal{P}.$$

We recall the Banach–Steinhaus theorem in the framework of multinormed spaces.

Theorem 2. *Suppose $\mathcal{P}$ is a complete multinormed space. Let $\{f_\tau\}_{\tau \in T}$, where T is a set of indices, be a family of continuous linear functionals on $\mathcal{P}$. We assume that for every $\xi \in \mathcal{P}$ the set $\{f_\tau(\xi) : \tau \in T\}$ is bounded in $\mathbf{C}$. Then there exists a constant $C < \infty$ and $k \in \mathbf{N}_0$ such that*

$$|f_\tau(\xi)| < C q_k(\xi) \qquad \text{for} \quad \xi \in \mathcal{P} \quad (\tau \in T).$$

Corollary 1. *Let $\mathcal{P}$ be as in Theorem 2. Let $\{f_\nu\}_{\nu=1}^{\infty}$ be a sequence of continuous linear functionals on $\mathcal{P}$. Suppose that for every $\xi \in \mathcal{P}$ the limit $f(\xi) \stackrel{\text{def}}{=} \lim_{\nu \to \infty} f_\nu(\xi)$ is finite. Then there exists a $C < \infty$ and $k \in \mathbf{N}_0$ such that*

$$|f(\xi)|, |f_\nu(\xi)| \le C\, q_k(\xi) \qquad \text{for} \quad \xi \in \mathcal{P} \quad (\nu = 1, 2, \ldots).$$

By $\mathcal{P}'$ we denote the space of continuous linear functionals on $\mathcal{P}$ with the topology of pointwise convergence. It is called the *dual space* of $\mathcal{P}$. Observe that it follows from Corollary 1 and Theorem 1 that the space $\mathcal{P}'$ is complete.

In applications we shall need the following theorem on the separate continuity of 2-*linear functionals*

Theorem 3 (cf. [Hö2] or [Tr]). *Let $\mathcal{P}_1$, $\mathcal{P}_2$ be two complete linear topological spaces with the topologies given by non-decreasing sequences of seminorms q_k^i ($i = 1,2$; $k = 0,1,\ldots$). Let f be a separately continuous 2-linear functional on $\mathcal{P}_1 \times \mathcal{P}_2$ i.e. such that for every fixed ξ_1, $f(\xi_1,\cdot) \in \mathcal{P}_2'$ and for every fixed ξ_2, $f(\cdot,\xi_2) \in \mathcal{P}_1'$. Then f is continuous as a mapping from $\mathcal{P}_1 \times \mathcal{P}_2$ into* C, *i.e. for some constants* $C < \infty$ and $k \in \mathbb{N}_0$

$$|f(\xi_1,\xi_2)| \le C\, q_k^1(\xi_1)\, q_k^2(\xi_2) \qquad \text{for} \quad \xi_1 \in \mathcal{P}_1, \quad \xi_2 \in \mathcal{P}_2.$$

2. Inductive and Projective Limits

Let $\{\mathcal{P}_\tau\}_{\tau\in T}$ be a family of multinormed vector spaces.

By the *inductive limit* $\mathcal{P} = \varinjlim_{\tau\in T} \mathcal{P}_\tau$ of the spaces $\mathcal{P}_\tau$ we understand the vector space $P = \bigcup_{\tau\in T} P_\tau$ with the convergence topology defined as follows:

> A sequence $\xi_s \in P$ ($s = 1,2,\ldots$) is convergent to zero if there exists a $\tau_0 \in T$ such that $\xi_s \in P_{\tau_0}$ ($s = 1,2,\ldots$) and for every P_τ such that all $\xi_s \in P_\tau$ we have $\lim_{s\to\infty} \xi_s = 0$ in $\mathcal{P}_\tau$.

By the *projective limit* $\mathcal{R} = \varprojlim_{\tau\in T} \mathcal{P}_\tau$ of the spaces $\mathcal{P}_\tau$ we understand the vector space $R = \bigcap_{\tau\in T} P_\tau$ with the following convergence topology:

> A sequence $\xi_s \in R$ ($s = 1,2,\ldots$) tends to zero if $\xi_s \to 0$ as $s \to \infty$ in every $\mathcal{P}_\tau$, $\tau \in T$.

Note that $f \in \mathcal{P}'$ (= the dual space of $\mathcal{P}$ with the pointwise topology) if and only if it is linear on P and its restriction to any $\mathcal{P}_\tau$ is in $(\mathcal{P}_\tau)'$.

From Corollary 1 we get:

Corollary 2. *Suppose $\mathcal{P}$ is an inductive limit of complete multinormed spaces $\mathcal{P}_\tau$. Let $f_\nu \in \mathcal{P}'$ and let the limit $f(\xi) = \lim_{\nu\to\infty} f_\nu(\xi)$ exist for every $\xi \in P$ and be finite. Then $f \in \mathcal{P}'$.*

In Section 3 we present examples of complex multinormed vector spaces and their inductive limits. Some of them are well-known from the theory of distributions.

3. Subspaces. The Hahn–Banach Theorem

We consider pairs $\mathcal{Q}, \mathcal{P}$ of topological vector spaces (not necessarily multinormed) with topologies given by convergence of sequences to zero. We write $\mathcal{Q} \subset \mathcal{P}$ if for the underlying sets we have $Q \subset P$ and the convergence to zero in $\mathcal{Q}$ implies the convergence to zero in $\mathcal{P}$. If $f \in \mathcal{P}'$ we denote by $Zf = f|_Q$ the restriction of f to Q.

Proposition 1. *Suppose $\mathcal{Q} \subset \mathcal{P}$ and the set Q is dense in $\mathcal{P}$. Then $Z(\mathcal{P}') \subset \mathcal{Q}'$ and the restriction mapping $Z\colon \mathcal{P}' \to \mathcal{Q}'$ is a homeomorphism onto the image. We identify $\mathcal{P}'$ with $Z(\mathcal{P}')$ under this bijection and write $\mathcal{P}' \subset \mathcal{Q}'$.*

Remark 1. Clearly if $\mathcal{Q} \subset \mathcal{P}$ and $f \in \mathcal{P}'$ then $f|_Q \in \mathcal{Q}'$. However to conclude that $\mathcal{P}' \subset \mathcal{Q}'$ it is essential to assume the density of Q in $\mathcal{P}$.

Theorem 4 (The Hahn–Banach extension theorem; [Ru1]). *Let $\mathcal{Q} \subset \mathcal{P}$ be topological vector spaces. Let q be a seminorm on $\mathcal{P}$ and $f \in \mathcal{Q}'$ be such that*

$$|f(\xi)| \le q(\xi) \qquad \text{for } \xi \in Q\,.$$

Then there exists a linear functional $\tilde{f}$ on $\mathcal{P}$ which extends f, i.e.

$$\tilde{f}(\xi) = f(\xi) \qquad \text{for } \xi \in Q$$

and satisfies the inequality

$$|\tilde{f}(\xi)| \le q(\xi) \qquad \text{for } \xi \in P\,.$$

Remark 2. The extension $\tilde{f}$ is in general, not unique. However if Q is dense in $\mathcal{P}$ the uniqueness holds. If, in addition, $\mathcal{P}$ is multinormed then the proof of the Hahn–Banach theorem becomes very simple.

Corollary 3 ([Tr]). *It follows from Theorem 4 that if $\mathcal{Q} \subset \mathcal{P}$ are multinormed vector spaces and $\xi_0 \in P$ is not in the closure of Q in $\mathcal{P}$ then there exists a functional $f \in \mathcal{P}'$ such that $f(\xi_0) = 1$ and $f(\xi) = 0$ for $\xi \in Q$.*

Corollary 4. *If $\mathcal{Q}$ is a topological subspace of a multinormed vector space $\mathcal{P}$ then $\mathcal{P}'$ is a subspace of $\mathcal{Q}'$ (see Proposition 1) if and only if Q is dense in $\mathcal{P}$. Equivalently, Q is not dense in $\mathcal{P}$ if and only if there exists an $f \in \mathcal{P}'$, $f \neq 0$ such that $f|_Q = 0$.*

Exercise

1. Let A be a subset of a vector space $\mathcal{P}$ with topology defined by a sequence $\{q_k\}_{k=1}^{\infty}$ of seminorms. Prove that a point ξ belongs to the closure of A if and only if, for every $\varepsilon > 0$ and $k \in \mathbf{N}_0$ there exists a point $\xi_{\varepsilon k} \in A$ such that $q_k(\xi_{\varepsilon k} - \xi) < \varepsilon$.

§3. A REVIEW OF BASIC FACTS IN THE THEORY OF DISTRIBUTIONS

1. Spaces D_K and $(D_K)'$

We denote by C_K^∞ the set of smooth functions on $\mathbf{R}^n$ with supports in a compact set K. Observe that the formula

$$|||\varphi|||_k = \sum_{|\alpha| \le k} \sup_{x \in K} \left| \left(\frac{\partial}{\partial x} \right)^\alpha \varphi(x) \right| \qquad \text{for } \varphi \in C_K^\infty \quad (k = 0, 1, \dots)$$

defines an increasing sequence of norms on C_K^∞. The set C_K^∞ equipped with the topology defined by this sequence of (semi)norms is denoted by D_K. The space D_K is complete (see e.g. [Sz1] or Proposition 5.1 for a similar proof).

Let $u \in (D_K)'$. The value of u on a function $\varphi \in C_K^\infty$ is denoted by $u[\varphi]$. It follows from Theorem 2.1 that a linear functional u on C_K^∞ belongs to $(D_K)'$ if and only if one of the following equivalent conditions holds:

Condition $\mathbf{W_1}$: If $\varphi_\nu \in C_K^\infty$ $(\nu = 1, 2, \dots)$ and $\lim_{\nu \to \infty} |||\varphi_\nu|||_k = 0$ $(k = 0, 1, \dots)$ then $\lim_{\nu \to \infty} u[\varphi_\nu] = 0$.

Condition $\mathbf{W_2}$: There exist constants $C < \infty$ and $p \in \mathbf{N}_0$ such that

$$|u[\varphi]| \le C |||\varphi|||_p \qquad \text{for } \varphi \in C_K^\infty .$$

Corollary 2.1 asserts that $(D_K)'$ equipped with the pointwise convergence topology is complete.

2. Spaces $D(A)$ and $D'(A)$

Let $A \subset \Omega \subset \mathbf{R}^n$ be such that Ω is open in $\mathbf{R}^n$ and A is relatively closed in Ω. We denote by $C_{(0)}^\infty(A)$ ($C^\infty(A)$, resp.) the space of restrictions to A of functions in $C_0^\infty(\Omega)$ ($C^\infty(\Omega)$, resp.). We equip $C_{(0)}^\infty(A)$ with the inductive limit topology as follows:

$$D(A) \stackrel{\text{def}}{=} \varinjlim_{K \subset \Omega} D_K|_A$$

where K ranges over all compact subsets of Ω and $D_K|_A$ denotes the space of restrictions to A of elements of D_K with the topology induced from D_K.

The dual space $D'(A)$ is called the *space of distributions* on A (see Exercises 1, 2, 3).

The convergence in the space D' will always be understood as a pointwise convergence. Hence by Corollary 2.2 the limit of a sequence of distributions is itself a distribution.

Note that if A is open then we take $A = \Omega$ and $D'(A)$ is the "usual" space of distributions on an open set (see Exercise 6).

In applications we take $A = (0, t] \subset \mathbf{R}^n_+$, or $A = [0, t] \subset \mathbf{R}^n$ with $t > 0$, $t \in \mathbf{R}^n$.

To verify that a linear functional u on $D(A)$ is a distribution it is enough to check (in view of the inductive limit topology and Condition W_1) that:

(i) for every sequence $\{\varphi_j\}_{j=1}^{\infty}$ of functions in $C_0^{\infty}(\Omega)$ with supports in a compact set $K \subset \Omega$ and such that $\lim_{j\to\infty} \left(\frac{\partial}{\partial x}\right)^{\alpha} \varphi_j = 0$ uniformly on Ω for $\alpha \in \mathbf{N}_0^n$, we have

$$u[\varphi_j|_A] \xrightarrow[j\to\infty]{} 0$$

or (in view of Condition W_2) that:

(ii) for every compact $K \subset \Omega$

$$|u[\varphi|_A]| \leq C(K) \sum_{|\alpha| \leq k(K)} \sup_{x \in K} \left| \left(\frac{\partial}{\partial x}\right)^{\alpha} \varphi(x) \right| \qquad \text{for } \varphi \in C_K^{\infty}$$

for some constants $C(K) < \infty$, $k(K) \in \mathbf{N}_0$.

If the set A is sufficiently regular then the inequality in (ii) can be improved so that the supremum is taken over the set A (see Theorem 1 below). We prove this fact in the case of a polyinterval $(0, t]$ (See Theorem 8.3 and Exercise 8.8).

Definition 1. A distribution u is said to be of *finite order on A* iff there exists a $k \in \mathbf{N}_0$ such that $k(K) \leq k$ for every $K \subset \Omega$ (see (ii) above). The smallest k satisfying this inequality is called the *order of the distribution* u on A.

As it is well known distributions in $D'(\Omega)$ enjoy the *localization property*: $u \in D'(\Omega)$ is zero on Ω if u is zero in a neighbourhood of every point in Ω. This property allows us to define the *support* of $u \in D'(\Omega)$ (denoted by $\operatorname{supp} u$) as the largest relatively closed subset of Ω such that u is zero on its complement in Ω.

Denote by $D'_A(\Omega)$ the space of distributions on Ω with support in A. We have the following characterization of $D'_A(\Omega)$:

Proposition 1. *The spaces $D'_A(\Omega)$ and $D'(A)$ are isomorphic (denoted $D'_A(\Omega) \simeq D'(A)$) under a canonical linear isomorphism L: $D'_A(\Omega) \to D'(A)$, given by the assignment*

$$D'_A(\Omega) \ni \tilde{u} \mapsto u \in D'(A),$$

where $u[\tilde{\varphi}|_A] = \tilde{u}[\tilde{\varphi}]$ for $\tilde{\varphi} \in C_0^\infty(\Omega)$ with $\tilde{u} \in D'_A(\Omega)$.

Proof: We start by showing that the definition of u is correct. To this end we have to verify that if $\tilde{\varphi} \in C_0^\infty(\Omega)$ and $\tilde{\varphi} = 0$ on A then $\tilde{u}[\tilde{\varphi}] = 0$. We take $\sigma \in C_0^\infty(\mathbf{R}^n)$, $\sigma = 1$ on $\operatorname{supp}\tilde{\varphi}$. Then $\tilde{\varphi} = \sigma\tilde{\varphi}$, $\tilde{u}[\tilde{\varphi}] = \sigma\tilde{u}[\tilde{\varphi}]$. Thus by Exercise 8 applied to $\sigma\tilde{u}$ we have $\tilde{u}[\tilde{\varphi}] = 0$.

It is clear that $u \in D'(A)$ and that the assignment $\tilde{u} \mapsto u$ is an isomorphism.

In view of this isomorphism we often identify u and $\tilde{u}$.

Theorem 1. *Let $u \in D'_K(\mathbf{R}^n)$ where K is a connected compact set in $\mathbf{R}^n$ such that any two points $x, y \in K$ can be joined by a rectifiable curve in K of length $\le \widetilde{C}|x-y|$, $\widetilde{C} < \infty$, then there exists a constant $C < \infty$ and $k \in \mathbf{N}_0$ such that*

$$|u[\psi]| \le C \sum_{|\alpha| \le k} \sup_{x \in K} \left| \left(\frac{\partial}{\partial x}\right)^\alpha \psi(x) \right| \qquad \text{for } \psi \in C^k(\mathbf{R}^n).$$

The proof of this theorem, based on Whitney extension theorem, is given in [Hö 2].

As examples of distribution we note:

(a) the Dirac delta $\delta_{(y)}$ at a point $y \in A$ defined as

$$\delta_{(y)}[\varphi] = \varphi(y) \qquad \text{for } \varphi \in C^\infty_{(0)}(A),$$

(b) locally integrable functions are imbedded into distributions by means of the identity

$$f[\varphi] = \int_A f(x)\varphi(x)\,dx \qquad \text{for } \varphi \in C^\infty_{(0)}(A)$$

if $f \in L^1_{\text{loc}}(A)$.

(c) We introduce a functional $\operatorname{Pf}\frac{1}{x}$ by the formula

$$\operatorname{Pf}\frac{1}{x}[\varphi] = \lim_{\varepsilon \to 0+} \int_{-\infty}^{-\varepsilon} + \int_{\varepsilon}^{+\infty} \frac{\varphi(x)}{x}\,dx \qquad \text{for } \varphi \in D(\mathbf{R}).$$

Let $\varphi \in C_0^\infty(\mathbf{R})$, $\varphi(x) = 0$ for $|x| \ge R$. Write $\varphi(x) = \varphi(0) + x\tilde{\varphi}(x)$ then

$$\left| \operatorname{Pf}\frac{1}{x}[\varphi] \right| = \left| \int_{-R}^{R} \tilde{\varphi}(x)\,dx \right| \le 2R \sup_{|x|<R} |\varphi'(x)|,$$

which shows that $\operatorname{Pf}\frac{1}{x}$ is a distribution on $\mathbf{R}$ of order 1. Clearly for $x \neq 0$ it coincides with the function $\frac{1}{x}$.

The spaces of distributions are closed under differentiation which is defined as (see Exercise 4)

$$(1)\qquad \left(\frac{\partial}{\partial x}\right)^{\alpha} u[\varphi] = (-1)^{|\alpha|} u\left[\left(\frac{\partial}{\partial x}\right)^{\alpha}\varphi\right] \qquad \text{for } u \in D'(A),\ \varphi \in D(A)$$

and under multiplication by $C^{\infty}(A)$-functions a defined as

$$au[\varphi] = u[a\varphi].$$

We shall prove now a theorem which characterizes distributions supported by a single point $\{y\}$.

Theorem 2. *$u \in D'_{\{y\}}(\mathbf{R}^n)$ if and only if it is a finite linear combination of the Dirac distribution $\delta_{(y)}$ and its derivatives.*

Proof: Let $u \in D'(\mathbf{R}^n)$, $\operatorname{supp} u = \{y\}$ and let $k \geq$ order of u. By the Taylor formula for $\psi \in C^{\infty}$ there exists a function $\sigma \in C^{\infty}$ such that

$$\psi(x) = \sum_{|\alpha|\leq k} \frac{1}{\alpha!}\left(\left(\frac{\partial}{\partial x}\right)^{\alpha}\psi\right)(y)(x-y)^{\alpha} + \sigma(x),$$

$$\left(\frac{\partial}{\partial x}\right)^{\alpha}\sigma(y) = 0 \qquad \text{for } |\alpha| \leq k.$$

Thus by Theorem 1 (or equivalently by Exercise 8) $u[\sigma] = 0$ and by the linearity of u we get

$$u[\psi] = \sum_{|\alpha|\leq k} a_{\alpha}\left(\left(\frac{\partial}{\partial x}\right)^{\alpha}\psi\right)(y), \qquad \text{where} \quad a_{\alpha} = u\left[\frac{(\cdot - y)^{\alpha}}{\alpha!}\right].$$

Since $\psi \in C^{\infty}$ was arbitrary the above formula denotes precisely that

$$u = \sum_{|\alpha|\leq k} (-1)^{|\alpha|} a_{\alpha}\left(\frac{\partial}{\partial x}\right)^{\alpha}\delta_{(y)}.$$

The second part of the theorem is obvious.

Certain distributions can be obtained as boundary values of holomorphic functions. Having in mind applications to Sections 13 and 15 we state the following theorem which also generalizes the classical theorem of Painlevé:

Theorem 3. *Let Ω be an open subset of* $\mathbf{R}$ *and* V *an open subset in* $\mathbf{C}$ *such that* $V \cap \mathbf{R} = \Omega$. *Let* $F \in \mathcal{O}(V \setminus \Omega)$ *be such that for every compact set* $K \subset \Omega$ *there exist constants* $N = N(K) \in \mathbf{N}_0$, $C = C(K) < \infty$ *such that*

$$(2) \quad |F(\alpha + i\beta)| \le C \cdot |\beta|^{-N} \qquad \text{for } \alpha \in K, \qquad 0 < |\beta| \le \eta, \quad \eta \text{ small enough.}$$

Then there exist distributions $F(\cdot \pm i0) \in D'(\Omega)$ *(called the* boundary values *of* F*) such that*

$$F(\cdot \pm i0)[\varphi] = \lim_{\varepsilon \to 0_+} \int_\Omega F(\alpha \pm i\varepsilon)\varphi(\alpha)\, d\alpha, \qquad \varphi \in C_0^\infty(\Omega).$$

Moreover if $F(\cdot + i0) = F(\cdot - i0)$ *then* F *extends holomorphically to* V.

We denote the difference $F(\cdot + i0) - F(\cdot - i0)$ by $b(F)$ and call the *jump of* F *across* $\mathbf{R}$.

Proof: We may assume that Ω is an open interval. Let K be a fixed closed subinterval of Ω. Choose $0 < \eta < 1$ so that the set

$$A = \{z = \alpha + i\beta \in \mathbf{C}\colon \alpha \in K,\ 0 < \beta \le \eta\} \subset V_+ = V \cap \{\operatorname{Im} z > 0\}$$

and (2) holds for $\alpha + i\beta \in A$. Let $\mathring{z}$ be a fixed point in A with $\operatorname{Im} \mathring{z} = \eta$. Define

$$J_+F(z) = \int_{\gamma_z} F(\theta)\, d\theta, \qquad z \in V_+,$$

where γ_z is the curve in V_+ joining $\mathring{z}$ to $z = \alpha + i\beta$ and consisting of two segments $[\mathring{z}, \alpha + i\eta]$ and $[\alpha + i\eta, \alpha + i\beta]$. Clearly $\frac{d}{dz} J_+F(z) = F(z)$ for $z \in V_+$ and from (2) and the fact that F is bounded on $K + i\eta$ we get for $z \in A$

$$|J_+F(z)| \le \left| \int_{\alpha+i\beta}^{\alpha+i\eta} F(\theta)\, d\theta + \int_{\alpha+i\eta}^{\mathring{z}} F(\theta)\, d\theta \right| \le \widetilde{C} + C \int_\beta^\eta t^{-N}\, dt \qquad \text{for } z \in A.$$

Hence there exists a constant $C_1(K)$ such that for $\alpha + i\beta \in A$

$$|J_+F(\alpha + i\beta)| \le \begin{cases} C_1 \beta^{-N+1} & \text{if } N > 1 \\ C_1\, |\ln \beta| & \text{if } N = 1. \end{cases}$$

Iterating the above operation $N + 1$ times we find $J_+^{N+1}F \in \mathcal{O}(V_+)$ such that $\frac{d^{N+1}}{dz^{N+1}} J_+^{N+1}F = F$ on V_+ and $J_+^{N+1}F(\alpha + i\varepsilon)$ converges uniformly, as $\varepsilon \to 0_+$, to a continuous function $F_+^{N+1}(\alpha)$ on K. Thus $J_+^{N+1}F$ extends to a continuous function on $\overline{A} = \{z \in \mathbf{C}\colon \alpha \in K,\ 0 \le \beta \le \eta\}$.

Let $\varphi \in C_0^\infty(\Omega)$, $\operatorname{supp}\varphi \subset K$. Integrating by parts we get from the Lebesgue theorem

$$\lim_{\varepsilon\to 0_+}\int_\Omega F(\alpha+i\varepsilon)\varphi(\alpha)\,d\alpha = \lim_{\varepsilon\to 0_+}(-1)^{N+1}\int_\Omega J_+^{N+1}F(\alpha+i\varepsilon)\Big(\frac{d}{d\alpha}\Big)^{N+1}\varphi(\alpha)\,d\alpha$$
$$= (-1)^{N+1}\int_\Omega F_+^{N+1}(\alpha)\Big(\frac{d}{d\alpha}\Big)^{N+1}\varphi(\alpha)\,d\alpha.$$

Thus the limit $F(\alpha+i0)$ exists and is a distribution. By symmetry one defines J_-F and $J_-^{N+1}F$ on $V_- = V\cap\{\operatorname{Im} z<0\}$ and one proves the existence of the limit $F(\alpha-i0)$. Now if $F(\alpha-i0)=F(\alpha+i0)$ it follows (see Exercise 10.1 in [Sz1]) that

$$F_+^{N+1}(\alpha) - F_-^{N+1}(\alpha) = w(\alpha)$$

where w is a polynomial of degree at most N. Hence by the classical Painlevé theorem [Ko] we infer that the function

$$F^{N+1}(z) = \begin{cases} J_+^{N+1}F(z) & \text{for } z\in V_+\\ F_+^{N+1}(\alpha) & \text{for } z=\alpha\\ J_-^{N+1}F(z)+w(z) & \text{for } z\in V_-\end{cases}$$

is holomorphic on V. Hence $\left(\frac{d}{dz}\right)^{N+1}F^{N+1}$ is the desired extension of $F\in\mathcal{O}(V\setminus\Omega)$ to V.

Example 1. With $\operatorname{Pf}\frac{1}{x}$ being the distribution of example (c) we have

$$\lim_{\varepsilon\to 0_+}\frac{1}{x\pm i\varepsilon} = \mp i\pi\delta_0 + \operatorname{Pf}\frac{1}{x}$$

and hence

$$\lim_{\varepsilon\to 0_+}\Big(\frac{1}{x+i\varepsilon}-\frac{1}{x-i\varepsilon}\Big) = -2\pi i\delta_0 \qquad \text{i.e.} \qquad \frac{1}{2\pi i}b\Big(\frac{-1}{z}\Big)=\delta_0.$$

Proof: Let $\varphi\in C_0^\infty(\mathbf{R})$, $\varphi(x)=0$ for $|x|>R$. Then

$$\lim_{\varepsilon\to 0_+}\int\frac{\varphi(x)}{x+i\varepsilon}\,dx = \lim_{\varepsilon\to 0_+}\int_{-R}^{R}\frac{x-i\varepsilon}{x^2+\varepsilon^2}\varphi(x)\,dx$$
$$=\varphi(0)\lim_{\varepsilon\to 0_+}\int_{-R}^{R}\frac{x-i\varepsilon}{x^2+\varepsilon^2}\,dx+$$
$$+\lim_{\varepsilon\to 0_+}\int_{-R}^{R}\frac{x-i\varepsilon}{x^2+\varepsilon^2}(\varphi(x)-\varphi(0))\,dx$$
$$=-2i\varphi(0)\lim_{\varepsilon\to 0_+}\operatorname{arctg}\frac{R}{\varepsilon}+\int_{-R}^{R}\frac{\varphi(x)-\varphi(0)}{x}\,dx$$
$$=-i\pi\varphi(0)+\operatorname{Pf}\frac{1}{x}[\varphi].$$

The proof for $\frac{1}{x-i\varepsilon}$ is analogous.

3. Spaces S and S'

By $S = S(\mathbf{R}^n)$ we denote the set of functions $\sigma \in C^\infty(\mathbf{R}^n)$ such that

$$(3) \qquad q_k(\sigma) \stackrel{\text{df}}{=} \sum_{|\alpha|\le k} \sup_{x\in\mathbf{R}^n} (1+\|x\|^2)^k \left|\left(\frac{\partial}{\partial x}\right)^\alpha \sigma(x)\right| \qquad (k=0,1,2,\dots)$$

is finite. A function $\sigma \in S$ is called a *rapidly decreasing function*. It is clear that

$$C_0^\infty \subset S \subset C^\infty$$

and the set S is indeed bigger then C_0^∞ since it contains e.g. the function $\mathbf{R}^n \ni x \mapsto e^{-\|x\|^2}$ which is not in C_0^∞.

It is easy to note that q_k is an increasing sequence of (semi) norms on S, and S endowed with the topology defined by the sequence of norms q_k (denoted by the same letter S) is complete (see e.g. [Sz1] Exercise 18.1). According to the general scheme we define the dual $S' = S'(\mathbf{R}^n)$ of $S(\mathbf{R}^n)$. It follows that if $u = \lim_{\nu\to\infty} u_\nu$ in S' then there exist constants $C < \infty$ and $k \in \mathbf{N}_0$ such that for all ν

$$|u[\sigma]|, |u_\nu[\sigma]| \le C \sum_{|\alpha|\le k} \sup\,(1+\|x\|^2)^k \left|\left(\frac{\partial}{\partial x}\right)^\alpha \sigma(x)\right| \qquad \text{for } \sigma \in S.$$

Further, the convergence in D implies the convergence in S which we write as $D \subset S$. Moreover, C_0^∞ is dense in S. These facts imply, in view of Corollary 2.4, that $S' \subset D'$ i.e. S' is a subspace of the space of distributions. We call it the space of *tempered distributions*. As in the case of $(D_K)'$, Corollary 2.1 implies that S' equipped with the pointwise topology is complete.

The following theorem characterizes tempered distributions.

Theorem 4. *A distribution $u \in D'(\mathbf{R}^n)$ is tempered if and only if it is a derivative $\left(\frac{\partial}{\partial x}\right)^\alpha$, for some $\alpha \in \mathbf{N}_0^n$, of a C^0-function slowly increasing at infinity i.e. of a continuous function f majorized by the function $x \mapsto C\left(1+\|x\|^2\right)^k$ for some $k \in \mathbf{N}$:*

$$u = \left(\frac{\partial}{\partial x}\right)^\alpha f.$$

The "if" part of Theorem 4 is clear: any C^0-function slowly increasing at infinity is a tempered distribution and so is any derivative of a tempered distribution. The "only if" part can be derived from the Hahn–Banach theorem (see for example [Tr] or Exercises 10–12).

We end this subsection with the famous "division theorem" which emphasizes the importance of the spaces S':

Theorem 5 ([Ł1], [Ł2], [Hö1]). *Let $u \in S'(\mathbf{R}^n)$ and P be a polynomial in $\mathbf{R}^n$. Then there exists a distribution $f \in S'(\mathbf{R}^n)$ such that $P \cdot f = u$ on $\mathbf{R}^n$.*

Some simple cases of the division problem are indicated in Exercises 14–16.

4. Spaces E and E'

By $E(\mathbf{R}^n)$ we denote the set of functions $\sigma \in C^\infty(\mathbf{R}^n)$ and define

$$q_k(\sigma) \overset{\text{def}}{=} \sum_{|\alpha| \le k} \sup_{\|x\| \le k} \left| \left(\frac{\partial}{\partial x}\right)^\alpha \sigma(x) \right| \qquad (k = 0, 1, 2, \dots).$$

It is easy to note that $\{q_k\}$ is an increasing sequence of seminorms on E. Hence the space E endowed with the topology defined by this sequence of seminorms (denoted by the same letter E) is complete (see e.g [Sz1], Theorem 7.2). Further C_0^∞ is dense in E and the convergence in D implies the convergence in E hence in view of Corollary 2.4 it follows that $E' \subset D'$. More precisely it can be shown that E' is the subspace of D' consisting of all distributions with bounded supports (cf. e.g. [Sz1], Theorem 7.1). We call it the space of *distributions with bounded support*. Clearly $E' \subset S'$ as topological spaces.

5. Substitution in Distributions. Homogeneous Distributions

Let $A \subset \Omega \subset \mathbf{R}^n$ be such that Ω is open in $\mathbf{R}^n$ and A is relatively closed in Ω. Let $x = f(y)$ be a one-to-one mapping of a $\Omega_1 \subset \mathbf{R}^n$ onto Ω, of class C^∞ with a non-vanishing *Jacobian* $Jf = \det \frac{\partial f}{\partial y}$. One easily checks that $A_1 = f^{-1}(A)$ is relatively closed in Ω_1 and that for every $u \in D'(A)$ the formal definition

$$(u \circ f)[\psi] = u\left[(\psi \circ f^{-1})|Jf^{-1}|\right] \qquad \text{for} \quad \psi \in C^\infty_{(0)}(A_1) \tag{4}$$

defines $u \circ f \in D'(A_1)$. Moreover the mapping $D'(A) \ni u \mapsto u \circ f \in D'(A_1)$ is an isomorphism of $D'(A)$ onto $D'(A_1)$.

In the case where u is a continuous function on A the distribution $u \circ f$ is a continuous function $A_1 \ni y \mapsto u(f(y))$.

Definition 2. A distribution $u \in D'(\mathbf{R}^n)$ is called *homogeneous of order* λ if and only if for every homothety $\mathbf{R}^n \ni y \mapsto f(y) = ry$ $(r > 0)$ we have

$$u \circ f = r^\lambda u.$$

It can be proved that a distribution $u \in D'(\mathbf{R}^n)$ is homogeneous of order λ if and only if u satisfies the *Euler equation* (see e.g. [Sz1] Exercise 9.7)

$$\sum_{j=1}^{n} x_j \frac{\partial}{\partial x_j} u = \lambda u.$$

Example (Homogeneous distributions on $\mathbf{R}$ cf. [G–S]). For $\lambda \in \mathbf{C}$ $\operatorname{Re}\lambda > -1$ we define x_+^λ as the regular distribution (=function)

$$x_+^\lambda[\varphi] = \int_0^\infty x^\lambda \varphi(x)\, dx \qquad \text{for } \varphi \in C_0^\infty(\mathbf{R}).$$

Subject to a suitable regularization at zero, this definition extends to all $\lambda \notin -\mathbf{N}$:

$$x_+^\lambda[\varphi] = \int_0^\infty x^\lambda \left(\varphi(x) - \sum_{i=0}^{n-1} \frac{\varphi^{(i)}(0)}{i!} x^i \right) dx \qquad \text{for } \varphi \in C_0^\infty(\mathbf{R})$$

if n is such that $-n-1 < \operatorname{Re}\lambda < -n$. By means of the substitution $x \mapsto -x$ we define the distributions x_-^λ. Observe that for $\lambda \notin \mathbf{N}$ $x_+^\lambda \in D'_{\overline{\mathbf{R}}_+}(\mathbf{R})$, $x_-^\lambda \in D'_{\overline{\mathbf{R}}_-}(\mathbf{R})$ are homogeneous of order λ. The distributions $x_\pm^\lambda$ do not exhaust the list of homogeneous distributions on $\mathbf{R}$ (see [G–S]). For example $\delta_0^{(n)}$ is homogeneous of order $-n-1$ and the distribution $\operatorname{Pf}\frac{1}{x}$ of example (c) is homogeneous of order -1.

6. Classical Order of a Distribution and Extendibility Theorems for Distributions

Let Ω be an open set in $\mathbf{R}^n$. By Definition 1 (cf. also Exercise 6) $u \in D'(\Omega)$ is of finite order $\leq k$ on Ω, $k \in \mathbf{N}_0$ if and only if for every compact $K \subset \Omega$ there exists a constant $C(K)$ such that

(5) $$|u[\varphi]| \leq C(K) |||\varphi|||_k \qquad \text{for } \varphi \in C_K^\infty,$$

where

(6) $$|||\varphi|||_k = \sum_{|\alpha| \leq k} \sup_{x \in \mathbf{R}^n} \left| \left(\frac{\partial}{\partial x}\right)^\alpha \varphi(x) \right| .$$

Observe that the distribution defined by a locally integrable function is of order zero and that every tempered distribution is of finite order. It is also clear that:

(7) $$u[\varphi] = \sum_{j=2}^{\infty} \left(\frac{d}{dx}\right)^j \varphi\left(\frac{1}{j}\right) \qquad \text{for } \varphi \in C_0^\infty((0,1))$$

defines $u \in D'((0,1))$. It can be proved (cf. Exercises 17 and 18) that u is of infinite order.

Proposition 2. *Let $\Omega_0 \subset \Omega \subset \mathbf{R}^n$ be open sets and $\overline{\Omega}_0$ a compact subset of Ω. If $u \in D'(\Omega)$ then there exist constants $C < +\infty$, $k \in \mathbf{N}_0$ such that*

$$(8) \qquad |u[\varphi]| \leq C |||\varphi|||_k \qquad \text{for} \quad \varphi \in C_0^\infty(\Omega_0).$$

Thus the restriction of u to Ω_0 is a distribution of order $\leq k$.

Proof: Since $u \in D'(\Omega)$, by Exercise 6 there exist constants $C = C(\overline{\Omega}_0)$ and $k = k\left(\overline{\Omega}_0\right)$ satisfying (8) as asserted.

Proposition 3. *Let u be a distribution on a bounded open set $\Omega_0 \subset \mathbf{R}^n$ and let $\Omega_1 \supset \overline{\Omega}_0$. Then the following conditions are equivalent:*

(i) *u is extendible to Ω_1,*

(ii) *there exist constants $C < +\infty$, $k \in \mathbf{N}_0$ satisfying (8),*

(iii) *the condition $C_0^\infty(\Omega_0) \ni \varphi_\nu \to 0$ in $D(\mathbf{R}^n)$ (equivalently $\varphi_\nu \in C_0^\infty(\Omega_0)$ $(\nu = 1, 2, \dots)$, $|||\varphi_\nu|||_k \underset{\nu \to \infty}{\longrightarrow} 0$ $(k = 0, 1, 2, \dots)$) implies $u[\varphi_\nu] \to 0$.*

Proof: The implication (i) $\Rightarrow$ (ii) is just Proposition 2. For the proof of (ii) $\Rightarrow$ (i) let u satisfy (8). Since $|||\cdot|||_k$ is a norm in $C_0^\infty(\Omega_1)$ the functional u can be extended by the Hahn–Banach theorem to a linear functional on $C_0^\infty(\Omega_1)$ continuous in this norm. The extended functional is a distribution on Ω_1. The implication (ii) $\Rightarrow$ (iii) is clear. The proof of (iii) $\Rightarrow$ (ii) follows from Theorem 2.1.

Proposition 3 implies immediately

Corollary 1. *A distribution u admitting an extension from a bounded open set Ω_0 to a set $\Omega_1 \supset \overline{\Omega}_0$ is also extendible to the whole of $\mathbf{R}^n$.*

By Proposition 2 every distribution on a bounded open set which is of infinite order cannot be extended to $\mathbf{R}^n$ hence Exercise 18 implies

Corollary 2. *There exists a distribution on a bounded open set which cannot be extended to $\mathbf{R}^n$.*

The next proposition concerns the extendibility to $\mathbf{R}^n$ of a distribution defined on any open set $\Omega \subset \mathbf{R}^n$.

Its proof requires the following *partition of unity*:

Lemma 1 ([Schw]). *There exists a locally finite covering of $\mathbf{R}^n$ by open bounded sets Ω_ν and functions $\alpha_\nu \in C_0^\infty(\Omega_\nu)$, $\nu \in A$ such that $\sum_{\nu \in A} \alpha_\nu \equiv 1$ in $\mathbf{R}^n$. (Note that for every $x \in \mathbf{R}^n$ only finitely many $\alpha_\nu(x)$ are different from zero).*

Proposition 4 ([L2], Proposition 1). *Let $u \in D'(\Omega)$. In order that u be extendible to $\mathbf{R}^n$ it is necessary and sufficient that u restricted to an arbitrary open set be extendible to $\mathbf{R}^n$.*

Proof: The condition is clearly necessary so it is enough to prove its sufficiency. By Lemma 1 there exists a locally finite covering of $\mathbf{R}^n$ by bounded open sets Ω_ν and a partition of unity $\alpha_\nu \in C_0^\infty(\Omega_\nu)$. Let u_ν be an extension to $\mathbf{R}^n$ of the restriction of u to $\Omega_\nu \cap \Omega$ (we take $u_\nu = 0$ if $\Omega_\nu \cap \Omega = \emptyset$). Since $\alpha_\nu u_\nu = \alpha_\nu u$ in Ω we have $\sum_\nu \alpha_\nu u_\nu = \sum_\nu \alpha_\nu u = u$ in Ω. On the other hand for every test function $\varphi \in D(\mathbf{R}^n)$ the sum $\sum_\nu \alpha_\nu u_\nu[\varphi]$ is finite which easily gives that $\sum_\nu \alpha_\nu u \in D'(\mathbf{R}^n)$.

We end this subsection with the theorem which states that distributions can be regarded as the derivatives of continuous functions.

Theorem 6. *Let u be a distribution on a bounded open set $\Omega \subset \mathbf{R}^n$ extendible to $\mathbf{R}^n$. Then there exist a multi-index α and a continuous function G (whose support may be chosen to be contained in any prescribed neighbourhood of $\overline{\Omega}$) such that $u = \left(\frac{\partial}{\partial x}\right)^\alpha G$ on Ω.*

The proof of this theorem (see e.g. [Sz1]) is sketched here in Exercises 20 and 21.

7. Convolution of Distributions

We begin with the definition of sequences $\{\eta_\nu\}$ of $C_0^\infty(\mathbf{R}^n)$-functions convergent to one in $\mathbf{R}^n$ (written as $\eta_\nu \to 1$ in $\mathbf{R}^n$) in the following sense (see [V2]):

1^0 for every compact set $K \subset \mathbf{R}^n$ there exists $N \in \mathbf{N}_0$ such that $\eta_\nu(x) = 1$ for $x \in K$ if $\nu \geq N$,

2^0 for every $\alpha \in \mathbf{N}_0^n$ there exists an $M_\alpha < \infty$ such that

$$\sup_{x \in \mathbf{R}^n} \left| \left(\frac{\partial}{\partial x}\right)^\alpha \eta_\nu \right| < M_\alpha \qquad (\nu = 1, 2, \ldots).$$

Let $u, v \in D'(\mathbf{R}^n)$ and let

$$\omega_\nu[\varphi] = u[v[\varphi(x+y)\eta_\nu(x,y)]] \qquad \text{for } \varphi \in D(\mathbf{R}^n) \qquad (\nu = 1, 2, \ldots), \tag{9}$$

where $\eta_\nu \to 1$ in $\mathbf{R}^{2n}$. It is easy to see that $\omega_\nu \in D'(\mathbf{R}^n)$ $(\nu = 1, 2, \ldots)$ and hence if $\{\omega_\nu\}$ is convergent its limit also belongs to $D'(\mathbf{R}^n)$.

Definition 3. We say that a distribution $\omega \in D'(\mathbf{R}^n)$ is the *convolution of distributions* $u, v \in D'(R^n)$ (denoted $\omega = u*v$) if it is the limit in $D'(\mathbf{R}^n)$ of distributions ω_ν $(\nu = 1, 2, \ldots)$ defined by (9).

Remark 1. The existence of $u * v$ implies the existence of $v * u$ and $u * v = v * u$.

Example 2. If $u, v \in L^1(\mathbf{R}^n)$ we get

$$(u * v)[\varphi] = \int u(x)v(y)\varphi(x+y)\,d(x,y) = \int \varphi(x)\Big(\int v(y)u(x-y)\,dy\Big)\,dx$$

i.e. $(u * v)(x) = \int v(y)u(x-y)\,dy$ which agrees with the clasical definition of the convolution.

Example 3. If $u \in E'(\mathbf{R}^n)$, $v \in D'(\mathbf{R}^n)$ then

$$(u * v)[\varphi] = u[v[\varphi(x+y)]] \qquad \text{for } \varphi \in D(\mathbf{R}^n).$$

Example 4. Assume one of the following conditions:

i) $\sigma \in C_0^\infty(\mathbf{R}^n)$, $u \in D'(\mathbf{R}^n)$, ii) $\sigma \in C^\infty(R^n)$, $u \in E'(R^n)$.

Then the convolution $u * \sigma$ is a C^∞-function defined by

$$(\sigma * u)(\xi) = u[\sigma(\xi - x)] \qquad \text{for } \xi \in \mathbf{R}^n.$$

The following simple properties of convolution will be used later on (here and later on we write δ for $\delta_{(0)}$): $u * \delta = \delta * u = u$ for $u \in D'(\mathbf{R}^n)$ and the Dirac delta distribution; if $u, v \in D'(\mathbf{R}^n)$ and $u * v$ exists, then for every $k = 1, \ldots, n$

$$\frac{\partial}{\partial x_k}(u * v) = \frac{\partial}{\partial x_k}u * v = u * \frac{\partial}{\partial x_k}v;$$

for every distribution $u \in D'(\mathbf{R}^n)$ and every differential operator

$$P\Big(\frac{\partial}{\partial x}\Big) = P\Big(\frac{\partial}{\partial x_1}, \ldots, \frac{\partial}{\partial x_n}\Big)$$

with constant coefficients we have

$$P\Big(\frac{\partial}{\partial x}\Big)u = P\Big(\frac{\partial}{\partial x}\Big)\delta * u.$$

8. Tensor Product of Distributions

Let $\varphi \in S(\mathbf{R}^n)$, $\psi \in S(\mathbf{R}^m)$. We denote by $\varphi \otimes \psi$ the function

$$\mathbf{R}^n \times \mathbf{R}^m \ni (x, y) \mapsto \varphi(x) \cdot \psi(y).$$

Now, given $u \in S'(R^n)$, $v \in S'(\mathbf{R}^m)$ we denote by $w = u \otimes v$ a unique element of $S'(\mathbf{R}^n \times \mathbf{R}^m)$ such that

$$w[\varphi \otimes \psi] = u[\varphi] \cdot v[\psi] \qquad \text{for } \varphi \in S(\mathbf{R}^n),\ \psi \in S(\mathbf{R}^m)$$

and call it the *tensor product* of u and v.

Exercises

1. Give an example showing that the space $(D_K)'$ defined in Subsection 1 is different from the space $D'(K)$ defined in Subsection 2 (K — a compact set).

2. Let $A \subset \Omega \subset \mathbf{R}^n$, A, Ω open and denote by K a compact set. Show that

$$\varinjlim_{K \subset \Omega} D_K|_A = \varinjlim_{K \subset A} D_K$$

if and only if $A = \Omega$.

3. Let $A \subset \Omega \subset \mathbf{R}^n$, Ω open. Show that

$$\varinjlim_{K \subset \Omega} D_K|_A = \big(\varinjlim_{K \subset \Omega} D_K\big)|_A.$$

4. Let A be relatively closed in an open set $\Omega \subset \mathbf{R}^n$. For $u \in D'(A)$, $\alpha \in \mathbf{N}_0^n$ let $\left(\frac{\partial}{\partial x}\right)^\alpha u$ be the derivative defined by (1). (If $A = \Omega$ formula (1) gives the "usual" distributional derivative of the "usual" distribution $u \in D'(\Omega)$).

Prove that $\left(\frac{\partial}{\partial x}\right)^\alpha L = L\left(\frac{\partial}{\partial x}\right)^\alpha$ where L is the isomorphism given by Proposition 1.

5. Let χ be a linear functional on $L^1((0,t))$ defined by

$$\chi[\varphi] = \int_0^t \varphi(x)\,dx \qquad \text{for } \varphi \in L^1((0,t)).$$

Prove that

(i) $\chi \in D'((0,t])$, $D'((0,t]) \sim D'_{(0,t]}(\mathbf{R}_+)$, $\frac{d\chi}{dx} = -\delta_{(t)}$,

(ii) $\chi \in D'([0,t])$, $D'([0,t]) \sim D'_{[0,t]}(\mathbf{R})$, $\frac{d\chi}{dx} = \delta_{(0)} - \delta_{(t)}$ (χ is a distribution on $\mathbf{R}$ defined by the function equal to 1 on $[0,t]$ and to zero outside),

(iii) $\chi \in (D_{[0,t]})'$, $\frac{d\chi}{dx} = 0$.

6. Let Ω be an open set in $\mathbf{R}^n$. Deduce from the definitions given in this section that

(a) $\varphi_\nu \to 0$ in $D(\Omega)$ iff there exists a compact set $K \subset \Omega$ such that $\varphi_\nu \in C_K^\infty$ and for any $\alpha \in \mathbf{N}_0^n$ $\lim_{j\to\infty} \left(\frac{\partial}{\partial x}\right)^\alpha \varphi_\nu = 0$ uniformly in Ω,

(b) $u \in D'(\Omega)$ iff u is linear on $C_0^\infty(\Omega)$ and for every compact $K \subset \Omega$ there exists a constant $C(K) < \infty$ and a non-negative integer $k(K)$ such that for any $\varphi \in C_K^\infty$

$$|u[\varphi]| \leq C(K) \sum_{|\alpha| \leq k(K)} \sup_{x \in \Omega} \left| \left(\frac{\partial}{\partial x}\right)^\alpha \varphi(x) \right|.$$

7. Let Y be the *Heaviside function* on $\mathbf{R}$: i.e. $Y(x) = 1$ for $x > 0$, $Y(x) = 0$ for $x < 0$. Show that $\frac{dY}{dx} = \delta_{(0)}$.

8. Suppose that $u \in D'_K(\mathbf{R}^n)$, K compact, is of order $\leq k$ and that $\sigma \in C^k(\mathbf{R}^n)$, $\left(\frac{\partial}{\partial x}\right)^\alpha \sigma(x) = 0$ for $|\alpha| \leq k$ and $x \in K$. Show that $u[\sigma] = 0$ (see e.g. [Sz1] Theorem 7.4).

HINT. Let $\psi \in C_0^\infty$, $\int \psi(x)dx = 1$, $\psi(x) \geq 0$ for $x \in \mathbf{R}^n$, $\operatorname{supp}\psi \subset \overline{B(0,1)}$. Denote $K_r = \bigcup_{x \in K} \overline{B(x,r)}$ and define

$$\chi_\varepsilon(x) = \varepsilon^{-n} \int_{K_{2\varepsilon}} \psi\left(\frac{x-y}{\varepsilon}\right) dy \qquad \text{for } x \in \mathbf{R}^n, \ 0 < \varepsilon < \varepsilon_0.$$

Prove that $\chi \in C_0^\infty$, $\chi_\varepsilon = 1$ on K_ε, $\operatorname{supp}\chi_\varepsilon \subset K_{3\varepsilon}$ and that for every $\beta \in \mathbf{N}_0^n$:

$$\sup_{x \in \mathbf{R}^n} \left| \left(\frac{\partial}{\partial x}\right)^\beta \chi_\varepsilon(x) \right| = 0\left(\varepsilon^{-|\beta|}\right) \qquad \text{as } \varepsilon \to 0.$$

Next show that for every function $\varphi \in C^k$: $u[\varphi] = u[\chi_\varepsilon \varphi]$ and that there exists a constant C such that

$$|u[\varphi]| \leq C \sum_{|\alpha| \leq k} \sup_{x \in K_{3\varepsilon}} \left| \left(\frac{\partial}{\partial x}\right)^\alpha (\chi_\varepsilon(x)\varphi(x)) \right| \qquad \text{for } 0 < \varepsilon \leq \varepsilon_0 .$$

Now it suffices to prove that for every $|\alpha| \leq k$ $\lim_{\varepsilon \to 0} \sup_{x \in K_{3\varepsilon}} \left| \left(\frac{\partial}{\partial x}\right)^\alpha (\chi_\varepsilon(x)\sigma(x)) \right| = 0$.

For this aim apply the Leibniz rule to $\chi_\varepsilon \sigma$ and the Taylor formula for $\left(\frac{\partial}{\partial x}\right)^\gamma \sigma$ ($|\gamma| \leq k$) to deduce that

$$\left(\frac{\partial}{\partial x}\right)^\gamma \sigma(x) = o(|x-y|^{k-|\gamma|}) \qquad \text{for } |\gamma| \leq k.$$

9. Let $u \in D'(\mathbf{R}^n)$ be a distribution of order $\leq k$. Show that $u[\sigma] = 0$ for every function $\sigma \in C_0^k(\mathbf{R}^n)$ such that $\left(\frac{\partial}{\partial x}\right)^\alpha \sigma(x) = 0$ for $|\alpha| \leq k$ and $x \in \operatorname{supp} u$.

10. Let $u \in S'(\mathbf{R}^n)$. Show that there exist $k \in \mathbf{N}_0$ and functions $h_\alpha \in L^\infty$ such that

$$u = \sum_{|\alpha| \leq k+n} (1 + \|x\|^2)^k \left(\frac{\partial}{\partial x}\right)^\alpha h_\alpha.$$

HINT. If $u \in S'(\mathbf{R}^n)$ then there exist $k \in \mathbf{N}_0$ and $C < \infty$ such that

$$|u[\varphi]| \le C \sum_{|\alpha|\le k} \sup \left|(1+\|x\|^2)^k \left(\frac{\partial}{\partial x}\right)^\alpha \varphi(x)\right| \qquad \text{for } \varphi \in C_0^\infty(\mathbf{R}^n).$$

Denote $\varphi_k(x) = \left(1+\|x\|^2\right)^k \varphi(x)$ for $x \in \mathbf{R}^n$. Observe that there exists a constant $C_1 = C_1(\alpha)$ such that

$$\left(1+\|x\|^2\right)^k \left|\left(\frac{\partial}{\partial x}\right)^\alpha \varphi(x)\right| \le C_1 \sum_{\beta\le\alpha} \left|\left(\frac{\partial}{\partial x}\right)^\beta \varphi_k(x)\right|$$

and hence with some constant C_2 we have

$$|u[\varphi]| \le C_2 \sum_{|\alpha|\le k} \sup \left|\left(\frac{\partial}{\partial x}\right)^\alpha \varphi_k\right| \le C_2 \sum_{|\alpha|\le k+n} \left\|\left(\frac{\partial}{\partial x}\right)^\alpha \varphi_k\right\|_{L_1}.$$

Apply the Hahn–Banach theorem and the Riesz theorem. (See the proof of Proposition 8.4).

11. Let $k \in \mathbf{N}_0$, $\beta \in \mathbf{N}_0^n$ and let g be a continuous function. Show that there exist polynomials p_γ such that

$$\left(1+\|x\|^2\right)^k \left(\frac{\partial}{\partial x}\right)^\beta g = \sum_{\gamma\le\beta} \left(\frac{\partial}{\partial x}\right)^\gamma (p_\gamma(x) g(x)).$$

12. Let $u \in S'(\mathbf{R}^n)$. Show that there exist a $C^0(\mathbf{R}^n)$-function f slowly increasing at infinity and $\alpha \in \mathbf{N}_0^n$ such that $u = \left(\frac{\partial}{\partial x}\right)^\alpha f$.

HINT. 1) Show that there exist $k \in \mathbf{N}_0$ and functions g_β, $|\beta| \le k+2n$ continuous, slowly increasing at infinity such that

$$u = \sum_{|\beta|\le k+2n} \left(1+\|x\|^2\right)^k \left(\frac{\partial}{\partial x}\right)^\beta g_\beta. \tag{10}$$

To this aim apply Exercise 10 and consider the functions

$$g_\alpha(x) = \int_0^{x_1} \cdots \int_0^{x_n} h_\alpha(y_1,\dots,y_n)\, dy_1 \dots dy_n \qquad \text{for } |\alpha| \le k+n.$$

2) Using Exercise 11 deduces from (10) that there exist continuous functions f_γ slowly increasing at infinity such that

$$u = \sum_{|\gamma|\le m} \left(\frac{\partial}{\partial x}\right)^\gamma f_\gamma.$$

3) Denoting $J_j f_\gamma = \int_0^{x_j} f_\gamma(x_1, \ldots, x_{j-1}, t, x_{j+1}, \ldots, x_n)\, dt \quad (j = 1, \ldots, n)$ put $f = \sum_{|\gamma|\le m} J_1^{m-\gamma_1} \cdots J_n^{m-\gamma_n} f_\gamma$ and prove that $u = \left(\frac{\partial}{\partial x}\right)^{\mathbf{m}} f$ where $\mathbf{m} = (m, \ldots, m)$.

13. Show that the transformation $\frac{\partial}{\partial x_k}\colon S' \to S'$ is surjective.

14. Prove that if the polynomial Q has zeros in $\mathbf{R}^n$ then the equation $Qv = 0$ has non-zero distributional solutions.

15. Let $g \in S'(\mathbf{R}^1)$. Show that there exist infinitely many distributions $v \in S'(\mathbf{R}^1)$ satisfying the equation $xv = g$. Any two of them differ by a multiple of the Dirac distribution δ.

16. Show that the general solution of the equation $(x_1^2 + \cdots + x_n^2)u = 0$ is the distribution $u = \sum_{|\alpha|\le 1} a_\alpha \left(\frac{\partial}{\partial x}\right)^\alpha \delta$ where a_α ($|\alpha| \le 1$, $\alpha \in \mathbf{N}_0^n$) are any constants and $\delta \in D'(\mathbf{R}^n)$ is the Dirac distribution.

17. Show that for every $\mathring{x} \in \mathbf{R}$, $r > 0$, $p \in \mathbf{N}$ there exists a function $f \in C^\infty(\mathbf{R})$ such that

$$f^{(q)}(\mathring{x}) = \begin{cases} r & \text{for } q = p \\ 0 & \text{for } q < p, \end{cases} \qquad \text{and} \quad \left|f^{(q)}(x)\right| < 1 \quad \text{for} \quad |x - \mathring{x}| < 1,\ q < p,$$

where $f^{(q)} = \left(\frac{d}{dx}\right)^q f$.

HINT. Let $g \in C_0^\infty((\mathring{x}-1, \mathring{x}+1))$ be a non-negative function such that $\int g(x)\, dx < 1$, $g(\mathring{x}) = r$. Define f by $f^{(p)}(x) = g(x)$, $f^{(q)}(\mathring{x}) = 0$ for $q < p$.

18. Let $u[\varphi] = \sum_{j=2}^{\infty} \varphi^{(j)}\left(\frac{1}{j}\right)$ for $\varphi \in C_0^\infty((0,1))$. Show that $u \in D'((0,1))$ and that u has infinite order.

OUTLINE OF THE PROOF. Suppose conversely that the distribution u is of order $\le m$ ($m \in \mathbf{N}_0$). Take $\mathring{x} = \frac{1}{m+1}$, $0 < \eta < \frac{1}{m+1} - \frac{1}{m+2}$, $K = [\mathring{x} - \eta, \mathring{x} + \eta]$ and let C be a constant such that $|u[\varphi]| \le C|||\varphi|||_m$ for $\varphi \in C_K^\infty$. Choose a function $\chi \in C_0^\infty((-1,1))$, $\chi = 1$ for $|x| \le \frac{1}{2}$. Let f_r be a function from Exercise 17 corresponding to $p = m+1$ and $\mathring{x} = \frac{1}{m+1}$. Take $\psi_r(x) = f_r(x)\chi\left(\frac{x-\mathring{x}}{\eta}\right)$ for $x \in \mathbf{R}$. Observe that $\psi_r \in C_K^\infty$, $u[\psi_r] = r$ and that $|u[\psi_r]| \le C|||\psi_r|||_m \le CM/\eta^n$ where M is a constant depending only on the function χ, and independent of r. Taking $r > CM/\eta^n$ we get a contradiction.

19. Give an example of a distribution of finite order on an open proper subset of $\mathbf{R}$, which is not extendible.

20. Let A be a bounded open cube. Let $H_k(A)$, $k = 1, \ldots, n$, ($H^\gamma(A), \gamma \in \mathbf{N}_n^0$ resp.) be the image of $C_0^\infty(A)$ under the differentiation $\frac{\partial}{\partial x_k}$ ($\left(\frac{\partial}{\partial x}\right)^\gamma$ resp.). Define operations: $J_k\colon H_k(A) \ni \psi \mapsto \varrho = J_k(\psi) \in C_0^\infty(A)$ where $k = 1, \ldots, n$, $\varrho(x) =$

$\int_{-\infty}^{x_k} \psi(x_1, \dots, x_{k-1}, t, x_{k+1}, \dots, x_n)\,dt$ for $x \in E^n$, and $J^\gamma = J_n^{\gamma_n} \dots J_1^{\gamma_1}\colon H^\gamma(A) \to C_0^\infty(A)$ where $J_k^0 =$ identity, $J_k^1 = J_k$, $J_k^p = J_k J_k^{p-1}$ for $p = 2, 3, \dots$ Show that

$$\varphi = J^\gamma \psi \in C_0^\infty(A), \qquad \left(\frac{\partial}{\partial x}\right)^\gamma J^\gamma \psi = \psi \qquad \text{for } \psi \in H^\gamma(A)$$

and that for any $m \in \mathbf{N}_0$, $\tilde{\gamma} = \mathbf{m} + \mathbf{1} \in \mathbf{N}_0^n$ and any $\psi \in H^{\tilde{\gamma}}(A)$ there exists a constant $C < \infty$ such that $|||\varphi|||_m \le C\,\|\psi\|_{L^1}$ where $\varphi = J^{\tilde{\gamma}}\psi$.

21. Prove Theorem 6.

HINT. Observe first that there exist a bounded open cube $A \supset \overline{\Omega}$ and constants $C > 0$, $m \in \mathbf{N}$ such that $|u[\varphi]| \le C|||\varphi|||_m$ for $\varphi \in C_0^\infty(A)$. Adopt the notation of Exercise 20 and consider the linear functional $w[\psi] = u[J^{\tilde{\gamma}}\psi]$ for $\psi \in H^{\tilde{\gamma}}(A)$. Show that there exists a constant $C^* > 0$ such that $|w[\psi]| \le C^*\,\|\psi\|_{L^1}$ for $\psi \in H^{\tilde{\gamma}}(A) \subset L^1(A)$. Apply the Hahn–Banach theorem and by the familiar theorem on the conjugate space to L^1 get $u = (-1)^{(m+1)n}\left(\frac{\partial}{\partial x}\right)^{\tilde{\gamma}} g$ on A where g is a bounded measurable function on A. Let Ω_ε be any neighbourhood of $\overline{\Omega}$, $\psi \in C_0^\infty(\Omega_\varepsilon)$, $\psi = 1$ for $x \in \overline{\Omega}$. Take a continuous function F such that $\frac{\partial}{\partial x} F = g$. Show that $G = (-1)^{(m+1)n} F\psi$ is the required function corresponding to the multi-index $\alpha = \tilde{\gamma} + \mathbf{1}$.

22. Let $u, v \in D'(\mathbf{R}^n)$, $A = \operatorname{supp} u$, $B = \operatorname{supp} v$. Suppose that for every compact set $H \subset \mathbf{R}^n$ the set

$$\widetilde{H} = \{(x, y)\colon x \in A,\ y \in B,\ x + y \in H\} \tag{11}$$

is bounded.

Show that the convolution $u * v$ exists and that the condition (11) is satisfied if one of the sets A, B is bounded (cf. Example 3).

Chapter II

MELLIN DISTRIBUTIONS AND THE MELLIN TRANSFORMATION

§4. THE FOURIER AND THE FOURIER–MELLIN TRANSFORMATIONS

1. The Fourier Transformation in S'

The *Fourier transform* $\mathcal{F}\sigma$ of a function $\sigma \in S(\mathbf{R}^n)$ is defined by

$$\mathcal{F}\sigma(\xi) = (2\pi)^{-\frac{n}{2}} \int_{\mathbf{R}^n} e^{-ix\xi}\sigma(x)\,dx \qquad \text{for } \xi \in \mathbf{R}^n.$$

The transformation $\mathcal{F}$ is an isomorphism of S onto S with the *inverse*

$$\mathcal{F}^{-1}\sigma(x) = (2\pi)^{-\frac{n}{2}} \int_{\mathbf{R}^n} e^{i\xi x}\sigma(\xi)\,d\xi \qquad \text{for } x \in \mathbf{R}^n.$$

Transformations $\mathcal{F}$ and $\mathcal{F}^{-1}$ preserve the L^2-norm of functions $\sigma \in S$, i.e. for every $\sigma \in S$ the following equalities, known as the *Parseval equalities* hold:

$$\|\mathcal{F}\sigma\|_{L^2} = \|\sigma\|_{L^2} = \|\mathcal{F}^{-1}\sigma\|_{L^2}.$$

By means of the relations

$$\begin{aligned} \mathcal{F}u[\sigma] &\overset{\text{def}}{=} u[\mathcal{F}\sigma] \\ \mathcal{F}^{-1}u[\sigma] &\overset{\text{def}}{=} u[\mathcal{F}^{-1}\sigma] \end{aligned} \qquad \text{for } u \in S',\ \sigma \in S$$

the Fourier transformation and the inverse Fourier transformation extend to isomorphism of S' onto S'. Below we recall basic theorems and spaces occurring in the theory of the Fourier transformation.

Definition 1. By $O_M(\mathbf{R}^n)$ we denote the vector space of the functions $m \in C^\infty(\mathbf{R}^n)$ such that for every $\beta \in \mathbf{N}_0^n$ there exist constants $C_\beta < \infty$ and $s_\beta > 0$ fulfilling the estimation

$$\left|\left(\frac{\partial}{\partial x}\right)^\beta m(x)\right| \leq C_\beta \left(1 + \|x\|^2\right)^{s_\beta} \qquad \text{for } x \in \mathbf{R}^n.$$

The elements of O_M are called smooth *functions slowly increasing at infinity.*

O_M is the class of *multipliers in* $S(\mathbf{R}^n)$ and consequently in $S'(\mathbf{R}^n)$. More precisely $m \in O_M$ if and only if the mapping $S \ni \psi \mapsto m \cdot \psi \in S$ is continuous.

Denote by O'_C the image of O_M under $\mathcal{F}$. Clearly $O'_C \subset S'$.

Now we recall the following *exchange formulas for the Fourier transformation* (cf. [Tr] or [Schw]).

Theorem 1. *If* $v \in S'$, $w \in O_M$, $u \in O'_C$ *then*

$$\mathcal{F}(wv) = (2\pi)^{-\frac{n}{2}}(\mathcal{F}w * \mathcal{F}v), \qquad \mathcal{F}(u * v) = (2\pi)^{\frac{n}{2}}\mathcal{F}u \cdot \mathcal{F}v.$$

The same equalities hold with $\mathcal{F}$ *replaced by* $\mathcal{F}^{-1}$. *If* $v \in S'$, $\sigma \in S$ *then* $\psi = v * \sigma \in O_M$ *and* $\psi(\xi) = v\left[\sigma(\xi - x)\right]$ *for* $\xi \in \mathbf{R}^n$. *Consequently* $v \cdot \sigma \in O'_C$.

The scale of the spaces $O_M^s \subset O_M$ introduced below will play an important role in Section 14.

Definition 2. Let $s \in \mathbf{R}$. We denote by O_M^s the class of functions $m \in C^\infty(\mathbf{R}^n)$ such that for every $\delta \in \mathbf{N}_0^n$ there exists a constant $C_\delta = C_\delta(m) < \infty$ such that

$$\left|\left(\frac{\partial}{\partial x}\right)^\delta m(x)\right| \leq C_\delta \langle x\rangle^s.$$

O_M^s is equipped with a natural topology of uniform convergence given by the smallest constants C_δ.

Theorem 2. (see e.g. [Sz1]). *The Fourier transform* $\mathcal{F}u$ *of a distribution* $u \in E'(\mathbf{R}^n)$ *is a function of class* O_M *given by*

$$(\mathcal{F}u)(\xi) = (2\pi)^{-\frac{n}{2}} u\left[e^{-ix\xi}\right] \qquad \text{for } \xi \in \mathbf{R}^n.$$

Similarly $\mathcal{F}^{-1}u \in O_M$ *and*

$$\left(\mathcal{F}^{-1}u\right)(x) = (2\pi)^{-\frac{n}{2}} u\left[e^{ix\xi}\right] \qquad \text{for } x \in \mathbf{R}^n.$$

We end this subsection with an S'-version of the *Schwartz kernel theorem*. We start with definitions:

Denote by $S'(\mathbf{R}^n, S'(\mathbf{R}^m))$ $n, m \in \mathbf{N}$ the space of $S'(\mathbf{R}^m)$-valued tempered distributions on $\mathbf{R}^n$ i.e. $T \in S'(\mathbf{R}^n, S'(\mathbf{R}^m))$ if for every $\varphi \in S(\mathbf{R}^n)$, $T\varphi \in S'(\mathbf{R}^m)$ and the assignment

$$S(\mathbf{R}^n) \ni \varphi \mapsto T\varphi \in S'(\mathbf{R}^m)$$

is linear and continuous. Then it follows from Theorem 2.3 that there exist constants $C < \infty$ and $k \in \mathbf{N}_0$ such that

$$\big|(T\varphi)[\psi]\big| \le C q_k(\varphi) q_k(\psi) \qquad \text{for } \varphi \in S(\mathbf{R}^n), \psi \in S(\mathbf{R}^m), \tag{1}$$

where q_k is the seminorm given by (3.3) in $\mathbf{R}^n$ and $\mathbf{R}^m$.

Theorem 3. *The spaces* $S'(\mathbf{R}^n, S'(\mathbf{R}^m))$ *and* $S'\left(\mathbf{R}^{n+m}\right)$ *are isomorphic in a canonical way. The isomorphism is given by the assignment*

$$S'\left(\mathbf{R}^{n+m}\right) \ni \widetilde{T} \mapsto T \in S'(\mathbf{R}^n, S'(\mathbf{R}^m)),$$

where

$$(T\varphi)[\psi] = \widetilde{T}[\varphi \otimes \psi] \qquad \text{for } \varphi \in S(\mathbf{R}^n), \psi \in S(\mathbf{R}^m) \tag{2}$$

and $\varphi \otimes \psi$ *denotes the function* $\mathbf{R}^n \times \mathbf{R}^m \ni (x, y) \mapsto \varphi(x) \cdot \psi(y)$.

Proof: Let $T \in S'(\mathbf{R}^n, S'(\mathbf{R}^m))$. We shall find a distribution $\widetilde{T} \in S'\left(\mathbf{R}^{n+m}\right)$ such that (2) holds. Observe that (by Exercise 1 and Hahn–Banach theorem) (1) holds for those functions φ and ψ for which the norm q_k is finite. Thus by Exercise 2 we have for some constant $0 < C < \infty$

$$\begin{aligned} g(\xi, \eta) &\stackrel{\text{def}}{=} \left| \left(T\left(\frac{e^{ix\xi}}{(1 + \|x\|^2)^{k+1}} \right) \right) \left[\frac{e^{iy\eta}}{(1 + \|y\|^2)^{k+1}} \right] \right| \\ &\le C(1 + \|\xi\|)^k (1 + \|\eta\|)^k \qquad \text{for } (\xi, \eta) \in \mathbf{R}^{n+m}. \end{aligned} \tag{3}$$

Let $\Phi \in S(\mathbf{R}^{n+m})$ and denote $\Psi(x, y) = (1 + \|x\|^2)^{k+1}(1 + \|y\|^2)^{k+1}\Phi(x, y)$ for $(x, y) \in \mathbf{R}^{n+m}$ (with k from (3)). $\widetilde{T}$ is defined explicitly as follows:

$$\widetilde{T}[\Phi] \stackrel{\text{def}}{=} (2\pi)^{-\frac{n+m}{2}} \int_{\mathbf{R}^{n+m}} g(\xi, \eta)(\mathcal{F}\Psi)(\xi, \eta)\, d\xi\, d\eta \tag{4}$$

where g is given by (3) and hence the integral in (4) makes sense and $\widetilde{T}$ is a tempered distribution. If $\Phi = \varphi \otimes \psi$ with $\varphi \in S(\mathbf{R}^n)$, $\psi \in S(\mathbf{R}^m)$ then by (4) we get

$$\widetilde{T}[\varphi \otimes \psi] = (2\pi)^{-\frac{n+m}{2}} \int_{\mathbf{R}^n} (\mathcal{F}((1+\|x\|^2)^k \varphi))(\xi) \times \\ \times \left\{ \int_{\mathbf{R}^m} g(\xi, \eta) \mathcal{F}((1+\|y\|^2)^k \psi(y))(\eta)\, d\eta \right\} d\xi,$$

and hence by Exercises 2–4 we get

$$\widetilde{T}[\varphi \otimes \psi] = (2\pi)^{-\frac{n}{2}} \int_{\mathbf{R}^n} (\mathcal{F}(1+\|x\|^2)^k \varphi))(\xi) \left(T_x \left(\frac{e^{ix\xi}}{(1+\|x\|^2)^k} \right) \right) [\psi]\, d\xi \\ = (T\varphi)[\psi].$$

2. The Fourier–Mellin Transformation in the Space of Mellin Distributions with Support in $\mathbf{R}^n_+$

Below we present basic properties of the Fourier transformation in logarithmic variables, which we call the Fourier–Mellin transformation.

Denote by $\mu\colon \mathbf{R}^n \to \mathbf{R}^n_+$ the diffeomorphism (see Exercise 5)

$$\mu(y) = e^{-y} \stackrel{\text{def}}{=} \left(e^{-y_1}, \dots, e^{-y_n}\right).$$

We define the space of *Mellin distributions on* $\mathbf{R}^n_+$ for every $\alpha \in \mathbf{R}^n$ as the dual of the space

$$\mathfrak{M}_\alpha = \mathfrak{M}_\alpha(\mathbf{R}^n_+) = \left\{\sigma \in C^\infty(\mathbf{R}^n_+) \colon \left(x^{\alpha+1}\sigma\right) \circ \mu \in S(\mathbf{R}^n)\right\}$$

with a natural topology in $\mathfrak{M}_\alpha$ induced from $S(\mathbf{R}^n)$. Note that $\mathfrak{M}'_\alpha(\mathbf{R}^n_+)$ is a subspace of $D'(\mathbf{R}^n_+)$ (cf. Exercise 6). Hence and from the definition of substitution in a distribution (cf. Section 3.5) follows

Remark 1. $u \in \mathfrak{M}'_\alpha(\mathbf{R}^n_+)$ if and only if $e^{\alpha y}(u \circ \mu) \in S'(\mathbf{R}^n)$.

Definition 3. Let $\alpha \in \mathbf{R}^n$. Define

$$\mathcal{M}^{\text{tr}}_\alpha\colon S \to \mathfrak{M}_\alpha, \qquad \left(\mathcal{M}^{\text{tr}}_\alpha\right)^{-1} \colon \mathfrak{M}_\alpha \to S,$$

$$\left(\mathcal{M}^{\text{tr}}_\alpha \psi\right)(x) = \int_{\mathbf{R}^n} x^{-\alpha - i\beta - 1} \psi(\beta)\, d\beta \qquad \text{for } \psi \in S,$$

$$\left(\left(\mathcal{M}^{\text{tr}}_\alpha\right)^{-1} \sigma\right)(\beta) = (2\pi)^n \int_{\mathbf{R}^n_+} x^{i\beta + \alpha} \sigma(x)\, dx \qquad \text{for } \sigma \in \mathfrak{M}_\alpha.$$

Proposition 1. *The transformations* $\mathcal{M}_\alpha^{\mathrm{tr}}$ *and* $(\mathcal{M}_\alpha^{\mathrm{tr}})^{-1}$ *are continuous and mutually inverse.*

The proof follows from the properties of the Fourier transformation on S in view of the relations

$$\begin{aligned} \mathcal{M}_\alpha^{\mathrm{tr}}\psi &= (2\pi)^{n/2}x^{-\alpha-1}(\mathcal{F}^{-1}\psi)\circ\mu^{-1} && \text{for } \psi\in S,\\ (\mathcal{M}_\alpha^{\mathrm{tr}})^{-1}\sigma &= (2\pi)^{-n/2}\mathcal{F}((x^{\alpha+1}\sigma)\circ\mu) && \text{for } \sigma\in\mathfrak{M}_\alpha. \end{aligned} \tag{5}$$

Definition 4. Let $u\in\mathfrak{M}'_\alpha$. The $\mathcal{M}_\alpha$ *Mellin transform* of u is the distribution $\mathcal{M}_\alpha u\in S'$ defined by the duality

$$\mathcal{M}_\alpha u[\psi] = u[\mathcal{M}_\alpha^{\mathrm{tr}}\psi] \qquad \text{for } \psi\in S,$$

or equivalently (see (5))

$$\mathcal{M}_\alpha u = (2\pi)^{n/2}\mathcal{F}^{-1}(e^{\alpha y}(u\circ\mu)). \tag{6}$$

If $T\in S'$ then the duality

$$(\mathcal{M}_\alpha)^{-1}T[\sigma] = T[(\mathcal{M}_\alpha^{\mathrm{tr}})^{-1}\sigma] \qquad \text{for } \sigma\in\mathfrak{M}_\alpha$$

defines a Mellin distribution $(\mathcal{M}_\alpha)^{-1}T\in\mathfrak{M}'_\alpha$.

Proposition 2. $\mathcal{M}_\alpha\colon \mathfrak{M}'_\alpha\to S'$ *is an isomorphism with inverse* $(\mathcal{M}_\alpha)^{-1}$.

The proof is obvious in view of Proposition 1.

Proposition 3. *If* $u\in\mathfrak{M}'_\alpha$ *then* $x_j\frac{\partial}{\partial x_j}u\in\mathfrak{M}'_\alpha$ *and*

$$\mathcal{M}_\alpha\Big(x_j\frac{\partial}{\partial x_j}u\Big) = (\alpha_j+i\beta_j)\mathcal{M}_\alpha u \qquad (j=1,\dots,n).$$

Proof: In the proof of the first part we apply Remark 1 and the identity

$$\Big(x\frac{\partial}{\partial x}^{\nu}u\Big)\circ\mu = (-1)^{|\nu|}\Big(\frac{\partial}{\partial y}\Big)^{\nu}(u\circ\mu) \qquad \text{for } u\in D'(\mathbf{R}^n_+) \text{ and } \nu\in\mathbf{N}_0^n. \tag{7}$$

The second part follows by Definition 4 and 3.

Similarly to the case of the Fourier transformation we have the following *exchange formula* for the $\mathcal{M}_\alpha$*Mellin transformation*

Proposition 4. *Let* $\psi\in\mathfrak{M}_{-1}$. *If* $u\in\mathfrak{M}'_\alpha(\mathbf{R}^n_+)$ *then*

$$(2\pi)^n\mathcal{M}_\alpha(\psi u) = \mathcal{M}_0\psi * \mathcal{M}_\alpha u,$$

where the right hand side is the convolution of a function $\mathcal{M}_0\psi\in S$ *with a tempered distribution* $\mathcal{M}_\alpha u$ *(i.e. can be written as* $\mathcal{M}_\alpha u[\mathcal{M}_0\psi(\beta-\gamma)]$*).*

Proof: Since $\psi \in \mathfrak{M}_{-1}$ it follows from the definition of $\mathfrak{M}_{-1}$ that $\psi \circ \mu \in S(\mathbf{R}^n)$ and by Remark 1 and (6) we get $\mathcal{M}_0\psi \in S$. As it was noted the assumption $u \in \mathfrak{M}'_\alpha$ means that $e^{\alpha y}(u \circ \mu) \in S'$. Hence

$$e^{\alpha u}((\psi u) \circ \mu) = \psi \circ \mu \cdot e^{\alpha y}(u \circ \mu) \in S'$$

since $\psi \circ \mu \in S$. Therefore $\psi u \in \mathfrak{M}'_\alpha$. From the definition of the transformation $\mathcal{M}_\alpha$

$$\mathcal{M}_\alpha(\psi u) = (2\pi)^{n/2}\mathcal{F}^{-1}(e^{\alpha y}((\psi u) \circ \mu)) = (2\pi)^{n/2}\mathcal{F}^{-1}(\psi \circ \mu \cdot e^{\alpha y}(u \circ \mu))$$

and by the exchange formula for the Fourier transformation we get

$$\mathcal{M}_\alpha(\psi u) = \mathcal{F}^{-1}(\psi \circ \mu) * \mathcal{F}^{-1}(e^{\alpha y}(u \circ \mu)) = \frac{1}{(2\pi)^n}\mathcal{M}_0\psi * \mathcal{M}_\alpha u,$$

which ends the proof.

Exercises

1. Let $k, p \in \mathbf{N}_0$. Denote by $S_k^p = S_k^p(\mathbf{R}^n)$ the space of functions $\sigma \in C^p(\mathbf{R}^n)$ such that for every $\varepsilon > 0$ there exists an $R > 0$ fulfilling the inequality

$$(1 + \|x\|^2)^k \left|\left(\frac{\partial}{\partial x}\right)^\alpha \sigma(x)\right| < \varepsilon$$

for $\|x\| > R$, $|\alpha| \leq p$. Define a norm q_k^p in S_k^p putting

$$q_k^p(\sigma) = \sum_{|\alpha| \leq p} \sup \left|(1 + \|x\|^2)^k \left(\frac{\partial}{\partial x}\right)^\alpha \sigma(x)\right| \qquad \text{for } \sigma \in S_k^p.$$

Show the following inclusions between the topological vector spaces D, S, S_k^p, S_l^q and their duals

$$\begin{matrix} D \subset S \subset S_l^q \subset S_k^p \\ (S_k^p)' \subset (S_l^q)' \subset S' \subset D' \end{matrix} \qquad \text{if } p, k, q, l \in \mathbf{N}_0, \quad p \leq q, \quad k \leq l.$$

Observe that $u \in S'$ iff there exists a $k \in \mathbf{N}_0$ such that $u \in (S^k)'$, where $S^k = S_k^k$. (See [Sz1], Subsection 18.2.)

2. Show that for every $\xi \in \mathbf{R}^n$ and $k \in \mathbf{N}$ the function $\mathbf{R}^n \ni x \mapsto e^{ix\xi}(1 + \|x\|^2)^{-k}$ belongs to the space S^{k-1} (see Exercise 1).

3. Let $T \in S'(\mathbf{R}^n, S'(\mathbf{R}^m))$. Show that there exists $k \in \mathbf{N}_0$ such that the assignment

$$S^k(\mathbf{R}^n) \ni \varphi \mapsto T\varphi \in (S^k)'(\mathbf{R}^m)$$

is linear and continuous.

4. Let $S^k = S^k_k$, $(S^k)' = (S^k_k)'$ (see Exercise 1). Let λ be a function defined on $\mathbf{R}^{n+m}$ satisfying the conditions:

(i) $\lambda(\eta, \cdot) \in S^k(\mathbf{R}^m)$ for $\eta \in \mathbf{R}^n$,

(ii) for any $\alpha \in \mathbf{N}_0^m$, $|\alpha| \leq k$, the function

$$\mathbf{R}^n \times \mathbf{R}^m \ni (\eta, y) \mapsto \frac{(1+\|y\|^2)^k}{(1+\|\eta\|)^{|\alpha|}} \left(\frac{\partial}{\partial y}\right)^\alpha \lambda(\eta, y)$$

is continuous and bounded in $\mathbf{R}^{n+m}$,

(iii) For any $\varepsilon > 0$, compact set $K \subset \mathbf{R}^n$ and $\alpha \in \mathbf{N}_0^m$ there exists $0 < \delta = \delta(\varepsilon, K, \alpha)$ such that

$$\sup_{y \in \mathbf{R}^m} (1+\|y\|^2)^k \left| \left(\frac{\partial}{\partial y}\right)^\alpha \lambda(\eta, y) - \left(\frac{\partial}{\partial y}\right)^\alpha \lambda(\tilde{\eta}, y) \right| < \varepsilon$$

$$\text{for } \eta, \tilde{\eta} \in K, \quad |\eta - \tilde{\eta}| < \delta.$$

Finally let $\sigma \in S(\mathbf{R}^n)$. Then if $u \in (S^k)'(\mathbf{R}^m)$

$$\int_{\mathbf{R}^n} u[\lambda(\eta, \cdot)]\sigma(\eta)\,d\eta = u\left[\int_{\mathbf{R}^n} \lambda(\eta, \cdot)\sigma(\eta)\,d\eta\right].$$

HINT. Assume first that $\sigma \in C_0^\infty(\mathbf{R}^n)$. Consider Riemann sums for the integral $J(y) = \int_{\mathbf{R}^n} \lambda(\eta, y)\sigma(\eta)\,d\eta$ and show that they converge to J in S^k (see Theorem 18.11 in [Sz1]). In the case of $\sigma \in S(\mathbf{R}^n)$ choose a sequence of functions $\chi_j \in C_0^\infty(\mathbf{R}^n)$ such that $\sigma \cdot \chi_j \to \sigma$ in $S(\mathbf{R}^n)$ and pass to the limit in the already proved formula for $\sigma\chi_j \in C_0^\infty(\mathbf{R}^n)$.

5. Prove that the transformation $\mu\colon \mathbf{R}^n \to \mathbf{R}^n_+$ defined in Section 2 is a diffeomorphism with an inverse $\mu^{-1}\colon \mathbf{R}^n_+ \to \mathbf{R}^n$, $\mu^{-1}(x) = -\ln x = (-\ln x_1, \dots, -\ln x_n)$ for $x = (x_1, \dots, x_n) \in \mathbf{R}^n_+$. Also prove that the operations of composition with μ and μ^{-1} are continuous

$$\circ\mu\colon D(\mathbf{R}^n_+) \to D(\mathbf{R}^n), \qquad \circ\mu^{-1}\colon D(\mathbf{R}^n) \to D(\mathbf{R}^n_+),$$

$$\circ\mu\colon D'(\mathbf{R}^n_+) \to D'(\mathbf{R}^n), \qquad \circ\mu^{-1}\colon D'(\mathbf{R}^n) \to D'(\mathbf{R}^n_+).$$

where for $u \in D'(\mathbf{R}^n_+)$ (cf. Section 3.5)

$$(u \circ \mu)[\varphi] = u[\varphi \circ \mu^{-1} \cdot x^{-1}] \qquad \text{for } \varphi \in D(\mathbf{R}^n),$$

and for $v \in D'(\mathbf{R}^n)$

$$(v \circ \mu^{-1})[\psi] = v[\psi \circ \mu \cdot \mathrm{e}^{-y}] \qquad \text{for } \psi \in D(\mathbf{R}^n_+)$$

(see also Exercise 5.9).

6. Prove that $C_0^\infty(\mathbf{R}^n_+)$ is dense in $\mathfrak{M}_\alpha$ for any $\alpha \in \mathbf{R}^n$.

7. Give an example of a function $\sigma \in \mathfrak{M}_\alpha(\mathbf{R}^n_+)$ with unbounded support.

8. Let $u \in \mathfrak{M}_0'(\mathbf{R}^n_+)$, $\psi \in \mathfrak{M}_{-\alpha-1}(\mathbf{R}^n_+)$. Show that $(2\pi)^n \mathcal{M}_\alpha(\psi u) = \mathcal{M}_\alpha \psi * \mathcal{M}_0 u$, $\mathcal{M}_\alpha \psi \in S(R^n)$, $\mathcal{M}_0 u \in S'(R^n)$.

§5. THE SPACES OF MELLIN DISTRIBUTIONS WITH SUPPORT IN A POLYINTERVAL

A natural setting for the local theory of the Mellin transformation are the spaces $M'_{(\omega)}((0,t])$ of Mellin distributions, defined below, whose definition is modelled upon that introduced by Zemanian [Z].

1. Spaces $M_a((0,t])$ and $M'_a((0,t])$

Let $a, b \in \mathbf{R}^n$. Recall that we write $a < b$ ($a \le b$ resp.) to denote that $a_j < b_j$ ($a_j \le b_j$ resp.) for $j = 1, \ldots, n$. Take an arbitrary $0 < t \in \mathbf{R}^n$. We denote by I the polyinterval

$$I = (0,t] = \{x \in \mathbf{R}^n \colon 0 < x \le t\}.$$

Recall (cf. 3.2) that $C^\infty(I)$ denotes the space of restrictions to I of smooth functions on $C^\infty(\mathbf{R}^n_+)$.

Definition 1. For $a \in \mathbf{R}^n$ we introduce the space

$$M_a = M_a(I) = \left\{\varphi \in C^\infty(I) \colon \sup_{x \in I} \left| x^{a+\alpha+1} \left(\frac{\partial}{\partial x}\right)^\alpha \varphi(x)\right| < \infty,\ \alpha \in \mathbf{N}_0^n\right\}$$

equipped with the convergence topology defined by a sequence of seminorms $\{\varrho_{a,\alpha}\}_{\alpha \in \mathbf{N}_0^n}$:

$$\varrho_{a,\alpha}(\varphi) = \sup_{x \in I} \left| x^{a+\alpha+1} \left(\frac{\partial}{\partial x}\right)^\alpha \varphi(x)\right|, \qquad \alpha \in \mathbf{N}_0^n.$$

As noted in Section 2 $\{\varrho_{a,\alpha}\}$ can be replaced by an increasing sequence of (semi)norms

$$q_k(\varphi) = \sum_{|\alpha|\leq k} \varrho_{a,\alpha}(\varphi),$$

without changing the topology in M_a.

Since the linear spaces spanned by the operators $x^\beta\left(\frac{\partial}{\partial x}\right)^\beta$ and $\left(x\frac{\partial}{\partial x}\right)^\beta$ coincide (see Exercise 1) it follows that Definition 1 may be replaced by the equivalent

Definition 1′. For $a \in \mathbf{R}^n$

$$M_a = \left\{\varphi \in C^\infty(I)\colon \sup_{x\in I}\left|x^{a+1}\left(x\frac{\partial}{\partial x}\right)^\alpha \varphi(x)\right| < \infty \text{ for } \alpha \in \mathbf{N}_0^n\right\}$$

with the topology defined by the sequence of seminorms[1]

$$\tilde{\varrho}_{a,\alpha}(\varphi) = \sup_{x\in I}\left|x^{a+1}\left(x\frac{\partial}{\partial x}\right)^\alpha \varphi(x)\right|, \qquad \alpha \in \mathbf{N}_0^n.$$

Note that the topology in $M_a(I)$ is defined in "inner" terms i.e. in terms of supremum on I, in contrast to the "outer" topology in terms of extensions, where the supremum is taken on larger sets (c.f. the definition of $D(A)$ in Subsection 3.2). To prove that those topologies coincide we shall construct a linear *continuous* extension mapping

$$\mathcal{E}\colon M_a((0,t]) \to M_a((0,\tilde{t}\,]) \qquad \text{for any } \tilde{t} > t.$$

We begin with a lemma which is a *parameter version* of the *Seeley extension theorem* from a half-line to the real line (see [Mel] and Exercise 14).

Let $a \in \mathbf{R}^n$, $x \in \mathbf{R}^n$, $t \in \mathbf{R}_+^n$ and write $a = (a_1, a')$, $x = (x_1, x')$, $t = (t_1, t')$ with $a' = (a_2, \ldots, a_n)$, $x' = (x_2, \ldots, x_n)$, $t' = (t_2, \ldots, t_n)$. Take $\varepsilon \in \mathbf{R}_+$ and denote

$$\widetilde{C}^\infty([0,\varepsilon)) = \Big\{\varphi \in C^\infty((0,\varepsilon))\colon \frac{d^j}{dx^j}\varphi \text{ extends continuously to } [0,\varepsilon)\ (j = 0,1,2,\ldots)\Big\},^{(2)}$$

$$\widetilde{C}^\infty([0,\varepsilon); M_{a'}((0,t'])) = \Big\{\varphi \in C^\infty((0,\varepsilon)\times(0,t'])\colon \sup_{(0,\varepsilon)\times(0,t']}\left|(x')^{a'+1}\left(x'\frac{\partial}{\partial x'}\right)^{\alpha'}\left(\frac{\partial}{\partial x_1}\right)^{\alpha_1}\varphi(x_1,x')\right| < \infty$$

$$\text{for } \alpha = (\alpha_1,\alpha') \in \mathbf{N}_0^n,\ \varphi(\,\cdot\,,x') \in \widetilde{C}^\infty([0,\varepsilon) \text{ for every } x' \in (0,t']\Big\}.$$

Analogously we define $C^\infty((-\varepsilon,\varepsilon); M_{a'}((0,t']))$ and $C^\infty_{(0,\frac{\varepsilon}{2}]}((0,\infty); M_{a'}((0,t']))$.

---

(1) See Exercise 2 for other examples of equivalent seminorms on M_a.

(2) In an analogous way we define $\tilde{C}^\infty((0,\varepsilon])$. The "tilde notation" is used exclusively in Lemma 1. It will result therefrom (see Exercise 4(ii)) that $\tilde{C}^\infty([0,\varepsilon)] = C^\infty([0,\varepsilon))$.

Lemma 1. *There exists a linear extension mapping*

$$\mathcal{E}^1 \colon \widetilde{C}^\infty([0,\varepsilon); M_{a'}((0,t'])) \to C^\infty((-\varepsilon,\varepsilon)); M_{a'}((0,t']))$$

such that for every $\alpha \in \mathbb{N}_0^n$ there exists a constant C_{α_1} such that

$$(1) \qquad \sup_{(-\varepsilon,\varepsilon)\times(0,t']} \left|(x')^{a'+1}\Big(x'\frac{\partial}{\partial x'}\Big)^{\alpha'}\Big(\frac{\partial}{\partial x_1}\Big)^{\alpha_1}(\mathcal{E}^1\varphi)(x_1,x')\right|$$
$$\leq C_{\alpha_1}\sum_{p=0}^{\alpha_1}\sup_{(0,\varepsilon)\times(0,t']}\left|(x')^{a'+1}\Big(x'\frac{\partial}{\partial x'}\Big)^{\alpha'}\Big(\frac{\partial}{\partial x_1}\Big)^{p}\varphi(x_1,x')\right|$$

for every $\varphi \in \widetilde{C}^\infty([0,\varepsilon); M_{a'}((0,t']))$.

Proof: Let $\chi \in C_0^\infty(\mathbb{R})$ be 1 in a neighbourhood of zero and $\chi(x_1) = 0$ for $|x_1| \geq \frac{\varepsilon}{2}$. Define

$$(2) \qquad (\widetilde{\mathcal{E}}^1\varphi)(x) = \sum_{l=0}^{\infty} a_l\varphi(3^l x_1, x')\chi(3^l x_1)$$

for $\varphi \in \widetilde{C}^\infty([0,\varepsilon); M_{a'}((0,t']))$, where $\{a_l\}$ is a sequence of real numbers. Observe that for each $x_1 > 0$ only finitely many terms on the right hand side of (2) are non zero and $(\widetilde{\mathcal{E}}^1\varphi)(x) = 0$ for $x_1 \geq \frac{\varepsilon}{2}$. Clearly

$$\widetilde{\mathcal{E}}^1 \colon \widetilde{C}^\infty([0,\varepsilon); M_{a'}((0,t'])) \to C^\infty_{(0,\frac{\varepsilon}{2}]}((0,\infty); M_{a'}((0,t']))$$

and this map is linear. We shall choose the sequence $\{a_l\}$ to satisfy

$$(3) \qquad \sum_{l=0}^{\infty} |a_l|\, 3^{lp} < \infty \qquad \text{for } p \in \mathbb{N}$$

and

$$(4) \qquad \sum_{l=0}^{\infty} a_l\, 3^{lp} = (-1)^p \qquad \text{for } p \in \mathbb{N}.$$

Assuming this for a moment we find by differentiating (2)

$$(5) \quad \Big(\frac{\partial}{\partial x_1}\Big)^p(\widetilde{\mathcal{E}}^1\varphi)(x_1,x') = \sum_{l=0}^{\infty} a_l\, 3^{lp}\left(\Big(\frac{\partial}{\partial s}\Big)^p(\chi(s)\varphi(s,x'))\right)_{s=3^l x_1} \qquad \text{for } p \in \mathbb{N}.$$

Thus from the properties of χ and from (3) we get the estimation

$$(6)\qquad \sup_{(0,\varepsilon)\times(0,t']}\left|(x')^{a'+1}\left(x'\frac{\partial}{\partial x'}\right)^{\alpha'}\left(\frac{\partial}{\partial x_1}\right)^{\alpha_1}(\widetilde{\mathcal{E}}^1\varphi)(x_1,x')\right|$$
$$\leq C_{\alpha_1}\sum_{p=0}^{\alpha_1}\sup_{(0,\varepsilon/2]\times(0,t']}\left|(x')^{a'+1}\left(x'\frac{\partial}{\partial x'}\right)^{\alpha'}\left(\frac{\partial}{\partial x_1}\right)^{p}\varphi(x_1,x')\right|,$$

with some constant $C_{\alpha_1}<\infty$, and on the other hand we find

$$\lim_{x_1\to 0+}\left(x'\frac{\partial}{\partial x'}\right)^{\alpha'}\left(\frac{\partial}{\partial x_1}\right)^{\alpha_1}(\widetilde{\mathcal{E}}^1\varphi)(x_1,x')=\sum_{l=0}^{\infty}a_l 3^{l\alpha_1}\left(x'\frac{\partial}{\partial x'}\right)^{\alpha'}\left(\left(\frac{\partial}{\partial x_1}\right)^{\alpha_1}\varphi\right)(0,x'),$$

where

$$\left(x'\frac{\partial}{\partial x'}\right)^{\alpha'}\left(\left(\frac{\partial}{\partial x_1}\right)^{\alpha_1}\varphi\right)(0,x')=\lim_{x_1\to 0+}\left(x'\frac{\partial}{\partial x'}\right)^{\alpha'}\left(\frac{\partial}{\partial x_1}\right)^{\alpha_1}\varphi(x_1,x').$$

Now by (4)

$$\lim_{x_1\to 0+}\left(x'\frac{\partial}{\partial x'}\right)^{\alpha'}\left(\left(\frac{\partial}{\partial x_1}\right)^{\alpha_1}(\widetilde{\mathcal{E}}^1\varphi)\right)(x_1,x')=(-1)^{\alpha_1}\left(x'\frac{\partial}{\partial x'}\right)^{\alpha'}\left(\left(\frac{\partial}{\partial x_1}\right)^{\alpha_1}\varphi\right)(0,x')$$

for every $\alpha\in\mathbb{N}_0^n$.

Put

$$\left(\left(x'\frac{\partial}{\partial x'}\right)^{\alpha'}\left(\frac{\partial}{\partial x_1}\right)^{\alpha_1}(\widetilde{\mathcal{E}}^1\varphi)\right)(0,x')=(-1)^{\alpha_1}\left(x'\frac{\partial}{\partial x'}\right)^{\alpha'}\left(\left(\frac{\partial}{\partial x_1}\right)^{\alpha_1}\varphi\right)(0,x')$$

for $\alpha\in\mathbb{N}_0^n$. The desired extension is obtained by taking

$$(\mathcal{E}^1\varphi)(x_1,x')=\begin{cases}\varphi(x_1,x') & \text{for } 0<x_1<\varepsilon,\ 0<x'\leq t'\\ (\widetilde{\mathcal{E}}^1\varphi)(-x_1,x') & \text{for } -\varepsilon<x_1\leq 0,\ 0<x'\leq t'\end{cases}$$

and (1) follows from (6) and the assumption that $\varphi\in\widetilde{C}^\infty([0,\varepsilon);M_{a'}((0,t']))$.

Finally to find the sequence $\{a_l\}$ satisfying (3) and (4) we note that the function

$$h(z)=\cos\left(\pi\left(\frac{3^z-1}{2}\right)\right)$$

is entire and $h(p)=(-1)^p$ for $p\in\mathbb{N}$. We take a_l $(l=0,1,\ldots)$ as the coefficients of the power series expansion of the function $\mathbb{C}\ni w\mapsto\cos\pi\left(\frac{w-1}{2}\right)$ i.e. $h(z)=\sum_{l=0}^{\infty}a_l(3^z)^l$.

Remark 1. $(\mathcal{E}^1\varphi)(x_1, x') = 0$ for $x_1 \leq -\frac{\varepsilon}{2}$ since $(\widetilde{\mathcal{E}}^1\varphi)(x) = 0$ for $x_1 \geq \frac{\varepsilon}{2}$.

Theorem 1. *Let $a \in \mathbf{R}^n$, $0 < t < \tilde{t} \in \mathbf{R}^n_+$. Then to every $0 < \varepsilon < \tilde{t} - t$, $\varepsilon < t$ there exists a linear extension mapping*

$$\mathcal{E}_\varepsilon \colon M_a((0,t]) \to M_a((0,\tilde{t}])$$

continuous in the respective topologies and such that for every $\varphi \in M_a((0,t])$ $(\mathcal{E}_\varepsilon\varphi)(x) = 0$ if $t_j + \varepsilon_j < x_j \leq \tilde{t}_j$ for some $1 \leq j \leq n$.

Proof: Let $\varphi \in M_a((0,t])$, choose $0 < \varepsilon < t$; $\varepsilon < \tilde{t} - t$ and observe that the function

$$(0, \varepsilon_1) \times (0, t'] \ni (x_1, x') \mapsto \tilde{\varphi}(x_1, x') = \varphi(t_1 - x_1, x')$$

belongs to $\widetilde{C}^\infty([0, \varepsilon_1); M_{a'}((0, t']))$. Thus by Lemma 1 and Remark 1

$$\mathcal{E}^1\tilde{\varphi} \in C^\infty((-\varepsilon_1, \varepsilon_1); M_{a'}((0, t'])), \qquad \mathcal{E}^1\tilde{\varphi}(x_1, x') = 0 \quad \text{for} \quad x_1 < -\frac{\varepsilon_1}{2}.$$

$$(\mathcal{E}^1\tilde{\varphi})(t_1 - x_1, x') = \tilde{\varphi}(t_1 - x_1, x') \qquad \text{for} \quad t_1 - \varepsilon_1 < x_1 < t_1, \ 0 < x' \leq t'.$$

Since $\tilde{\varphi}(t_1 - x_1, x') = \varphi(x_1, x')$ for $t_1 - \varepsilon_1 < x_1 < t_1$, $0 < x' < t'$ we get

$$\varphi(x_1, x') = (\mathcal{E}^1\tilde{\varphi})(t_1 - x_1, x') \qquad \text{for} \quad t_1 - \varepsilon_1 < x_1 < t_1, \quad 0 < x' \leq t',$$

which yields the correctness of the following definition:

$$(\mathcal{E}_{\varepsilon_1}\varphi)(x_1, x') = \begin{cases} \varphi(x_1, x') & \text{for} \ \ 0 < x_1 < t_1, \ \ 0 < x' \leq t' \\ (\mathcal{E}^1\tilde{\varphi})(t_1 - x_1, x') & \text{for} \ \ t_1 - \varepsilon_1 < x_1 < t_1 + \varepsilon_1, \ \ 0 < x' \leq t'. \end{cases}$$

It is clear that $\mathcal{E}^1\tilde{\varphi}$ is an extension of $\tilde{\varphi}$ to $M_a((0, (t_1 + \varepsilon_1, t')])$ and in fact to $M_a((0, (\tilde{t}_1, t')])$ since

$$(\mathcal{E}^1\tilde{\varphi})(x_1, x') = 0$$

for $x_1 < -\frac{\varepsilon_1}{2}$. The continuity of $\mathcal{E}_{\varepsilon_1}$ follows from the continuity of $\mathcal{E}^1$.

If $n \geq 2$ we iterate the above procedure starting with $\mathcal{E}_{\varepsilon_1}\varphi$ defined above instead of φ and $(\tilde{t}_1, t')$ instead of t.

Proposition 1. *The space $M_a(I)$ is complete.*

Proof: Let $\{\varphi_j\}_{j=1}^{\infty}$ be a Cauchy sequence in $M_a(I)$. Take $\tilde{t} > t$, $0 < \varepsilon < \tilde{t} - t$, $\varepsilon < t$ and a continuous extension map $\mathcal{E}_\varepsilon$ from Theorem 1. Let $\tilde{\varphi}_j = \mathcal{E}_\varepsilon \varphi_j$ ($j = 1, 2, \ldots$). By the continuity of $\mathcal{E}_\varepsilon$ $\{\tilde{\varphi}_j\}_{j=1}^{\infty}$ is also a Cauchy sequence in $M_a((0, \tilde{t}])$. This means that for every $\alpha \in \mathbf{N}_0^n$ the sequence $\{x^{a+\alpha+1}(\frac{\partial}{\partial x})^\alpha \tilde{\varphi}_j\}_{j=1}^{\infty}$ satisfies the Cauchy condition for the uniform convergence on $\widetilde{I} = (0, \tilde{t}]$. Thus there exist functions h_α ($\alpha \in \mathbf{N}_0^n$) continuous on $\widetilde{I}$ vanishing near the boundary $\widetilde{I} \setminus I$ and such that

$$\sup_{x\in\tilde{I}} \left| x^{a+\alpha+1}\left(\frac{\partial}{\partial x}\right)^\alpha \tilde{\varphi}_j - h_\alpha \right| \underset{j\to\infty}{\longrightarrow} 0. \tag{7}$$

Hence, for every compact set $K \subset \widetilde{I}$

$$\sup_{x\in\tilde{I}} \left| x^{a+\alpha+1}\left(\left(\frac{\partial}{\partial x}\right)^\alpha \tilde{\varphi}_j - \frac{h_\alpha}{x^{a+\alpha+1}}\right)\right| \geq C_{\alpha K} \max_{x\in K} \left| \left(\frac{\partial}{\partial x}\right)^\alpha \tilde{\varphi}_j - \frac{h_\alpha}{x^{a+\alpha+1}}\right|,$$

where $C_{\alpha K} = \min_{x\in K} x^{a+\alpha+1} > 0$.

This in view of (7) implies that

$$\lim_{j\to\infty} \left(\left(\frac{\partial}{\partial x}\right)^\alpha \tilde{\varphi}_j\right) = \frac{h_\alpha}{x^{a+\alpha+1}} \quad \text{almost uniformly on } \widetilde{I}. \tag{8}$$

Denote $\tilde{\varphi}(x) = h_0/x^{a+1}$, $x \in \operatorname{Int} \widetilde{I}$ and $\varphi = \tilde{\varphi}|_I$. Then $\varphi \in C^\infty(I)$ and from (8) $(\frac{\partial}{\partial x})^\alpha \tilde{\varphi} = h_\alpha / x^{a+\alpha+1}$ for $x \in \operatorname{Int} \widetilde{I}$, $\alpha \in \mathbf{N}_0^n$ and consequently

$$\sup_{x\in I} \left| x^{a+\alpha+1}\left(\frac{\partial}{\partial x}\right)^\alpha \varphi_j - h_\alpha \right| \underset{j\to\infty}{\longrightarrow} 0,$$

which proves that $\varphi_j \to \varphi$ in $M_a(I)$.

Example 1. Let f_z be a function depending on a complex parameter $z \in \mathbf{C}^n$ defined by

$$I \ni x \mapsto f_z(x) = x^{-z-1}.$$

Let $a \in \mathbf{R}^n$. The function f_z is in $M_a(I)$ if and only if $\operatorname{Re} z \leq a$. Indeed for a fixed $\alpha \in \mathbf{N}_0^n$ there exists a polynomial w_α of degree $|\alpha|$ such that

$$\varrho_{a,\alpha}(f_z) = |w_\alpha(z)| \sup_{x\in I} x^{a-\operatorname{Re} z},$$

and $\varrho_{a,\alpha}(f_z)$ is finite if and only if $\operatorname{Re} z \leq a$.

Recall that for $x \in \mathbf{R}_+^n$ $\ln x$ denotes the function $\ln x = (\ln x_1, \ldots, \ln x_n)$ and consequently for $\alpha \in \mathbf{N}_0^n$, $(\ln x)^\alpha = (\ln x_1)^{\alpha_1} \cdot \ldots \cdot (\ln x_n)^{\alpha_n}$.

Example 2. Let for $z \in \mathbf{C}^n$ g_z be the function $I \ni x \mapsto g_z(x) = (\ln x)^\beta x^{-z-1}$ for some $\beta \in \mathbf{N}_0^n$. The function g_z is in $M_a(I)$ if and only if $\operatorname{Re} z_j < a_j$ if $\beta_j > 0$ and $\operatorname{Re} z_j \le a_j$ if $\beta_j = 0$. This is seen immediately from the following two facts:

(i) For arbitrary α, $\beta \in \mathbf{N}_0^n$ there exist polynomials $w_{\alpha\beta\gamma}$ such that

$$\left(\frac{\partial}{\partial x}\right)^\alpha \left((\ln x)^\beta x^{-z-1}\right) = x^{-z-\alpha-1} \sum_{\gamma \le \beta} (\ln x)^\gamma w_{\alpha\beta\gamma}(z),$$

(ii) $x_j^{a_j - \operatorname{Re} z_j} |\ln x_j|^{\beta_j} = O(x_j)$ as $x_j \to 0$ if and only if

$$\operatorname{Re} z_j \le a_j \text{ if } \beta_j = 0 \qquad \text{and} \qquad \operatorname{Re} z_j < a_j \text{ if } \beta_j > 0.$$

Other examples of functions belonging to M_a are given in Exercise 3.

Returning to the spaces M_a we note the following topological inclusions:

$$D(I) \subset M_a(I) \subset M_b(I)$$

valid if $a, b \in \mathbf{R}^n$ $a \le b$, where $D(I)$ was defined in Section 3.2 for $A = I$.

Next, it follows from Theorem 2.1 that $u \in M_a'(I)$ — the dual space of $M_a(I)$, if and only if u is linear and for some $m \in \mathbf{N}_0$ and $C < \infty$

$$(9) \qquad \left|u[\varphi]\right| \le C \sum_{|\alpha| \le m} \sup_{x \in I} \left| x^{a+\alpha+1} \left(\frac{\partial}{\partial x}\right)^\alpha \varphi(x)\right| \qquad \text{for } \varphi \in M_a(I).$$

We then say that u is of *Mellin order* less or equal to m. For a fixed $m \in \mathbf{N}_0$ the space of the elements of $M_a'(I)$ whose Mellin order is not greater than m can be defined as a dual of the space $M_a^m(I)$ of complex functions $\varphi \in C^m(I)$ for which the norm $\sum_{|\alpha| \le m} \varrho_{a,\alpha}$ is finite. The dual space of $M_a^m(I)$ is denoted by $M_a'^m(I)$. Note that it follows from (9) that

$$M_a'(I) = \bigcup_{m \in \mathbf{N}_0} M_a'^m(I).$$

Proposition 2. *The set $C_{(0)}^\infty(I)$ is not dense in $M_a(I)$ and consequently $M_a'(I)$ is not a subspace of the space of distributions $D'(I)$ for any $a \in \mathbf{R}^n$.*

Proof: On account of Corollary 2.4 it is enough to construct a non-zero functional $u \in M_a'(I)$ whose restriction to $C_{(0)}^\infty(I)$ is zero.

Consider a subspace B_a of M_a defined as

$$B_a = \left\{\varphi \in M_a \colon \lim_{x \to 0} x^{a+1} \varphi(x) \text{ exists and is finite}\right\}.$$

We define on B_a a linear functional $\tilde{u}$; $\tilde{u}[\varphi] = \lim_{x\to 0} x^{a+1}\varphi(x)$. The estimation $|x^{a+1}\varphi(x)| \le \varrho_{a,0}(\varphi)$ for $x \in I$ implies that $|\tilde{u}[\varphi]| \le \varrho_{a,0}(\varphi)$. Now from the Hahn–Banach Theorem 2.4 it follows that $\tilde{u}$ extends to an $u \in M_a'(I)$ such that $|u[\varphi]| \le \varrho_{a,0}(\varphi)$ for $\varphi \in M_a(I)$. Clearly u is non-zero since $u[f_a] = 1$ where $f_a(x) = x^{-a-1}$ for $x \in I$. However for $\varphi \in C^\infty_{(0)}(I)$ we have $u[\varphi] = \tilde{u}[\varphi] = \lim_{x\to 0} x^{a+1}\varphi(x) = 0$, which is what we wanted.

In Exercise 4 (cf. also Exercises 5 and 6) the reader will find a direct proof that $C^\infty_{(0)}(I)$ is not dense in $M_a(I)$.

In Subsection 2 we introduce a subspace $M_{(a)} \subset M_a$ such that $C^\infty_{(0)}(I)$ is dense in $M_{(a)}$.

2. Spaces $M_{(\omega)}((0,t])$ and $M'_{(\omega)}((0,t])$

As in Subsection 1 we take $I = (0,t]$ for some $t \in \mathbf{R}^n_+$. Let $\omega \in (\mathbf{R} \cup \{\infty\})^n$. We define the function space $M_{(\omega)}(I)$ as the inductive limit

$$M_{(\omega)}(I) = \varinjlim_{a<\omega} M_a(I).$$

It follows from the definition of the inductive topology that a sequence $\{\varphi_\nu\}$, $\varphi_\nu \in M_{(\omega)}$ $(\nu = 1,2,\ldots)$ converges to zero in $M_{(\omega)}$ if there exists $a < \omega$ such that $\varphi_\nu \in M_a$ $(\nu = 1,2,\ldots)$ and for every such a the sequence $\{\varphi_\nu\}$ is convergent to zero in M_a.

The following topological inclusion is clear:

$$D(I) \subset M_{(\omega)}(I). \tag{10}$$

Moreover we shall prove:

Proposition 3. *The set $C^\infty_{(0)}(I)$ is dense in $M_{(\omega)}(I)$.*

Proof: Take an arbitrary $\varphi \in M_{(\omega)}$. Thus there exist $a < b < \omega$ such that $\varphi \in M_a(I) \subset M_b(I)$. To prove the proposition it is enough to construct a sequence $\{\varphi_\nu\}_{\nu=1}^\infty$ of $C^\infty_{(0)}(I)$ functions such that $\varphi_\nu \to \varphi$ in M_b[(1)]. We introduce auxiliary functions $f_i \in C^\infty(\mathbf{R}_+)$, $0 \le f_i(s) \le 1$, $s \in \mathbf{R}_+$ $(i = 1,\ldots,n)$

$$f_i(s) = \begin{cases} 0 & \text{for } 0 < s < \frac{1}{4}t_i \\ 1 & \text{for } \frac{1}{2}t_i < s < \infty, \end{cases}$$

---

(1) One does not have convergence in M_a since $C^\infty_{(0)}(I)$ in not dense there.

and we define

$$\psi_\nu(x) = f_1(\nu x_1)\cdot\ldots\cdot f_n(\nu x_n) \qquad \text{for } x \in I \qquad (\nu = 1,2,\ldots).$$

Let $A_\nu = \{x \in I\colon 0 < x_j < \frac{1}{2\nu}t_j \text{ for some } j,\ 1 \le j \le n\}$. We check that

(11) $\psi_\nu \in C^\infty(I), \quad 0 \le \psi_\nu(x) \le 1, \quad \psi_\nu = 1 \text{ on } I\setminus A_\nu \text{ and } \sup_{x\in A_\nu}|x^{b-a}| \underset{\nu\to\infty}{\longrightarrow} 0.$

We shall prove that the sequence

$$\varphi_\nu(x) = \psi_\nu(x)\varphi(x) \qquad \text{for } x \in I$$

has the desired properties.

Fix arbitrarily $\alpha \in \mathbb{N}_0^n$. Let c be a constant such that

(12) $$\left|f_i^{(k)}(s)\right| \le c \qquad \text{for } k \le \alpha_i,\ s > 0\ (i = 1,\ldots,n).$$

It follows from the Leibniz formula and (11) that

(13) $$\begin{aligned}\varrho_{b,\alpha}(\varphi_\nu - \varphi) &= \sup_{x\in I}\left|x^{b+\alpha+1}\left(\frac{\partial}{\partial x}\right)^\alpha(\varphi(x)(\psi_\nu(x)-1))\right|\\ &\le \sup_{x\in I}\left|x^{b+\alpha+1}\left(\frac{\partial}{\partial x}\right)^\alpha\varphi(x)\cdot(\psi_\nu(x)-1)\right| +\\ &\quad+\sum_{\substack{\beta\le\alpha\\ \beta\ne\alpha}}\binom{\alpha}{\beta}\sup_{x\in I}\left|x^{a+\beta+1}\left(\frac{\partial}{\partial x}\right)^\beta\varphi(x)\left(\frac{\partial}{\partial x}\right)^{\alpha-\beta}(\psi_\nu(x)-1)x^{b-a+\alpha-\beta}\right|\\ &\le \varrho_{a,\alpha}(\varphi)\sup_{x\in A_\nu}\left|x^{b-a}(\psi_\nu(x)-1)\right| +\\ &\quad+\sum_{\substack{\beta\le\alpha\\ \beta\ne\alpha}}\binom{\alpha}{\beta}\varrho_{a,\beta}(\varphi)\sup_{x\in I}\left|\left(\frac{\partial}{\partial x}\right)^{\alpha-\beta}(\psi_\nu(x)-1)x^{b-a+\alpha-\beta}\right|.\end{aligned}$$

It is seen from (11) that the first summand above tends to zero as $\nu\to\infty$ (which proves that $\varrho_{b,0}(\varphi_\nu-\varphi)\to 0$). If $\alpha\ne 0$ then in each of the remaining summands there is a j, $1\le j\le n$ with $\beta_j<\alpha_j$. The assumptions on f_i imply that

(14) $$x_k^{\alpha_k-\beta_k}\left(\frac{\partial}{\partial x_k}\right)^{\alpha_k-\beta_k}(\psi_\nu(x)-1) = 0 \qquad \text{if } \alpha_k > \beta_k \text{ and } x_k > \frac{1}{2\nu}t_k,$$

which in view of the relations (11) and (12) leads to

$$\left|x_k^{\alpha_k-\beta_k}\left(\frac{\partial}{\partial x_k}\right)^{\alpha_k-\beta_k}(\psi_\nu(x)-1)\right| \le \max\left\{1, c^n\left(\frac{t_k}{2}\right)^{\alpha_k-\beta_k}\right\} \qquad \text{for } x \in I.$$

Hence there exists a constant $c^* < \infty$ such that

$$\sup_{x\in I}\left|x^{\alpha-\beta}\left(\frac{\partial}{\partial x}\right)^{\alpha-\beta}(\psi_\nu(x)-1)\right| \le c^*. \tag{15}$$

Assuming that $\beta_j < \alpha_j$ we have by (14) and (15) the estimates

$$\sup_{x\in I}\left|x^{b-a+\alpha-\beta}\left(\frac{\partial}{\partial x}\right)^{\alpha-\beta}(\psi_\nu(x)-1)\right| = \sup_{\substack{x\in I\\ x_j<\frac{1}{2\nu}t_j}}\left|x^{b-a+\alpha-\beta}\left(\frac{\partial}{\partial x}\right)^{\alpha-\beta}(\psi_\nu(x)-1)\right|$$

$$\le c^* \sup_{\substack{x\in I\\ x_j\le\frac{1}{2\nu}t_j}}\left|x^{b-a}\right| \le c^*\left|\frac{t_j}{2\nu}\right|^{b_j-a_j}\prod_{\substack{i=1\\ i\neq j}}^{n} t_i^{b_i-a_i},$$

which imply that the left hand side tends to zero as $\nu \to \infty$, which proves the proposition.

By $M'_{(\omega)} = M'_{(\omega)}(I)$, $\omega \in (\mathbf{R}\cup\{\infty\})^n$ we denote as usual the dual space of $M_{(\omega)}$ endowed with the pointwise convergence topology. By (10), Proposition 3 and Proposition 2.1 we derive that $M'_{(\omega)}(I)$ is a subspace of $D'(I)$. Therefore the elements of $M'_{(\omega)}$ are called *Mellin distributions*. The totality of Mellin distributions is denoted by $M'(I)$:

$$M'(I) = \bigcup_{\omega\in(\mathbf{R}\cup\{\infty\})^n} M'_{(\omega)}(I) = \bigcup_{\omega\in\mathbf{R}^n} M'_{(\omega)}(I).$$

Definition 2. We say that $u \in D'(I)$ ($u \in D'_I(\mathbf{R}^n_+)$)) is a *Mellin distribution* if there exists an $\omega \in (\mathbf{R}\cup\{\infty\})^n$ such that u extends to $M_{(\omega)}$ as a continuous linear functional on $M_{(\omega)}$.

Remark 1. Observe that a functional $u \in D'(I)$ may extend to different spaces $M_{(\omega)}$, however for a given ω if an extension exists then it is unique due to the density of $C^\infty_{(0)}(I)$ in $M_{(\omega)}$. Moreover the extensions are compatible in the sense that for any extensions u^ω, $u^{\tilde\omega}$ of u to $M_{(\omega)}$ and $M_{(\tilde\omega)}$ respectively

$$u^\omega[\psi] = u^{\tilde\omega}[\psi] = u^{\mathring{\omega}}[\psi] \qquad \text{for} \quad \psi \in M_{(\mathring{\omega})}$$

where $\mathring{\omega} = (\min(\omega_1,\tilde\omega_1),\dots,\min(\omega_n,\tilde\omega_n))$.

Definition 2 may be reformulated equivalently as

Definition 2′. We say that $u \in D'(I)$ ($u \in D'_I(\mathbf{R}^n_+)$) is a Mellin distribution if there exists an $a \in \mathbf{R}^n$ such that u extends (clearly, non-uniquely) to a linear continuous functional on M_a.

Indeed if $u \in M'_{(\omega)}$ then for any $a < \omega$, $u \in M'_a$ since $M'_{(\omega)} \subset M'_a$. On the other hand given a $u \in M'_a$ it follows from the topological inclusion, $M_{(\omega)} \subset M_a$ if $\omega < a$, that $u \in M'_{(\omega)}$.

We shall yet introduce a space $M_{[a]}(I)$ (considered in detail in Subsection 6.2)

$$M_{[a]}(I) = \varprojlim_{b>a} M_b(I)$$

equipped with the projective limit topology: $\varphi_\nu \to 0$ in $M_{[a]}$ if and only if $\varphi_\nu \to 0$ in every space M_b for $b > a$.

Let $\alpha, a, b \in \mathbf{R}^n$, $\omega \in (\mathbf{R} \cup \{\infty\})^n$, $a < b < \omega$. The topological inclusions for the spaces M_a given in Subsection 1 can now be completed (see Exercise 7 for the proof) as follows:

$$\begin{gathered} D(I) \subset M_{(a)}(I) \subset M_a(I) \subset M_{[a]}(I) \subset M_b(I) \subset M_{(\omega)}(I), \\ M'_{(\omega)}(I) \subset D'(I). \end{gathered} \tag{16}$$

Also note that it follows from (16) and the definitions of the spaces $M_{[a]}$, $M_{(a)}$ that

$$M_{(\omega)} = \varinjlim_{c<\omega} M_{(c)}, \qquad M_{[a]} = \varprojlim_{b>a} M_{(b)}. \tag{17}$$

As it was shown in Subsection 1 M'_a is not a subspace of $D'(I)$. Nevertheless, due to the topological inclusion $D \subset M_a$, the restriction of a functional $u \in M'_a$ to $D(I)$ belongs to $D'(I)$. The same is also valid for a pair of spaces $M'_{(a)}$, M'_a. Also note that $\mathfrak{M}_a \not\subset \mathfrak{M}_b$ (see Exercise 10).

We shall distinguish now another important dense subset of $M_{(\omega)}(I)$ different from $C^\infty_{(0)}(I)$. We start with the notation:

For $a \in \mathbf{R}^n$ we denote $X_a = \operatorname{span}\{x^{-b-1}\}_{b \le a}$ the space of finite linear combinations of elements of the set $\{x^{-b-1}\}_{b \le a}$ with complex coefficients, $X_{(\omega)} = \bigcup_{a<\omega} X_a$. From Example 1 $X_a \subset M_a$ and $X_{(\omega)} \subset M_{(\omega)}$. We have

Proposition 4. *$X_{(\omega)}$ is dense in $M_{(\omega)}$.*

The proof is immediate from the following

Lemma 2. *Let $a, b \in \mathbf{R}^n$, $b < a$. Then M_b is contained in the closure of X_a in M_a.*

Proof: Let $\varphi \in M_b$. It follows from Exercise 2.1 that it is enough to show that for every $\varepsilon > 0$ and $m \in \mathbf{N}_0$ there exists a $\psi \in X_a$ such that

$$\varrho_m(\varphi - \psi) < \varepsilon, \tag{18}$$

where $\varrho_m = \sum_{|\alpha| \le m} \varrho_{a,\alpha}$ with $\varrho_{a,\alpha}$ given in Definition 1. Since $\varphi \in M_{(\omega)}$ if $\omega > b$ and the space $C^\infty_{(0)}(I)$ is dense in $M_{(\omega)}$ for any ω, again by Exercise 2.1, there exists $\tilde{\psi} \in C^\infty_{(0)}(I)$ such that $\varrho_m(\varphi - \tilde{\psi}) < \frac{\varepsilon}{2}$.

To prove (18) we thus need find a $\psi \in X_a$ such that $\varrho_m(\psi - \tilde{\psi}) < \frac{\varepsilon}{2}$. Since $\tilde{\psi}$ is in $C^\infty_{(0)}(I)$, the function $\bar{I} \ni x \mapsto x^{a+1}\tilde{\psi}(x)$ is smooth and from the Weierstrass theorem for every $\delta > 0$ one can find a polynomial w_δ with the property

$$\sup_{x \in I} \left| \left(\frac{\partial}{\partial x}\right)^\beta (x^{a+1}\tilde{\psi}(x) - w_\delta(x)) \right| < \delta \qquad \text{for } \beta \le \alpha.$$

We shall prove that subject to an appropriate choice of δ, the function $\psi(x) = x^{-a-1}w_\delta(x)$ has the desired property. Indeed $\psi \in X_a$ and by Exercise 1 there exist constants $c_{a,\alpha,\beta}$ such that

$$\varrho_m(\psi - \tilde{\psi}) \le \sum_{|\alpha| \le m} \sum_{\beta \le \alpha} c_{a,\alpha,\beta} \sup_{x \in I} \left| \left(\frac{\partial}{\partial x}\right)^\beta (x^{a+1}(\psi - \tilde{\psi})) \right| t^\beta \le c\delta$$

for $c = \sum_{|\alpha| \le m} \sum_{\beta \le \alpha} c_{a,\alpha,\beta} t^\beta$. This ends the proof.

In the same way that locally integrable functions were identified with distributions we can identify functions in $M_a(I)$ with distributions in $M'_{(-a-1)}$. Indeed given an $f \in M_a$ we define

$$f[\psi] = \int_I f(x)\psi(x)\,dx \qquad \text{for } \psi \in M_c \text{ with } c < -a-1.$$

Note that the above formal definition makes sense and defines $f \in M'_c$. Due to the arbitrariness of $c < -a-1$ we have $f \in M'_{(-a-1)} \subset D'(I)$. Thus

$$M_a \subset M'_{(-a-1)}. \tag{19}$$

Analogously

$$M_{[a]} \subset M'_{(-a-1)}.$$

We shall complete inclusions (16) with those between the spaces $M_{(\omega)}$, M_ω, $M'_{(\omega)}$ and the spaces considered in Subsection 4.2. Denote by $\mathfrak{M}_\alpha(I)$ the space of restrictions to I of functions in $\mathfrak{M}_\alpha(\mathbf{R}^n_+)$ with the topology induced from $\mathfrak{M}_\alpha(\mathbf{R}^n_+)$. It is clear that

$$C^\infty_{(0)}(I) \subset \mathfrak{M}_\alpha(I) \qquad \text{for } \alpha \in \mathbf{R}^n. \tag{20}$$

We shall prove now

Proposition 5. *Let $\alpha \in \mathbf{R}^n$, $I = (0,t]$. Then for every $\omega > \alpha$ the topological inclusion*

$$\mathfrak{M}_\alpha(I) \subset M_{(\omega)}(I) \tag{21}$$

holds and the set of functions in $\mathfrak{M}_\alpha(I)$ is dense in $M_{(\omega)}(I)$. Hence for every $\alpha < \omega$ $M'_{(\omega)}(I)$ is a subspace of $\mathfrak{M}'_\alpha(I)$:

$$M'_{(\omega)}(I) \subset \mathfrak{M}'_\alpha(I) \qquad \text{for every } \alpha < \omega. \tag{22}$$

Proof: Let $\tilde{\sigma} \in \mathfrak{M}_\alpha(\mathbf{R}^n_+)$ i.e. $\psi \overset{\text{def}}{=} (x^{\alpha+1}\tilde{\sigma}) \circ \mu \in S(\mathbf{R}^n)$ ($\mu(y) = e^{-y}$ for $y \in \mathbf{R}^n$). Suppose $\omega > \alpha$ and let $\sigma = \tilde{\sigma}|_I$, $I_1 = \mu^{-1}(I)$ (cf. Exercise 9). By the Leibniz rule and formula (4.7) there exist constants $C = C(\beta)$ and $\tilde{C} = C(\beta, t)$ such that (cf. Definition 5.1′)

$$\begin{aligned}
\tilde{\varrho}_{\omega,\beta}(\sigma) &= \sup_{x\in I}\left| x^{\omega+1}\left(x\frac{\partial}{\partial x}\right)^\beta (x^{-\alpha-1}(\psi\circ\mu^{-1}))\right| \\
&\le C\sum_{\gamma\le\beta}\sup_{x\in I}\left|x^{\omega-\alpha}\left(\left(\frac{\partial}{\partial y}\right)^\gamma\psi\right)\circ\mu^{-1}\right| = C\sum_{\gamma\le\beta}\sup_{y\in I_1}\left|e^{(\alpha-\omega)y}\left(\frac{\partial}{\partial y}\right)^\gamma\psi\right| \\
&\le \tilde{C}\sum_{\gamma\le\beta}\sup_{y\in\mathbf{R}^n}\left|\left(\frac{\partial}{\partial y}\right)^\gamma\psi\right| < \infty,
\end{aligned}$$

since $\omega > \alpha$ i.e. $\sigma \in M_\omega(I)$. In the same way, taking $\tilde{\sigma}_\nu \to 0$ in $\mathfrak{M}_\alpha(I)$, $\psi_\nu = (x^{\alpha+1}\tilde{\sigma}_\nu)\circ\mu$ we obtain: $\tilde{\varrho}_{\omega,\beta}(\sigma_\nu) \le \tilde{C}\sum_{\gamma\le\beta}\sup_{y\in\mathbf{R}^n}\left|\left(\frac{\partial}{\partial y}\right)^\gamma\psi_\nu\right|$, hence $\sigma_\nu \to 0$ in $M_\omega(I)$ for every $\omega > \alpha$. In this way we have proved the topological inclusion $\mathfrak{M}_\alpha(I) \subset M_{[\alpha]}(I)$ which in view of (16) implies the desired inclusion (21). By (20), (21) and Proposition 3 the set $\mathfrak{M}_\alpha(I)$ is dense in $M_{(\omega)}(I)$ for every $\omega > \alpha$.

Exercises

1. Denote by span X the space of finite linear combinations with complex coefficients of elements of a space X. Let $c \in \mathbf{R}^n$, $\alpha, \beta \in \mathbf{N}_0^n$. Prove that the following spaces of operators coincide: $\operatorname{span}\{x^{c+\beta}(\frac{\partial}{\partial x})^\beta\}_{\beta\le\alpha}$, $\operatorname{span}\{(x\frac{\partial}{\partial x})^\beta(x^c\cdot)\}_{\beta\le\alpha}$, $\operatorname{span}\{x^c(\frac{\partial}{\partial x}x)^\beta\}_{\beta\le\alpha}$, $\operatorname{span}\{x^\beta(\frac{\partial}{\partial x})^\beta(x^c\cdot)\}_{\beta\le\alpha}$.

2. Prove that the system $\{\tilde{\tilde{\varrho}}_{a,\alpha}\}_{\alpha\in\mathbf{N}_0^n}$

$$\tilde{\tilde{\varrho}}_{a,\alpha}(\varphi) = \sup_{x\in I}\left|x^{a+1}\left(\frac{\partial}{\partial x}\right)^\alpha (x^\alpha\varphi(x))\right|, \qquad \alpha \in \mathbf{N}_0^n$$

constitutes a system of seminorms on $M_a(I)$ equivalent to $\{\varrho_{a,\alpha}\}_{\alpha\in\mathbf{N}}$.

3. Prove that

(i) The function $I \ni x \mapsto (\ln x)^1$ belongs to $M_{[-1]}$,

(ii) If $0 < r < 1$ then the function $(0, r] \ni x \mapsto \cos(\frac{1}{\ln x})$ belongs to $M_{-1} \backslash C^\infty([0, r])$.

4. Let $\varepsilon \in \mathbf{R}_+$. Show that:

(i) $\widetilde{C}^\infty([0, \varepsilon))$ can be equivalently defined by

$$\left\{\varphi \in C^\infty((0, \varepsilon)) \colon \sup_{x \in (0,\varepsilon)} \left|\frac{d^j}{dx^j}\varphi(x)\right| < \infty \text{ for every } j \in \mathbf{N}_0\right\} :$$

(ii) $\widetilde{C}^\infty([0, \varepsilon)) = C^\infty([0, \varepsilon))$.

5. Let $a \in \mathbf{R}^n$ and f_a be the function from Example 1. Show that $\varrho_{a,0}(f_a - \psi) \geq 1$ for every $\psi \in C^\infty_{(0)}(I)$.

6. Let $a, b \in \mathbf{R}^n$, $a < b$. Show that:

(i) the set of functions in M_a is not dense in the topological space M_b (hence M'_b is not a subspace of M'_a);

(ii) the set $M_{(a)}$ is not dense in the topological space M_a,

(iii) the set M_a is dense in $M_{(b)}$ for every $b > a$,

(iv) $M'_{(b)}$ is a subspace of M'_a for every $b > a$,

(v) $M'_{(\omega)} = \bigcap_{a<\omega} M'_a$.

7. Prove the topological inclusion $M_{(a)} \subset M_a$.

8. Let f_z be the function from Example 1. Show that $f_z \in M_{(\omega)}$ if and only if $\operatorname{Re} z < \omega$.

9. (Cf. Exercise 4.5 and Subsection 3.5). Let $I = (0, t]$, $t \in \mathbf{R}^n_+$, and $\mu(y) = e^{-y}$ for $y \in \mathbf{R}^n$. Put $I_1 = \mu^{-1}(I) = \{y \in \mathbf{R}^n : y \geq -\ln t\}$. Show the continuity of the following operations of composition with μ and μ^{-1}:

$$\circ\mu\colon D(I) \to D(I_1), \qquad \circ\mu\colon D'(I) \to D'(I_1),$$
$$\circ\mu^{-1}\colon D(I_1) \to D(I), \qquad \circ\mu^{-1}\colon D'(I_1) \to D'(I).$$

10. Let α, $\beta \in \mathbf{R}^n$, $\alpha < \beta$. Show that

$$\{\sigma\colon \sigma \in \mathfrak{M}_\alpha,\ \operatorname{supp}\sigma \subset I\} \subset \{\sigma\colon \sigma \in \mathfrak{M}_\beta,\ \operatorname{supp}\sigma \subset I\} \subset \mathfrak{M}_\beta,$$

but $\mathfrak{M}_\alpha \not\subset \mathfrak{M}_\beta$ and $\mathfrak{M}'_\beta \not\subset \mathfrak{M}'_\alpha$.

11. Let $\omega \in (\mathbf{R} \cup \{-\infty\})^n$ and let f be a measurable function on I such that $\int_I x^b\,|f(x)|\,dx < \infty$ for every $b > \omega$. Show that $f \in M'_{(-\omega-1)}$ is of Mellin order ≤ 0. In particular if $f \in L^\infty$ on I then $f \in M'_{(0)}$ is of Mellin order ≤ 0.

12. Give an example of a $C^\infty(I)$ function f which does not belong to any $M_a(I)$ nevertheless $f \in M'_{(0)}(I)$.

13. Let $I = (0,t]$, $t \in \mathbf{R}_+$, let $\theta \in \mathbf{C}$, $\operatorname{Re}\theta < 0$ and let $c \in \mathbf{R}$. Show that the function $I \ni x \mapsto (-\ln x + c)^\theta$ belongs to $M'_{(0)}$ if $t < e^c$. (See Example 7.6).

14. Prove that there exists a linear *continuous* extension mapping

$$\mathcal{E}\colon \widetilde{C}^\infty([a,b]) \to C_0^\infty((a^*,b^*)), \qquad a^* < a < b < b^*,$$

(see [Mel], Seeley extension theorem; cf. Theorem 1).

§6. OPERATIONS OF MULTIPLICATION AND DIFFERENTIATION IN THE SPACE OF MELLIN DISTRIBUTIONS

1. Multiplication and Differentiation in M_a, $M_{(\omega)}$ and Their Duals

Proposition 1. *Let $\beta \in \mathbf{C}^n$. The operation of (pointwise) multiplication by x^β*

(1) $$x^\beta\colon M_a \to M_{a-\operatorname{Re}\beta}, \qquad x^\beta\colon M_{(\omega)} \to M_{(\omega-\operatorname{Re}\beta)}$$

are continuous.

Proof: Suppose $\varphi_\nu \to 0$ in M_a. Choose arbitrarily $\alpha \in \mathbf{N}_0^n$ and note that for suitable constants C_λ we have the estimate

$$\sup_{x\in I}\left|x^{a-\operatorname{Re}\beta+\alpha+1}\left(\frac{\partial}{\partial x}\right)^\alpha (x^\beta\varphi_\nu)\right| \le \sum_{\lambda\le\alpha} C_\lambda \sup_{x\in I}\left|x^{a-\operatorname{Re}\beta+\alpha+1}\left(\frac{\partial}{\partial x}\right)^\lambda \varphi_\nu\cdot x^{\beta-\alpha+\lambda}\right|$$
$$\le \sum_{\lambda\le\alpha} C_\lambda \varrho_{a,\lambda}(\varphi_\nu).$$

Hence it follows that $\varrho_{a-\operatorname{Re}\beta,\alpha}(x^\beta\varphi_\nu) \to 0$ which proves the continuity of the first mapping in (1).

Concerning the continuity of the second mapping in (1), we assume $\varphi_\nu \to 0$ in $M_{(\omega)}$. Then there exists an $a < \omega$ such that $\varphi_\nu \to 0$ in M_a. In view of the part which we have just proven $x^\beta\varphi_\nu \to 0$ in $M_{a-\operatorname{Re}\beta}$ and since $a < \omega$ the convergence also holds in $M_{(\omega-\operatorname{Re}\beta)}$.

It follows from the proposition above that the formula

$$x^\beta u[\varphi] = u[x^\beta\varphi] \qquad \text{for } \varphi \in M_{(\omega+\operatorname{Re}\beta)},\ u \in M'_{(\omega)}$$

defines a continuous operation

(2) $$x^\beta\colon M'_{(\omega)} \to M'_{(\omega+\operatorname{Re}\beta)}.$$

Proposition 2. *For $\alpha \in \mathbf{N}_0^n$ the differentiation operations are continuous on the following spaces:*

$$\left(\frac{\partial}{\partial x}\right)^\alpha : M_a \to M_{a+\alpha}, \qquad \left(\frac{\partial}{\partial x}\right)^\alpha : M_{(\omega)} \to M_{(\omega+\alpha)}.$$

The simple proof is left to the reader.

Proposition 2 implies immediately that the formal definition

$$\left(\frac{\partial}{\partial x}\right)^\alpha u[\varphi] = (-1)^{|\alpha|} u\left[\left(\frac{\partial}{\partial x}\right)^\alpha \varphi\right] \qquad \text{for } \varphi \in M_{(\omega-\alpha)},\ u \in M'_{(\omega)}$$

defines a continuous operation

$$\left(\frac{\partial}{\partial x}\right)^\alpha : M'_{(\omega)} \to M'_{(\omega-\alpha)},$$

which is a restriction of the distributional derivative in $D'(I) \simeq D'_I(\mathbf{R}^n_+)$. More precisely if u is a Mellin distribution we consider it as a distribution in $D'(I)$ and by Proposition 3.1 as a distribution in $D'(\mathbf{R}^n_+)$ with support in I. We compute the derivative $(\frac{\partial}{\partial x})^\alpha u$ in $D'(\mathbf{R}^n_+)$ and return to M' under inverse isomorphisms (see also Exercises 1, 2, 3).

From Propositions 1 and 2 we readily get

Corollary 1. *The mappings*

$$\left(x\frac{\partial}{\partial x}\right)^\alpha : M_a \to M_a, \qquad \left(x\frac{\partial}{\partial x}\right)^\alpha : M_{(\omega)} \to M_{(\omega)},$$

$$\left(x\frac{\partial}{\partial x}\right)^\alpha : M'_a \to M'_a, \qquad \left(x\frac{\partial}{\partial x}\right)^\alpha : M'_{(\omega)} \to M'_{(\omega)}$$

are continuous.

2. Mellin Multipliers

Definition 1. By a Mellin *multiplier* we understand a function $m \in C^\infty(I)$ such that the multiplication by m is a continuous mapping

$$m\colon M_{(\omega)} \to M_{(\omega)}$$

for every $\omega \in (\mathbf{R} \cup \{\infty\})^n$.

Proposition 3. *A function $m \in C^\infty(I)$ is a Mellin multiplier if and only if the multiplication by m is continuous in the spaces*

$$m\colon M_{(0)} \to M_{(0)}. \tag{3}$$

Proof: The necessity of (3) is obvious. To prove the sufficiency take an arbitrary $\omega \in (\mathbf{R} \cup \{\infty\})^n$ and a sequence $\psi_\nu \to 0$ in $M_{(\omega)}$. By (5.17) there exists $\tilde{\omega} \in \mathbf{R}^n$, $\tilde{\omega} < \omega$, such that $\psi_\nu \to 0$ in $M_{(\tilde{\omega})}$. Let $\varphi_\nu = x^{\tilde{\omega}}\psi_\nu$. By Proposition 1 $\varphi_\nu \to 0$ in $M_{(0)}$ hence by assumption (3) $m\varphi_\nu \to 0$ in $M_{(0)}$. Again from Proposition 1 we infer that $m\psi_\nu \to 0$ in $M_{(\tilde{\omega})}$ and consequently in $M_{(\omega)}$ since $\tilde{\omega} \leq \omega$.

By duality a multiplier m defines a continuous operation

$$m\colon M'_{(\omega)} \to M'_{(\omega)}$$

for every $\omega \in (\mathbf{R} \cup \{\infty\})^n$.

In Section 5.2 we introduced the spaces $M_{[a]}$ for $a \in \mathbf{R}^n$. Now we are interested in the space $M_{[-1]}$:

$$M_{[-1]} = \varprojlim_{\varepsilon>0} M_{-1+\varepsilon} = \varprojlim_{\varepsilon>0} M_{(-1+\varepsilon)}.$$

Formula (5.6) gives the topological inclusions

$$M_{(-1)} \subset M_{-1} \subset M_{[-1]}. \tag{4}$$

Proposition 4. *The space of Mellin multipliers coincides with the space* $M_{[-1]}$.

Proof: Let $m \in M_{[-1]}$. We shall prove that the mapping (3) is continuous. Choose arbitrarily $\varphi_\nu \to 0$ in $M_{(0)}$. There is an $a < 0$ such that $\varphi_\nu \to 0$ in M_a. Let $a < c < 0$, $\alpha \in \mathbf{N}_0^n$. As in the proof of Proposition 1 for suitable constants C_λ

$$\varrho_{c,\alpha}(m\varphi_\nu) \leq \sum_{\lambda \leq \alpha} C_\lambda\, \varrho_{a,\lambda}(\varphi_\nu)\varrho_{-1+c-a,\alpha-\lambda}(m).$$

By assumption $m \in M_{-1+c-a}$ (since $c - a > 0$) hence $\varrho_{-1+c-a,\alpha-\lambda}(m) < \infty$. Since $\varphi_\nu \to 0$ in M_a it follows that $\varrho_{a,\lambda}(\varphi_\nu) \to 0$ for every $\lambda \in \mathbf{N}_0^n$. Thus the estimate above implies that $\varrho_{c,\alpha}(m\varphi_\nu) \to 0$ and due to the arbitrariness of $\alpha \in \mathbf{N}_0^n$, $m\varphi_\nu$ tends to 0 in M_c and consequently in $M_{(0)}$ since $c < 0$.

To prove the converse assume that m is a Mellin multiplier. Let $\varepsilon \in \mathbf{R}_+^n$. By Example 5.1 the function $I \ni x \to \varphi(x) = x^{-1+\varepsilon}$ belongs to $M_{-\varepsilon}$ and therefore to $M_{(0)}$. m being a Mellin multiplier we get $x^{-1+\varepsilon}m \in M_{(0)}$. Thus by Proposition 1 $m \in M_{(-1+\varepsilon)}$. Since $0 < \varepsilon$ is arbitrary this implies $m \in M_{[-1]}$.

Remark 1. M_{-1} is a proper subset of $M_{[-1]}$. Indeed, by (4) $M_{-1} \subset M_{[-1]}$ and by Exercise 5.3 the function $(\ln x)^1$ defined for $x \in I$ belongs to $M_{[-1]}$. However $(\ln x)^1 \notin M_{-1}$ because $\sup_{x\in I} \left|(\ln x)^1\right| = \infty$.

Below we present examples of Mellin multipliers.

Example 1. As noted in Remark 1 the function

$$I \ni x \mapsto (\ln x)^{\mathbf{1}}$$

is a Mellin multiplier in view of Proposition 4.

Example 2. (cf. Exercise 8). Let $\alpha_i \in \mathbf{C}$, $\beta_i \in \mathbf{R}$, $i = 1, \ldots, n$ and suppose

$$m(x) = x_1^{\alpha_1} \cdot \ldots \cdot x_n^{\alpha_n} \Big/ \sum_{i=1}^{n} x_i^{\beta_i} \qquad \text{for } x \in I$$

satisfies the condition: $\sup_{x \in I} |m(x)| < \infty$. Then $m \in M_{-\mathbf{1}}$ and consequently is a Mellin multiplier.

Proof: By assumption $\varrho_{-\mathbf{1},0}(m) = \sup_{x \in I} |m(x)| < \infty$. Denote $\mathcal{K}(x) = \sum_{i=1}^{n} x_i^{\beta_i}$. Then

$$x_j \frac{\partial}{\partial x_j} m(x) = m(x)(\alpha_j - \beta_j x_j^{\beta_j} / \mathcal{K}(x)) \qquad (j = 1, 2, \ldots, n).$$

By induction one proves that for $|\gamma| = k$, $(k = 1, 2, \ldots, n)$ there exist constants $C(\gamma)$, $C_{i_1,\ldots,i_s}(\gamma)$, $1 \leq i_j \leq n$ $(j = 1, \ldots, s; s = 1, \ldots, k)$ such that

$$x^\gamma \left(\frac{\partial}{\partial x}\right)^\gamma m(x) = m(x) \Big\{ C(\gamma) + \sum_{s=1}^{k} \sum_{i_1,\ldots,i_s} C_{i_1,\ldots,i_s}(\gamma) x_{i_1}^{\beta_{i_1}} \cdots x_{i_s}^{\beta_{i_s}} / (\mathcal{K}(x))^s \Big\}.$$

Note that $\left| x_i^{\beta_i} / \mathcal{K}(x) \right| \leq 1$, $(i = 1, \ldots, n)$ and it follows from above that for some constant $C_1 < \infty$, $\left| x^\gamma (\frac{\partial}{\partial x})^\gamma m(x) \right| < C_1 |m(x)|$. Since by assumption $\sup_{x \in I} |m(x)|$ is finite this implies that $m \in M_{-\mathbf{1}}$.

Example 3. Any smooth function on $\bar{I}$ (i.e. a restriction to $\bar{I}$ of a smooth function on $\mathbf{R}^n$) belongs to $M_{-\mathbf{1}}$ and consequently is a Mellin multiplier (cf. Exercise 5.3).

Remark 2. In Definition 1 of a Mellin multiplier one cannot replace $M_{(\omega)}$ by M_ω. Indeed taking $m(x) = (\ln x)^{\mathbf{1}}$, $\varphi(x) = x^{-a-\mathbf{1}}$ for $x \in I$ we observe that m is a Mellin multiplier, $\varphi \in M_a$ but $m\varphi \notin M_a$ since $\sup_{x \in I} \left| x^{a+\mathbf{1}} \varphi(x) m(x) \right| = \sup_{x \in I} \left| (\ln x)^{\mathbf{1}} \right| = \infty$.

Proposition 5. *The space of Mellin multipliers is closed under multiplication.*

In fact if m_1, m_2 are Mellin multipliers then $m_1 \in M_{(-1+\varepsilon)}$ for every $\varepsilon > 0$ and also $m_2 m_1 \in M_{(-1+\varepsilon)}$ for $\varepsilon > 0$ which means that $m_2 m_1 \in M_{[-1]}$.

Proposition 6. *Let $\varphi = x^b m$ where $b \in \mathbf{C}$ and m is a Mellin multiplier. Then for any $u \in M'_{(a)}$ the following Leibniz rule holds in $M'_{(a+\operatorname{Re} b)}$:*

$$\left(x_j \frac{\partial}{\partial x_j}\right)(\varphi u) = x_j \frac{\partial}{\partial x_j}\varphi \cdot u + \varphi \cdot x_j \frac{\partial}{\partial x_j} u$$

(cf. Exercise 9).

Proposition 7. *Let $u \in M'_{(\omega)}(I)$, $\omega \in (\mathbf{R} \cup \{+\infty\})^n$, $m_0, m_j \in M_{[-1]}(I)$ $(j = 1, 2, \ldots)$, $m_j \to m_0$ in the projective limit topology of the space $M_{[-1]}$ (see Subsection 5.2). Then*

$$m_j u \to m_0 u \qquad \text{in } M'_{(\omega)}(I).$$

Proof: We may assume that $m_0 = 0$. Take arbitrarily $\varphi \in M_{(\omega)}(I)$. Hence $\varphi \in M_b(I)$ for some $b < \omega$ and there exists $\delta > 0$ such that $b + \delta < \omega$. We shall prove that $m_j\varphi \to 0$ in $M_{b+\delta}$ and thus $m_j\varphi \to 0$ in $M_{(\omega)}(I)$. To this end choose arbitrarily $\alpha \in \mathbf{N}_0^n$ and observe (see Definition 1′ and Exercise 9) that

$$\begin{aligned}
\tilde{\varrho}_{b+\delta,\alpha}(m_j\varphi) &= \sup_{x\in I}\left|x^{b+\delta+1}\left(x\frac{\partial}{\partial x}\right)^{\alpha}(m_j\varphi)\right| \\
&\le \sum_{0\le\beta\le\alpha}\binom{\alpha}{\beta}\sup_{x\in I}\left|x^{b+1}\left(x\frac{\partial}{\partial x}\right)^{\beta}\varphi\right|\cdot\sup_{x\in I}\left|x^{\delta}\left(x\frac{\partial}{\partial x}\right)^{\alpha-\beta}(m_j)\right| \\
&= \sum_{0\le\beta\le\alpha}\binom{\alpha}{\beta}\tilde{\varrho}_{b,\beta}(\varphi)\cdot\tilde{\varrho}_{-1+\delta,\alpha-\beta}(m_j)
\end{aligned}$$

and the right hand side tends to 0 as $j \to \infty$. Now it is clear that $\lim_{j\to\infty} m_j u[\varphi] = \lim_{j\to\infty} u[m_j\varphi] = 0$.

Exercises

1. Let $\chi(x) = 1$ for $x \in I = (0, t]$, $t \in \mathbf{R}_+$ and let $\tilde{\chi}(x) = 1$ for $0 < x \le t$ and $\tilde{\chi}(x) = 0$ if $x > t$. Show that

(i) $\chi \in M'_{(0)}(I) \subset D'(I)$, $\frac{d\chi}{dx} \in M'_{-1}(I) \subset D'(I)$, $\frac{d\chi}{dx} = -\delta(t)$ where $\delta_{(t)}[\varphi] = \varphi(t)$ for $\varphi \in M_{-1}(I)$ hence for $\varphi \in D(I)$,

(ii) $\tilde{\chi} \in D'_I(\mathbf{R}_+)$, $\frac{d\tilde{\chi}}{dx} = -\delta_{(t)}$ where $\delta_{(t)}[\tilde{\varphi}] = \tilde{\varphi}(t)$ for $\tilde{\varphi} \in D(\mathbf{R}_+)$,

(iii) $L\tilde{\chi} = \chi$, $\frac{d}{dx}L\tilde{\chi} = L\frac{d}{dx}\tilde{\chi}$ where L is the isomorphism from Proposition 3.1.

2. Let $\varphi \in M'_{(b)}$ be a function of class $C^k(I)$, $I = (0,t]$, $t \in \mathbf{R}_+$, $k \in \mathbf{N}$. Let $\tilde{\varphi}(x) = \varphi(x)$ for $x \in I$, $\tilde{\varphi}(x) = 0$ for $x > t$. Denote by $\varphi^{(j)}$ $(j = 1,\dots,k)$ the classical derivative of φ on I ($\varphi^{(1)} = \varphi'$). Show (see Exercise 3.5) that

(i) $\dfrac{d\varphi}{dx} = \varphi' - \varphi(t)\delta_{(t)}$ in $M'_{(b-1)}$,

(ii) $\dfrac{d\tilde{\varphi}}{dx} = \varphi' - \varphi(t)\delta_{(t)}$ in $D'(\mathbf{R}_+)$,

(iii) $\dfrac{d^k\varphi}{dx^k} = \varphi^{(k)}$ if and only if $\varphi^{(j)}(t) = 0$ for $j = 0,1,\dots,k-1$.

3. Let $\varphi \in M_a(I)$, $a \in \mathbf{R}^n$. Fix $m \in \mathbf{N}$ and j, $1 \le j \le n$. Let $\varphi_{j0}(x) = \varphi(x)$, $\varphi_{jp}(x) = \left(x_j\frac{\partial}{\partial x_j}\right)^p\varphi(x)$ for $x \in I$ $(p = 1,\dots,m)$. Show that

$$\left(x_j\frac{\partial}{\partial x_j}\right)^p \varphi = \varphi_{jp} \qquad \text{in } M'_{(-a-1)} \qquad (p = 1,\dots,m)$$

if and only if

$$\varphi_{jp}(x_1,\dots,x_{j-1},t_j,x_{j+1},\dots,x_n) = 0 \qquad (p = 0,1,\dots,m-1)$$

for $0 < x_\nu < t_\nu$, $(\nu = 1,\dots,j-1,j+1,\dots,n)$.

4. Let $a \in \mathbf{R}^n$, $m \in C^\infty(I)$. Then the following conditions are equivalent

(i) The mapping m: $M_{(\omega)} \to M_{(\omega+a+1)}$ is continuous for every $\omega \in (\mathbf{R} \cup \{\infty\})^n$,

(ii) $m \in M_{[a]}$.

5. Prove that the bilinear mapping

$$M_{[\omega_1]} \times M_{[\omega_1]} \ni (\varphi,\psi) \mapsto \varphi\psi \in M_{[\omega_1+\omega_2+1]}$$

is continuous in the sense that if $\varphi_\nu \to \varphi$ in $M_{[\omega_1]}$ and $\psi_\nu \to \psi$ in $M_{[\omega_2]}$ then $\varphi_\nu\psi_\nu \to \varphi\psi$ in $M_{[\omega_1+\omega_2+1]}$.

6. Prove that every Mellin multiplier is contained in $\bigcap\limits_{\varepsilon>0} M'_{(-\varepsilon)}$.

7. Prove that if m is a Mellin multiplier then $\left(x\frac{\partial}{\partial x}\right)^\alpha m$ is also so.

8. Investigate if the assumption $\beta_i \in \mathbf{R}$ in Example 2 may be replaced by $\beta_i \in \mathbf{C}$ $(i = 1,\dots,n)$.

9. Let m be a Mellin multiplier, $u \in M'_{(a)}$. Prove the Leibniz rules:

$$\left(x_j\frac{\partial}{\partial x_j}\right)(mu) = m\left(x_j\frac{\partial}{\partial x_j}u\right) + \left(x_j\frac{\partial}{\partial x_j}m\right)u \qquad \text{in } M'_{(a)},$$

$$\left(\frac{\partial}{\partial x_j}\right)(mu) = m\frac{\partial}{\partial x_j}u + \left(\frac{\partial}{\partial x_j}\right)mu \qquad \text{in } M'_{(a-e^j)},$$

where e^j was defined in Section 1.

10. Show that the function $g(x) = \frac{1}{x^2} \sin \frac{1}{x}$ for $0 < x \le t$ belongs to $M'_{(-1)}$ although the integral $\int_0^t \frac{1}{x^2} \sin \frac{1}{x} \sigma(x)\, dx$ absolutely convergent for $\sigma \in M_{(-2)}$ is not absolutely convergent for a suitable $\sigma \in M_{(-1)}$.

§7. THE MELLIN TRANSFORMATION IN THE SPACE OF MELLIN DISTRIBUTIONS

The Mellin transform of a bounded measurable function f on $\mathbf{R}_+$ (vanishing at infinity) was defined in the classical form as the integral

$$\int_0^\infty f(x) x^{z-1}\, dx \qquad \text{for} \quad \operatorname{Re} z > 0. \tag{1}$$

The importance of Mellin transforms of certain functions was recognized already by Euler (without, of course, mentioning the Mellin transformation itself). For instance the Γ function

$$\Gamma(z) = \int_0^\infty x^{z-1} e^{-x}\, dx$$

is the one-dimensional Mellin transform of e^{-x}.

The Mellin transformation used in this book differs from the classical one by the change of variable $z \mapsto -z$. Moreover we consider a generalization of (1) to n dimensions by integrating over $\mathbf{R}^n_+$ i.e.

$$\mathcal{M}f(z) = \int_{\mathbf{R}^n_+} f(x) x^{-z-1}\, dx \qquad \text{for} \quad \operatorname{Re} z < 0 \tag{2}$$

for f bounded, measurable and supported by a bounded polyinterval $I = (0, t]$ for some $t \in \mathbf{R}^n_+$. The integration is in fact over a bounded set I, which is important from the point of view of the local analysis of the Mellin transforms. The change of sign allows to establish a natural relation with the theory of (generalized) asymptotic expansions (see Subsection 14.1 and Appendix). We wish to underline the fact that we are always interested in the holomorphic extensions of the function $\mathcal{M}f$ defined by (2). Also we want to extend the Mellin transformation to distributions in $D'(I)$. This amounts to evaluating distributions on test functions $I \ni x \mapsto x^{-z-1}$. Clearly not all distributions in $D'(I)$ admit such evaluation. This is however possible for Mellin distributions $M'_{(\omega)}(I)$ since the space of test functions $M_{(\omega)}$ is "spanned" by the functions $x \mapsto x^{-z-1}$ for $\operatorname{Re} z < \omega$ (see Proposition 5.4). Here are the details:

1. The Mellin transformation in the space of Mellin distributions and its relations with the Fourier–Laplace transformation

Definition 1. Let $u \in M'_{(\omega)}(I)$ for some $\omega \in (\mathbf{R} \cup \{\infty\})^n$ and $I = (0, t]$, $t > 0$. We define the *Mellin transform* of u

$$\mathcal{M}u(z) = u[x^{-z-1}] \qquad \text{for} \quad \operatorname{Re} z < \omega. \tag{3}$$

It follows from Remark 5.1 that the Mellin transform $\mathcal{M}u$ of a Mellin distribution u is well-defined on the set Ω_u

$$\Omega_u = \bigcup_{\{a:\, u \in M'_{(a)}\}} \{\operatorname{Re} z < a\}.$$

Theorem 1. *Let $u \in D'(I)$ be a Mellin distribution, $I = (0, t]$, $t \in \mathbf{R}^n_+$. Then the Mellin transform $\mathcal{M}u$ is a holomorphic function on Ω_u and*

$$\frac{\partial}{\partial z_i} \mathcal{M}u(z) = u[-\ln x_i \cdot x^{-z-1}], \qquad i = 1, \ldots, n, \quad z \in \Omega_u. \tag{4}$$

Proof: First note that by Hartogs theorem (cf. [Hö3]), in order to prove the holomorphicity of $\mathcal{M}u$, it is enough to verify (4). For the proof we take $i = 1$ and choose arbitrarily a point $\mathring{z} \in \Omega_u$. Then there exists $a \in (\mathbf{R} \cup \{\infty\})^n$ such that $u \in M'_{(a)}$ and $\operatorname{Re} \mathring{z} < a$. Take $\alpha \in \mathbf{R}^n$ and $r > 0$ so that $\operatorname{Re} \mathring{z}_1 + r < \alpha_1 < a_1$. Let $\Delta z = (\Delta z_1, 0, \ldots, 0)$ and $|\Delta z_1| < r_1 < r$ (for some $r_1 > 0$). We write $z = (z_1, z')$ where $z' = (z_2, \ldots, z_n)$. Then $\operatorname{Re}(\mathring{z} + \Delta z) < \alpha < a$, hence $\mathring{z} + \Delta z \in \Omega_u$. To prove that $\frac{\partial \mathcal{M}u}{\partial z_1}(\mathring{z})$ exists and (4) holds it is enough to verify that

$$\mathcal{J}_{\Delta z} = \frac{\mathcal{M}u(\mathring{z} + \Delta z) - \mathcal{M}u(\mathring{z})}{\Delta z_1} - u[-\ln x_1 \cdot x^{-\mathring{z}-1}] \to 0 \qquad \text{as} \quad \Delta z_1 \to 0.$$

From the definition of $\mathcal{M}u$ we get

$$\mathcal{J}_{\Delta z} = u[\psi_{\Delta z}],$$

where

$$\psi_{\Delta z} = \frac{1}{\Delta z_1}\left(x^{-\mathring{z}-\Delta z-1} - x^{-\mathring{z}-1}\right) - \left(\frac{\partial}{\partial z_1} x^{-z-1}\right)\Big|_{z=\mathring{z}}$$

is a function in M_α as it is easily seen from the inequalities: $\operatorname{Re} \mathring{z} < \alpha$, $\operatorname{Re}(\mathring{z}+\Delta z) < \alpha$, which imply the relations $x^{-\mathring{z}-1} \in M_\alpha$, $x^{-\mathring{z}-\Delta z-1} \in M_\alpha$, $-\ln x_1 \cdot x^{-\mathring{z}-1} \in M_\alpha$. Since $u \in M'_{(a)} \subset M'_\alpha$, all we have to do is to prove that $\psi_{\Delta z} \to 0$ in M_α as $\Delta z_1 \to 0$, i.e. that for $\beta \in \mathbf{N}_0^n$

$$\sup_{x \in I} \left| x^{\alpha+\beta+1} \left(\frac{\partial}{\partial x}\right)^{\beta} \psi_{\Delta z} \right| \to 0 \qquad \text{as} \quad \Delta z_1 \to 0. \tag{5}$$

To compute $\left(\frac{\partial}{\partial x}\right)^{\beta}\psi_{\Delta z}$ note first that

$$(-1)^{\beta_j}\left(\frac{\partial}{\partial x_j}\right)^{\beta_j}(x_j^{-z_j-1}) = H_j(z_j, x_j) \qquad (j = 1, \dots, n)$$

with

$$H_j(z_j, x_j) = (z_j + 1)\cdot \ldots \cdot (z_j + \beta_j)x_j^{-z_j-1-\beta_j} \qquad (j = 1, \dots, n).$$

Denote $x' = (x_2, \dots, x_n)$ and $H'(z', x') = H_2(z_2, x_2)\cdot \ldots \cdot H_n(z_n, x_n)$. Commuting $\frac{\partial}{\partial x_1}$ with $\frac{\partial}{\partial z_1}$ we get

$$(6)\quad (-1)^{|\beta|}\left(\frac{\partial}{\partial x}\right)^{\beta}\psi_{\Delta z}(x) = \frac{1}{\Delta z_1}\Big(H_1(\mathring{z}_1 + \Delta z_1, x_1) - H_1(\mathring{z}_1, x_1)\Big)\, H'(\mathring{z}', x') - \\ - \Big(\frac{\partial}{\partial z_1}H_1(z_1, x_1)\Big)\Big|_{z_1 = \mathring{z}_1} H'(\mathring{z}', x').$$

We modify the right-hand side of (6) by applying to H_1 the Cauchy formula in the variable z_1. To this end choose a circle γ of radius r and center $\mathring{z}_1$ in the plane $z' = \mathring{z}'$. We note that for $|z_1 - \mathring{z}_1| < r$ we have

$$H_1(z_1, x_1) = \frac{1}{2\pi i}\int_\gamma \frac{H_1(\zeta_1, x_1)}{\zeta_1 - z_1}\, d\zeta_1, \qquad \frac{\partial}{\partial z_1}H_1(z_1, x_1) = \frac{1}{2\pi i}\int_\gamma \frac{H_1(\zeta_1, x_1)}{(\zeta_1 - z_1)^2}\, d\zeta_1.$$

Hence from (6) we find

$$(7)\qquad (-1)^{\beta}\left(\frac{\partial}{\partial x}\right)^{\beta}\psi_{\Delta z}(x) = \frac{1}{2\pi i}\Delta z_1 \int_\gamma \frac{H_1(\zeta_1, x_1)H'(\mathring{z}', x')}{(\zeta_1 - \mathring{z}_1 - \Delta z_1)(\zeta_1 - \mathring{z}_1)^2}\, d\zeta_1.$$

Observe that

$$|\zeta_1 - \mathring{z}_1 - \Delta z_1| \ge |\zeta_1 - \mathring{z}_1| - |\Delta z_1| > r - r_1 > 0.$$

Let $\beta \in \mathbb{N}_0^n$. Recalling the form of the functions H_1 and H' we get by (7) the estimate

$$(8)\qquad \left|x^{\alpha+\beta+1}\left(\frac{\partial}{\partial x}\right)^{\beta}\psi_{\Delta z}\right| \le A\,|\Delta z_1| \int_\gamma x_1^{\alpha_1 - \mathrm{Re}\,\zeta_1}\, d\zeta_1 \cdot (x')^{\alpha' - \mathrm{Re}\,\mathring{z}'},$$

where

$$A = A(r, r_1, \beta, \mathring{z}_1) = \frac{1}{2\pi}\frac{1}{(r - r_1)r^2}(r + (|\mathring{z}_1 + 1|)\cdots(r + |\mathring{z}_1 + \beta_1|)\times \\ \times |\mathring{z}_2 + 1|\cdot \ldots \cdot |\mathring{z}_2 + \beta_2|\cdot \ldots \cdot |\mathring{z}_n + 1|\cdot \ldots \cdot |\mathring{z}_n + \beta_n|.$$

Note that

$$\sup_{0 < x_1 \le t_1} \int_\gamma x_1^{\alpha_1 - \mathrm{Re}\,\zeta_1}\, d\zeta_1 \le 2\pi r \max(1, t_1^{r + \alpha_1 - \mathrm{Re}\,\mathring{z}_1})$$

combined with

$$\sup_{x \in I}\left|x^{\alpha+\beta+1}\left(\frac{\partial}{\partial x}\right)^{\beta}\psi_{\Delta z}\right| \le A\Delta z_1 \cdot \sup_{x \in I}\left(\int_\gamma x_1^{\alpha_1 - \mathrm{Re}\,\zeta_1}\, d\zeta_1 \cdot (x')^{\alpha' - \mathrm{Re}\,\mathring{z}'}\right)$$

given by (8), clearly imply (5).

It follows from Theorem 1 that the Mellin transform of a Mellin distribution u is holomorphic on the set where it makes sense. In the sequel by the Mellin transform we shall understand any holomorphic extension of the function $\mathcal{M}u(z)$ given by (3).

The Mellin transformation so defined possesses the following important *operational properties*:

If $u \in M'_{(\omega)}$, $\alpha \in \mathbf{C}^n$ then

$$\mathcal{M}(x^\alpha u)(z) = (\mathcal{M}u)(z-\alpha) \qquad \text{for} \quad \operatorname{Re} z < \omega + \operatorname{Re}\alpha. \tag{9}$$

If $u \in M'_{(\omega)}$, $\gamma \in \mathbf{N}_0^n$ then (see Exercise 2)

$$\mathcal{M}\Big(\Big(\frac{\partial}{\partial x}\Big)^\gamma u\Big)(z) = (z^\gamma + 1)(\mathcal{M}u)(z+\gamma) \qquad \text{for} \quad \operatorname{Re} z < \omega - \gamma, \quad |\gamma| = 1, \tag{10}$$

$$\mathcal{M}\Big(\Big(x\frac{\partial}{\partial x}\Big)^\gamma u\Big)(z) = z^\gamma \mathcal{M}u(z) \qquad \text{for} \quad \operatorname{Re} z < \omega. \tag{11}$$

The proofs make use of the properties of multiplication and differentiation in spaces of Mellin distributions discussed in Subsection 6.1.

Now we proceed to discuss the relations between the Mellin transformation $\mathcal{M}$ and the transformation $\mathcal{M}_\alpha$ examined in Subsection 4.2

Proposition 1. *Let* $u \in M'_{(\omega)}(I)$. *Then* $\mathcal{M}u$ *is holomorphic for* $\operatorname{Re} z < \omega$ *and* $u \in \mathfrak{M}'_\alpha(I)$ *for every* $\alpha < \omega$. *The tempered distribution* $\mathcal{M}_\alpha u$ *is a function:*

$$(\mathcal{M}_\alpha u)(\beta) = \mathcal{M}u(\alpha + i\beta) = (u \circ \mu)[e^{(\alpha+i\beta)y}] \qquad \textit{for} \quad \beta \in \mathbf{R}^n. \tag{12}$$

Moreover $\mathcal{M}_\alpha \colon M'_{(\omega)} \to S'$ *is continuous for* $\alpha < \omega$.

Proof: Let $u \in M'_{(\omega)}(I)$. By Theorem 1 $\mathcal{M}u$ is holomorphic for $\operatorname{Re} z < \omega$. By Proposition 5.5 $u \in \mathfrak{M}'_\alpha(I)$ for every $\alpha < \omega$, hence (cf. Definition 4.4) $\mathcal{M}_\alpha u \in S'$. To see that $\mathcal{M}_\alpha u$ is a function take arbitrarily $\psi \in S$ and observe that by (4.5) and Definition 4.4 we get

$$\mathcal{M}_\alpha u[\psi] = (2\pi)^{\frac{n}{2}} u[x^{-\alpha-1}(\mathcal{F}^{-1}\psi) \circ \mu^{-1}]. \tag{13}$$

Hence by the definition of the inverse Fourier transformation in S, definition (3) of Mellin transformation and Exercise 3 we get

$$\begin{aligned}\mathcal{M}_\alpha u[\psi] &= u\Big[x^{-\alpha-1}\int_{\mathbf{R}^n} x^{-i\beta}\psi(\beta)\,d\beta\Big] = \int_{\mathbf{R}^n} \psi(\beta) u\big[x^{-\alpha-1-i\beta}\big]\,d\beta \\ &= \int_{\mathbf{R}^n} \psi(\beta)\mathcal{M}u(\alpha+i\beta)\,d\beta,\end{aligned}$$

which means that $\mathcal{M}_\alpha u(\beta) = \mathcal{M}u(\alpha + i\beta)$ for $\beta \in \mathbf{R}^n$. On the other hand by the definition of μ and of the substitution in distribution (cf. Subsection 3.5) we derive in view of (13) and Exercise 3 that

$$\mathcal{M}_\alpha u[\psi] = (2\pi)^{\frac{n}{2}} u\left[(e^{\alpha y}\mathcal{F}^{-1}\psi)\circ\mu^{-1}\cdot\frac{1}{x^1}\right] = (2\pi)^{\frac{n}{2}}(u\circ\mu)[e^{\alpha y}\mathcal{F}^{-1}\psi]$$

$$= (u\circ\mu)\left[\int_{\mathbf{R}^n} e^{(\alpha+i\beta)y}\psi(\beta)\,d\beta\right] = \int_{\mathbf{R}^n}\psi(\beta)(u\circ\mu)[e^{(\alpha+i\beta)y}]\,d\beta$$

which means that $\mathcal{M}_\alpha u(\beta) = (u\circ\mu)[e^{(\alpha+i\beta)y}]$ for $\beta \in \mathbf{R}^n$.

To prove the last assertion take a sequence $u_\nu \to 0$ in $M'_{(\omega)}$. By Proposition 5.5 $u_\nu \to 0$ in $\mathfrak{M}'_\alpha(I)$ for $\alpha < \omega$, hence by Proposition 4.2 $\mathcal{M}_\alpha u_\nu \to 0$ in S'.

From Proposition 1 and Theorem 1 we derive

Corollary 1. *Let $u \in M'_{(\omega)}(I)$, $I = (0,t]$. Then $(\mathcal{M}u)(z) = (u\circ\mu)[e^{zy}]$ is holomorphic for* $\operatorname{Re} z < \omega$.

Now we proceed to discuss the relations between the Mellin and the Fourier–Laplace transformation. We recall the definition of the latter.

If $v \in E'(\mathbf{R}^n)$ by Theorem 4.2

$$(\mathcal{F}^{-1}v)(\beta) = (2\pi)^{-\frac{n}{2}} v[e^{iy\beta}] \qquad \text{for } \beta \in \mathbf{R}^n.$$

We define

$$(\mathcal{L}v)(i\beta) = v[e^{iy\beta}] \qquad \text{for } \beta \in \mathbf{R}^n$$

hence

$$(\mathcal{L}v)(i\beta) = (2\pi)^{\frac{n}{2}}(\mathcal{F}^{-1}v)(\beta) \qquad \text{for } \beta \in \mathbf{R}^n.$$

In general for distributions supported by an infinite polyinterval $\{y \in \mathbf{R}^n\colon h_0 < y\}$, for some $h_0 \in \mathbf{R}^n$, we define the (right-sided) *Laplace transform* $\mathcal{L}v$ of v by the formula

$$\mathcal{L}v(z) = v[e^{zy}] \tag{14}$$

for those z (if any) for which it makes sense.

Let $v = u\circ\mu$ where $u \in M'_{(\omega)}$, $\omega \in \mathbf{R}^n$, $\mu(y) = e^{-y}$ for $y \in \mathbf{R}^n$.

By Corollary 1 we get

Corollary 2. *Let $u \in M'_{(\omega)}(I)$, $\omega \in \mathbf{R}^n$, $I = (0,t]$. Then $\mathcal{L}(u\circ\mu)(z) = \mathcal{M}u(z)$ is holomorphic for* $\operatorname{Re} z < \omega$.

From Proposition 1 we derive now two propositions which will turn useful in Section 13:

Proposition 2. *Let* $\varphi \in C_0^\infty(\mathbf{R}^n)$, $\operatorname{supp}\varphi \cap \mathbf{R}^n_+ \subset (0,t] = I$. *Then for every* $\varepsilon \in \mathbf{R}^n_+$ $x^\varepsilon\varphi \in \mathfrak{M}_{-1}$ *and we have the equality:*

$$\mathcal{M}_0(x^\varepsilon\varphi)(\beta) = \mathcal{M}(x^\varepsilon\varphi)(i\beta) \qquad \text{for } \beta \in \mathbf{R}^n. \tag{15}$$

The above function belongs to $S(\mathbf{R}^n)$.

Proof: Clearly $\varphi \in M_{-1}$ hence $\varphi \in M'_{(0)}$ and $x^\varepsilon\varphi \in M'_{(\varepsilon)}$. By Proposition 1 $x^\varepsilon\varphi \in \mathfrak{M}'_0$ and (15) holds. Note that $(x^\varepsilon\varphi)\circ\mu = e^{-\varepsilon y}\varphi(e^{-y}) \in S(\mathbf{R}^n)$ which by the definition of $\mathfrak{M}_{-1}$ means that $x^\varepsilon\varphi \in \mathfrak{M}_{-1}$ and by the definition of $\mathfrak{M}'_0$ $x^\varepsilon\varphi \in \mathfrak{M}'_0$. Thus $\mathcal{M}_0(x^\varepsilon\varphi) = (2\pi)^{\frac{n}{2}}\mathcal{F}^{-1}(x^\varepsilon\varphi\circ\mu) \in S(\mathbf{R}^n)$.

Proposition 3. *Let* φ *be as in Proposition 2 and let* $u \in M'_{(\omega)}$, $\omega \in (\mathbf{R}\cup\{+\infty\})^n$. *Then for every* $\varepsilon \in \mathbf{R}^n_+$

$$\mathcal{M}(\varphi u)(\alpha+i\beta) = (2\pi)^{-n}\int_{\mathbf{R}^n} \mathcal{M}u(\alpha+\varepsilon+i\gamma)\mathcal{M}\varphi(-\varepsilon+i(\beta-\gamma))\,d\gamma \tag{16}$$
$$\text{for } \alpha < \omega-\varepsilon, \quad \beta \in \mathbf{R}^n.$$

Proof: Since $\varphi \in M_{-1} \subset M_{[-1]}$ is a Mellin multiplier $\varphi u \in M'_{(\omega)}$ and by Proposition 1 we have for $\alpha < \omega$, $\varphi u \in \mathfrak{M}'_\alpha$ and

$$\mathcal{M}_\alpha(\varphi u)(\beta) = \mathcal{M}(\varphi u)(\alpha+i\beta) \qquad \text{for } \beta \in \mathbf{R}^n. \tag{17}$$

Since $x^{-\varepsilon}u \in M'_{(\omega-\varepsilon)}$ it follows from Proposition 1 that $x^{-\varepsilon}u \in \mathfrak{M}'_\alpha$ for $\alpha < \omega-\varepsilon$ and

$$\mathcal{M}_\alpha(x^{-\varepsilon}u)(\beta) = \mathcal{M}(x^{-\varepsilon}u)(\alpha+i\beta) = \mathcal{M}u(\alpha+\varepsilon+i\beta) \qquad \text{for } \beta \in \mathbf{R}^n. \tag{18}$$

By (17), Proposition 2, Proposition 4.4, (15) and (18) we get for $\alpha < \omega-\varepsilon$, $\beta \in \mathbf{R}^n$

$$\begin{aligned}\mathcal{M}(\varphi u)(\alpha+i\beta) &= \mathcal{M}_\alpha(x^\varepsilon\varphi\cdot x^{-\varepsilon}u)(\beta) = (2\pi)^{-n}(\mathcal{M}_0(x^\varepsilon\varphi)*\mathcal{M}_\alpha(x^{-\varepsilon}u))(\beta)\\ &= (2\pi)^{-n}(\mathcal{M}\varphi(-\varepsilon+i\,\cdot\,)*\mathcal{M}u(\alpha+\varepsilon+i\,\cdot\,))(\beta).\end{aligned}$$

Hence follows (16).

Remark 1. Let φ, u be as in Proposition 3. Take $\varepsilon > 0$ and a such that $a+\varepsilon < \omega$. Then

$$\mathcal{M}(\varphi u)(\alpha+i\beta) = (2\pi)^{-n}\int_{\mathbf{R}^n}\mathcal{M}u(a+\varepsilon+i\gamma)\mathcal{M}\varphi(\alpha-a-\varepsilon+i(\beta-\gamma))\,d\gamma \qquad \text{for } \alpha \le a.$$

In fact $x^{-\varepsilon-a}u \in \mathfrak{M}'_0$, $x^{\varepsilon+a}\varphi \in \mathfrak{M}_{-\alpha-1}$ if $\alpha \le a$ and by (17) and Exercise 4.8

$$\begin{aligned}\mathcal{M}(\varphi u)(\alpha+i\beta) &= \mathcal{M}_\alpha(x^{\varepsilon+a}\varphi \cdot x^{-\varepsilon-a}u)(\beta)\\ &= (2\pi)^{-n}(\mathcal{M}_\alpha(x^{\varepsilon+a}\varphi) * \mathcal{M}_0(x^{-\varepsilon-a}u))(\beta).\end{aligned}$$

2. Examples of Mellin Transforms of Some Functions

We start this Subsection with examples of Mellin transforms of some important functions supported by an interval $I = (0,t] \subset \mathbf{R}_+$. It turns out that in the cases considered below the Mellin transforms defined by (2) on a half-plane $\operatorname{Re} z < 0$ extend to meromorphic functions on $\mathbf{C}$ or to functions holomorphic outside the half-axis $[0,+\infty)$.

Below, we denote by χ_I the characteristic function of the interval I.

Example 1. Let $j \in \mathbf{N}_0$, $f_I(x) = \chi_I \cdot x^j$ then formula (2) defines

$$(\mathcal{M}f_I)(z) = \frac{t^{j-z}}{j-z} \quad \text{for} \quad \operatorname{Re} z < j. \tag{19}$$

The right-hand side of (19) is a meromorphic function on the complex plane $\mathbf{C}$ with a simple pole at the point $z = j$ with residuum -1. For $j = 0$ (19) gives $\mathcal{M}(\chi_I)(z) = -t^{-z}/z$ for $z \neq 0$.

Example 2. Let $f = \sum_{j=0}^{\infty} a_j x^j$ be an analytic function with the radius of convergence greater than t. Then by Example 1

$$(\mathcal{M}\chi_I f)(z) = t^{-z}\sum_{j=0}^{\infty} a_j t^j \frac{1}{j-z}$$

is a meromorphic function with simple poles at the points $j = 0,1,\ldots$, with $\operatorname{Res}_{z=j}(\mathcal{M}\chi_I f)(z) = -a_j$ $(j = 0,1,\ldots)$. Moreover for z such that $|z-j| > \varepsilon > 0$ for $j = 0,1,\ldots$,

$$|(M\chi_I f)(z)| \le C_\varepsilon t^{-\operatorname{Re} z}$$

with some constant $C_\varepsilon < \infty$.

Example 3. Let $l \in \mathbf{N}_0$, $\varrho \in \mathbf{C}$, $f_I(x) = \chi_I \cdot x^\varrho(\ln x)^l$. Suppose $\operatorname{Re} z < \operatorname{Re}\varrho$. Then

$$\begin{aligned}(\mathcal{M}f_I)(z) &= \int_0^t x^{\varrho-z-1}(\ln x)^l\,dx = (-1)^l\frac{d^l}{dz^l}\lim_{\varepsilon\to 0+}\int_\varepsilon^t x^{\varrho-z-1}\,dx\\ &= (-1)^l t^\varrho \frac{d^l}{dz^l}\left(\frac{t^{-z}}{\varrho-z}\right).\end{aligned}$$

Thus the Mellin transform of f_I is a meromorphic function

$$(\mathcal{M}f_I)(z) = (-1)^l t^{\varrho} \frac{d^l}{dz^l}\left(\frac{t^{-z}}{\varrho - z}\right)$$

which has a pole of order $l+1$ at the point $z = \varrho$.

Example 4. Consider

$$f(x) = \sum_{j=0}^{\infty}\sum_{l=0}^{k} c_{lj} x^{\lambda+j} (\ln x)^l \qquad \text{for } 0 < x \leq t$$

and suppose that the series $\sum\limits_{j=0}^{\infty} c_{lj} x^j$ $(l = 0, 1, \ldots, k)$ have the radii of convergence greater than t. By Example 3

$$(\mathcal{M}\chi_I f)(z) = \sum_{j=0}^{\infty}\sum_{l=0}^{k} c_{lj} (-1)^l t^{\lambda+j} \frac{d^l}{dz^l}\left(\frac{t^{-z}}{\lambda + j - z}\right),$$

i.e. $\mathcal{M}\chi_I f$ is meromorphic on $\mathbf{C}$ with poles at points $z = \lambda + j$ $(j = 0, 1, \ldots)$ of order at most $k+1$.

Example 5. If $f \in C^k([0,t])$ then

$$\mathcal{M}f(z) = \sum_{j=0}^{k-1} c_j \frac{t^{j-z}}{j-z} + N(z)$$

where

$$c_j = \frac{f^{(j)}(0)}{j!} \qquad (j = 0, 1, \ldots, k-1)$$

and N is holomorphic for $\operatorname{Re} z < k$.

Proof: By the Taylor formula

$$f(x) = \sum_{j=0}^{k-1} \frac{f^{(j)}(0)}{j!} x^j + R(x) x^k$$

for $0 \leq x \leq t$ with a continuous function R. Since $\mathcal{M}(R(x)x^k)$ is holomorphic on the half-plane $\operatorname{Re} z < k$ the assertion follows from Example 1.

Example 6. Let $c \in \mathbf{R}$, $\theta \in \mathbf{C}$, $I = (0,t]$, $0 < t < e^c$. If $\operatorname{Re}\theta < k$, $k \in \mathbf{N}_0$ then (cf. Exercise 5.13)

$$(20) \quad \mathcal{M}(\chi_I(x)\cdot(-\ln x + c)^\theta)(z) = \frac{(-1)^k}{\Gamma(-\theta+k)} \int_0^\infty \varrho^{-\theta+k-1} \frac{d^k}{d\varrho^k}\left(\frac{e^{-c\varrho}t^{\varrho-z}}{\varrho - z}\right) d\varrho$$

for $z \in \mathbf{C} \setminus [0,+\infty)$.

In the formula (20) the right-hand side is a holomorphic function on $\mathbf{C} \setminus [0,+\infty)$.

Proof: By definition $\Gamma(-\theta) = \int_0^\infty \xi^{-\theta-1} e^{-\xi}\, d\xi$, $\operatorname{Re}\theta < 0$ (see [Wa]).

By substitution $\xi = -\varrho \ln(xe^{-c})$, $(x < e^c)$ we easily derive the formula

$$(21) \quad (-\ln x + c)^\theta = \frac{1}{\Gamma(-\theta)} \int_0^\infty \varrho^{-\theta-1} e^{-c\varrho} x^\varrho \, d\varrho \qquad \text{for } 0 < x < e^c, \quad \operatorname{Re}\theta < 0$$

which implies that

$$\mathcal{M}(\chi_I(x)\cdot(-\ln x + c)^\theta)(z) = \frac{1}{\Gamma(-\theta)} \int_0^t x^{-z-1} \left\{ \int_0^\infty x^\varrho \varrho^{-\theta-1} e^{-c\varrho}\, d\varrho \right\} dx.$$

Observe that for $\operatorname{Re} z < \varrho$

$$\int_0^t x^{\varrho - z - 1}\, dx = \frac{t^{\varrho - z}}{\varrho - z}.$$

Hence by the Fubini–Tonelli theorem we get for $\operatorname{Re} z < 0$

$$(22) \qquad \mathcal{M}(\chi_I(x)\cdot(-\ln x + c)^\theta)(z) = \frac{1}{\Gamma(-\theta)} \int_0^\infty \varrho^{-\theta-1} e^{-c\varrho} \frac{t^{\varrho-z}}{\varrho - z}\, d\varrho.$$

The last integral is absolutely convergent in $\mathbf{C} \setminus [0,\infty)$ hence (20) holds for $k = 0$ and the Mellin transform of the function $(0,t] \ni x \mapsto (c - \ln x)^\theta$ $(t < e^c$, $\operatorname{Re}\theta < 0)$ is holomorphic in the set $\mathbf{C} \setminus [0,+\infty)$. Now, consider the function in (21) as a function of the complex variable θ. It is holomorphic in the half-plane $\operatorname{Re}\theta < 0$ (here x is a fixed number $0 < x < e^c$). Integrating by parts the right-hand side of (21) we get for $\operatorname{Re}\theta < 0$ the formula

$$\frac{1}{\Gamma(-\theta)} \int_0^\infty \varrho^{-\theta-1} e^{(-c+\ln x)\varrho}\, d\varrho = \frac{1}{\theta\Gamma(-\theta)} \int_0^\infty \varrho^{-\theta} \frac{d}{d\varrho} e^{(-c+\ln x)\varrho}\, d\varrho.$$

We note that the expression on the right is holomorphic in a larger set $\operatorname{Re}\theta < 1$. This together with (21) implies that

$$(-\ln x + c)^\theta = \frac{1}{\theta\Gamma(-\theta)} \int_0^\infty \varrho^{-\theta} \frac{d}{d\varrho} e^{(-c+\ln x)\varrho}\, d\varrho \qquad \text{for } \operatorname{Re}\theta < 1.$$

Proceeding by induction we derive that for an arbitrary $k \in \mathbb{N}_0$, $0 < x < e^c$

$$(23)\ (-\ln x + c)^\theta = \frac{1}{\theta(\theta-1)\cdots(\theta-k+1)\Gamma(-\theta)} \int_0^\infty \varrho^{-\theta+k-1} \frac{d^k}{d\varrho^k} e^{(-c+\ln x)\varrho}\, d\varrho$$
$$= \frac{(-1)^k}{\Gamma(-\theta+k)} \int_0^\infty \varrho^{-\theta+k-1} \frac{d^k}{d\varrho^k}(e^{-c\varrho} x^\varrho)\, d\varrho$$

is holomorphic for $\operatorname{Re}\theta < k$. Now (20) follows from (23) in a similar way that (20) for $k = 0$ followed from (21).

Example 7 (see also Exercise 14). Let $c \in \mathbb{R}$, $0 < t < e^c$ and let $\gamma = -\Gamma'(1)$ be the *Euler constant*. Then

$$\mathcal{M}(\chi_I(x) \cdot \ln(-\ln x + c)) = \frac{\gamma}{z} t^{-z} + \int_0^\infty \ln \varrho \frac{d}{d\varrho}\left(\frac{e^{-c\varrho} t^{\varrho - z}}{\varrho - z}\right) d\varrho$$

is a holomorphic function on $\mathbb{C} \setminus [0, \infty)$.

Proof: Let $t < e^c$. Since

$$\ln(-\ln x + c) = \frac{\partial}{\partial\theta}(-\ln x + c)^\theta \Big|_{\theta=0},$$

it follows from (20) with $k = 1$ that

$$\mathcal{M}(\chi_I(x) \cdot \ln(-\ln x + c))(z) = \frac{\partial}{\partial\theta} \mathcal{M}(\chi_I \cdot (-\ln x + c)^\theta)(z) \Big|_{\theta=0}$$
$$= \int_0^\infty \frac{\partial}{\partial\theta}\left(\frac{-\varrho^{-\theta}}{\Gamma(-\theta+1)}\right)\Big|_{\theta=0} \cdot \frac{d}{d\varrho}\left(\frac{e^{-c\varrho} t^{\varrho-z}}{\varrho - z}\right) d\varrho$$
$$= \int_0^\infty (\ln \varrho + \gamma) \frac{d}{d\varrho}\left(\frac{e^{-c\varrho} t^{\varrho-z}}{\varrho - z}\right) d\varrho$$
$$= \frac{\gamma}{z} t^{-z} + \int_0^\infty \ln \varrho \frac{d}{d\varrho}\left(\frac{e^{-c\varrho} t^{\varrho-z}}{\varrho - z}\right) d\varrho,$$

since

$$\frac{\partial}{\partial\theta}\left(\frac{\varrho^{-\theta}}{\theta\Gamma(-\theta)}\right)\Big|_{\theta=0} = \ln \varrho + \gamma.$$

Next we continue with examples of Mellin transforms of functions in several variables. We start with the following "division theorem" for smooth functions (see e.g. [Hö2]) which for simplicity we state in two dimensions

Lemma 1. *Let $g(x_1, x_2)$ be a smooth function in a neighbourhood of $0 \in \mathbf{R}^2$. Then for every $r \in \mathbf{N}$*

$$g(x_1,x_2) = \sum_{j=0}^{r-1} x_1^j n_j(x_2) + x_1^r \sum_{j=0}^{r-1} x_2^j s_j(x_1) + x_1^r x_2^r p(x_1,x_2), \tag{24}$$

where n_j, s_j are smooth functions in a neighbourhood of $0 \in \mathbf{R}$ for $j = 0, \ldots, r-1$ and p is a smooth function about $0 \in \mathbf{R}^2$.

Proof: Applying to g the Taylor formula with respect to x_1 (with x_2 fixed) we arrive at

$$g(x_1,x_2) = \sum_{j=0}^{r-1} x_1^j n_j(x_2) + x_1^r q(x_1,x_2)$$

where $n_j(x_2) = \frac{1}{j!}\frac{\partial^j}{\partial x_1^j} g(x_1,x_2)\big|_{x_1=0}$ $(j = 0, 1, \ldots, r-1)$ is smooth about $0 \in \mathbf{R}$ and $q(x_1,x_2) = \frac{1}{(r-1)!}\int_0^1 \frac{\partial^r g}{\partial x_1^r}\big|_{(tx_1,x_2)}(1-t)^{r-1}\,dt$ is smooth about $0 \in \mathbf{R}^2$. Again by the same formula applied to q in variable x_2 we get the desired expression for g.

Example 8 (The Mellin transform of a smooth function). Let φ be a smooth function in a neighbourhood of a polyinterval $\bar{I} = \{0 \le x \le t\} \subset \mathbf{R}^n$ and let χ_I be the characteristic function of I. Then for every $r \in \mathbf{N}_0$

$$\mathcal{M}(\chi_I\varphi)(z) - \sum_{\substack{\alpha \in \mathbf{N}_0^n \\ \alpha < \mathbf{r}}} c_\alpha \frac{t^{\alpha - z}}{(\alpha - z)^1} = \sum_{k=1}^{n} N_k(z),$$

where $\mathbf{r} = (r, \ldots, r) \in \mathbf{N}_0^n$, $c_\alpha = \frac{\partial^\alpha \varphi}{\partial x^\alpha}(0)$ and N_k is holomorphic in the set $\{z \in \mathbf{C}^n\colon z_l \ne j,\ j = 0, 1, \ldots, r-1 \text{ and } l \ne k\} \cap \{z \in \mathbf{C}^n\colon \operatorname{Re} z < \mathbf{r}\}$.

Proof: For simplicity of notation we assume $n = 2$. We write φ in the form (24)

$$\chi_I\varphi(x_1,x_2) = \chi_I \sum_{j=0}^{r-1} x_1^j n_j(x_2) + \chi_I x_1^r \sum_{j=0}^{r-1} x_2^j s_j(x_1) + \chi_I x_1^r x_2^r p(x_1,x_2) \tag{25}$$

where n_j, s_j and p are as in Lemma 1. Let χ^1 (resp. χ^2) be the characteristic function of the interval $(0, t_1] \subset \mathbf{R}_+$ (resp. $(0, t_2] \subset \mathbf{R}_+$). Then $\chi_I(x_1, x_2) = \chi^1(x_1) \cdot \chi^2(x_2)$ and by computing the Mellin transform of (25) and using the operational properties (9) we get

$$\mathcal{M}(\chi_I\varphi)(z) = \sum_{j=0}^{r-1} \frac{t_1^{j-z_1}}{j - z_1}\mathcal{M}^1(\chi^2 n_j)(z_2) + \sum_{j=0}^{r-1} \frac{t_2^{j-z_2}}{j - z_2}\mathcal{M}^1(\chi^1 s_j)(z_1 - r) +$$
$$+ \mathcal{M}(\chi_I p)(z_1 - r, z_2 - r)$$

since for the one-dimensional Mellin transformation $\mathcal{M}^1$ we have by (19)

$$\mathcal{M}^1(\chi^\nu(x_\nu)x_\nu^j)(z_\nu) = \frac{t_\nu^{j-z_\nu}}{j-z_\nu}, \qquad \nu = 1,2.$$

Now, $\mathcal{M}(\chi_I p)(z_1 - r, z_2 - r)$ is holomorphic on $\{z \in \mathbf{C}^2\colon \operatorname{Re} z < \mathbf{r}\}$ (since p is bounded and hence in $M'_{(0)}$) and $\mathcal{M}^1(\chi^1 s_j)(z_1 - r)$ is holomorphic for $\operatorname{Re} z_1 < r$ (for the same reason). Next we know by Example 5 and Lemma 1 that

$$\mathcal{M}^1(\chi^2 n_j)(z_2) = \sum_{l=0}^{r-1} c_{jl}\frac{t_2^{l-z_2}}{l-z_2} + h_j(z_2)$$

where

$$c_{jl} = \frac{1}{l!}\frac{d^l n_j(x_2)}{dx_2^l}\bigg|_{x_2=0} = \frac{1}{j!l!}\frac{\partial^j\partial^l\varphi(0)}{\partial x_1^j \partial x_2^l}$$

and $h_j(z_2)$ is holomorphic for $\operatorname{Re} z_2 < r$. This ends the proof.

Example 9. Let $a, b \in \mathbf{R}$, $0 \le a < b$ and let $\chi_{a,b}$ be the characteristic function of the set $\Delta_{a,b} = \{(x_1,x_2) \in \mathbf{R}^2\colon 0 < x_1 \le 1,\ ax_1 < x_2 < bx_1\}$. Then

$$\mathcal{M}\chi_{a,b}(z) = \frac{1}{z_1+z_2}\frac{1}{z_2}(b^{-z_2} - a^{-z_2}) \qquad \text{for } z_2 \ne 0,\ z_1 + z_2 \ne 0. \tag{26}$$

Observe that if $a > 0$ then $\mathcal{M}\chi_{a,b}$ is holomorphic outside $\{(z_1,z_2)\colon z_1 + z_2 = 0\}$. Indeed the function $f(z_2) \stackrel{\text{def}}{=} b^{-z_2} - a^{-z_2}$ has $f(0) = 0$ hence the function $\mathbf{C} \ni z_2 \mapsto (b^{-z_2} - a^{-z_2})/z_2$ is entire.

Example 10. Let $\theta \in \mathbf{C}$, $0 < b < 1$ and let $\chi_{0,b}$ be as in Example 9. Then

$$\mathcal{M}((x_1+x_2)^\theta \chi_{0,b}(x_1,x_2))(z) = \frac{1}{z_1+z_2-\theta}\sum_{j=0}^{\infty}\binom{\theta}{j}\frac{b^{j-z_2}}{z_2-j} \tag{27}$$

and the series is absolutely convergent for $z_2 \ne 0, 1, \ldots$. Thus the Mellin transform is holomorphic on $\mathbf{C}^2 \setminus \{z \in \mathbf{C}^2\colon z_2 = j\ (j = 0,1,\ldots) \text{ or } z_1 + z_2 = \theta\}$.

Proof: We have

$$\chi_{0,b}\cdot(x_1+x_2)^\theta = \chi_{0,b}\sum_{j=0}^{\infty}\binom{\theta}{j}x_1^{\theta-j}x_2^j,$$

$$\binom{\theta}{j} \stackrel{\text{def}}{=} \frac{\theta(\theta-1)\cdot\ldots\cdot(\theta-j+1)}{j!} \quad \text{for } j \ge 1, \qquad \binom{\theta}{0} \stackrel{\text{def}}{=} 1$$

and the series is absolutely uniformly convergent since on $\operatorname{supp}\chi_{0,b}$, $\left|\frac{x_2}{x_1}\right| \le b < 1$. The result now follows from (26) and the operational rules (9) for the Mellin transformation.

Example 11. Let $\theta \in \mathbf{R}$. Let χ be the characteristic function of the square $\{0 < x \leq 1\}$. Then

$$(28) \qquad \mathcal{M}(\chi \cdot (x_1 + x_2)^\theta)(z) \cong \frac{1}{z_1 + z_2 - \theta}\left(\sum_{j=0}^{\infty}\binom{\theta}{j}\frac{1}{z_2 - j} + \sum_{l=0}^{\infty}\binom{\theta}{l}\frac{1}{z_1 - l}\right),$$

where the sign $\cong$ means that for every $\varrho = (\varrho_1, \varrho_2) \in \mathbf{N}_0^2$

$$(z_1 + z_2 - \theta)\mathcal{M}(\chi \cdot (x_1 + x_2)^\theta)(z) - \sum_{j=0}^{\varrho_2}\binom{\theta}{j}\frac{1}{z_2 - j} - \sum_{l=0}^{\varrho_1}\binom{\theta}{l}\frac{1}{z_1 - l}$$

is holomorphic for $\operatorname{Re} z < \varrho + 1$.

For $\theta \geq -1$ the sign $\cong$ can be replaced by equality. Outside singularities the series on the right in (28) are absolutely convergent for $\theta > -1$, conditionally convergent for $\theta = -1$ and divergent for $\theta < -1$.

Proof: Formula (28) follows from Example 10, Exercise 9, their analogues for the variables x_1, x_2 interchanged and the observation that for $j = 0, 1, \ldots$

$$\frac{b^{j - z_2}}{z_2 - j} = \frac{1}{z_2 - j} + H_{j,b}(z_2)$$

where $H_{j,b}$ is a holomorphic function on $\mathbf{C}$. To discuss the convergence of the series in (28) we use the following estimation:

For $\theta \in \mathbf{C}$ there exist constants $C_1(\theta)$, $C_2(\theta)$ such that

$$C_1(\theta) j^{-\operatorname{Re}\theta - 1} \leq \left|\binom{\theta}{j}\right| \leq C_2(\theta) j^{-\operatorname{Re}\theta - 1}.$$

Moreover if $\theta \notin \mathbf{N}_0$, $C_1(\theta)$ can be taken greater than zero. The proof of the estimation follows easily from the Stirling formula

$$\Gamma(z) e^z z^{-z + \frac{1}{2}} \to \sqrt{2\pi} \qquad \text{as} \quad \operatorname{Re} z \to +\infty$$

since

$$\binom{\theta}{j} = (-1)^j \frac{\Gamma(j - \theta)}{\Gamma(-\theta)\Gamma(j + 1)}$$

and $\lim_{j \to \infty} (1 + \theta/j)^j = e^\theta$.

It follows from the above estimation that the series in (28) is absolutely convergent if $\theta > -1$ (here $\theta \in \mathbf{R}$). Consider now the case $\theta = -1$. We have $\binom{-1}{j} = (-1)^j$. Write $z_2 = \alpha_2 + i\beta_2$. Then

$$\operatorname{Re}\frac{1}{z_2 - j} = \frac{\alpha_2 - j}{(\alpha_2 - j)^2 + \beta_2^2}, \qquad \operatorname{Im}\frac{1}{z_2 - j} = \frac{-\beta_2}{(\alpha_2 - j)^2 + \beta_2^2}.$$

Thus $\operatorname{Re}\frac{1}{z_2-j} \to 0$, $\operatorname{Im}\frac{1}{z_2-j} \to 0$ as $j \to \infty$. Moreover $\operatorname{sgn}\left(\operatorname{Im}\frac{1}{z_2-j}\right) = -\operatorname{sgn}\beta_2$ and for large j, $\operatorname{sgn}\left(\operatorname{Re}\frac{1}{z_2-j}\right) = -1$. Thus the series $\sum_{j=1}^{\infty}\binom{-1}{j}\frac{1}{z_2-j}$ is conditionally convergent i.e. the real and the imaginary parts are so and from the Abel theorem it follows that there exist the radial limits

$$\lim_{\substack{b\to 1\\ b<1}} \sum_{j=0}^{\infty}\binom{-1}{j}\frac{b^j}{z_2-j} = \sum_{j=0}^{\infty}\binom{-1}{j}\frac{1}{z_2-j}. \tag{29}$$

The divergence of the series for $\theta < -1$ follows by a similar argument since $\binom{\theta}{2k} > 0$ and $\binom{\theta}{2k+1} < 0$ for $k \in \mathbf{N}_0$ by using the estimation on $\left|\binom{\theta}{j}\right|$ from below. To see that $\cong$ in (28) can be replaced by equality for $\theta \geq -1$ observe that (for instance by introducing polar coordinates analogously to the proof of Exercise 9) for $\operatorname{Re} z_2 < 0$ and $\operatorname{Re}(z_1+z_2) < 0$

$$\mathcal{M}(\chi_{0,1}(x_1,x_2)\cdot(x_1+x_2)^{\theta})(z) = \lim_{\substack{b\to 1\\ b<1}} \mathcal{M}(\chi_{0,b}(x_1,x_2)\cdot(x_1+x_2)^{\theta})(z) \tag{30}$$

where $\chi_{0,b}$ is the characteristic function of Example 9. Now the passage to the limit with $b \to 1$ in (30) is legitimate for $\theta > -1$ since the series in (28) is absolutely convergent and for $\theta = -1$ this is stated by (29). This together with an analogous result for x_1 and x_2 interchanged ends the proof.

The many dimensional Mellin transforms, in particular corresponding to Example 11 are given in [Ly 2].

3. Mellin Transforms of Certain Cut-Off Functions

The simplest cut-off function is $\chi_{(0,t]}$ — the characteristic function of the polyinterval $(0,t]$, $t \in \mathbf{R}^n_+$. By iterating Example 1 we find that

$$\mathcal{M}(\chi_{(0,t]})(z) = (-1)^n \frac{t^{-z}}{z^1} \qquad \text{for } z \in \mathbf{C}^n,\ z^1 \neq 0.$$

It will be useful for further applications to use smooth cut-off functions whose Mellin transforms have more convenient properties.

3.1. One-dimensional Smooth Cut-Off Functions

Proposition 4. *Let χ be a restriction to $\mathbf{R}_+$ of some function $\tilde{\chi} \in C_0^{\infty}(\mathbf{R})$, $\tilde{\chi} \equiv 1$ in a neighbourhood of zero. Let $0 < \tilde{t} < t$ be such that $\chi(x) = 1$ for $0 < x \leq \tilde{t}$, $\operatorname{supp}\chi \subset I = (0,t]$. Then*

(i) $\quad (\mathcal{M}\chi)(z) = -\frac{1}{z} + \widetilde{G}(z)$ *with* $\widetilde{G} \in \mathcal{O}(\mathbf{C})$,

(ii) *the function* $\mathbf{R} \ni \beta \mapsto (\alpha + i\beta)\mathcal{M}\chi(\alpha + i\beta)$ *is in* $S(\mathbf{R})$ *locally uniformly with respect to* α.[(1)]

In particular for any $p \in \mathbf{N}$, $\varepsilon > 0$ *and any bounded set* V *in* $\mathbf{R}$ *there exists a constant* $C_{\varepsilon,p,V}$ *such that*

$$|\mathcal{M}\chi(\alpha + i\beta)| \leq \frac{C_{\varepsilon,p,V}}{\langle\beta\rangle^p} \qquad \text{for } |\alpha + i\beta| > \varepsilon, \quad \alpha \in V. \tag{31}$$

Proof: Denote $\kappa = \chi \circ \mu$, $\mu(y) = e^{-y}$ and observe that $\frac{d\kappa}{dy} \in C_0^\infty(\mathbf{R})$ and by (4.7) $\frac{d\kappa}{dy} = -(x\frac{d\chi}{dx}) \circ \mu$. Hence[(2)]

$$\begin{aligned} \mathcal{L}\left(\frac{d\kappa}{dy}\right)(z) &= \int_{\mathbf{R}} e^{zy} \frac{d\kappa}{dy}(y)\, dy \in \mathcal{O}(\mathbf{C}), \\ \mathcal{L}\left(\frac{d\kappa}{dy}\right)(z) &= -\int_{-\infty}^{\infty} e^{zy} \left(x\frac{d\chi}{dx}\right) \circ \mu\, dy = -\mathcal{M}\left(x\frac{d\chi}{dx}\right)(z) = -z\mathcal{M}\chi(z). \end{aligned} \tag{32}$$

Note that

$$(z(\mathcal{M}\chi)(z))\big|_{z=0} = -\mathcal{L}\left(\frac{d\kappa}{dy}\right)(0) = -\int_{-\ln t}^{-\ln \bar{t}} \frac{d\kappa}{dy} dy = -1,$$

which together with (32) yields assertion (i).

Fix $\alpha \in \mathbf{R}$ and observe that by (32)

$$(\alpha + i\beta)(\mathcal{M}\chi)(\alpha + i\beta) = -\int_{-\infty}^{+\infty} e^{\alpha y} \frac{d\kappa}{dy} \cdot e^{i\beta y}\, dy = \sqrt{2\pi}\mathcal{F}^{-1}\left(e^{\alpha y}\frac{d\kappa}{dy}\right)(\beta).$$

Since $e^{\alpha y}\frac{d\kappa}{dy} \in C_0^\infty(\mathbf{R}) \subset S(\mathbf{R})$ the above function is also in $S(\mathbf{R})$. To end the proof of (ii) denote by V a bounded interval of $\mathbf{R}$ and define

$$\sigma_\alpha(\beta) = \int_{\mathbf{R}} e^{\alpha y + i\beta y} \frac{d\kappa}{dy}\, dy \qquad \text{for } \beta \in \mathbf{R}, \quad \alpha \in V.$$

Applying the standard technique for proving that $\sigma_\alpha \in S(\mathbf{R})$ and taking into account that $\frac{d\kappa}{dy}$ has bounded support we claim that to every $k, h \in \mathbf{N}_0$ there exists a constant $C_{h,k,V}$ (independent of the choice of $\alpha \in V$) such that

$$\sup_\beta \left|\beta^k \left(\frac{d}{d\beta}\right)^h (\alpha + i\beta)\mathcal{M}\chi(\alpha + i\beta)\right| < C_{h,k,V} < \infty. \tag{33}$$

---

(1) A family $\sigma_\alpha \in S(\mathbf{R}^n)$ is in $S(\mathbf{R}^n)$ locally uniformly with respect to α if $q_\nu(\sigma_\alpha) < C_\nu$ for $\nu = 1, 2, \ldots$, where C_ν are locally independent of α and q_ν are the seminorms (given by (3.3)) defining the topology in S.

(2) Note that if $f \in C_0^\infty(\mathbf{R}^n)$ then $\mathcal{L}f(\zeta) = \int_{\mathbf{R}^n} e^{\zeta y} f(y)\, dy$ is an entire function in $\mathbf{C}^n$.

Taking $h = 0$, $k = p$, $C_{0,p,V} = C_{p,V}$ we get

$$\begin{aligned} &\sup_\beta |\beta^p(\alpha+i\beta)\mathcal{M}\chi(\alpha+i\beta)| \le C_{p,V} \\ &\sup_\beta |(\alpha+i\beta)\mathcal{M}\chi(\alpha+i\beta)| < C_{0,V} \end{aligned} \qquad \text{for } \alpha \in V$$

hence

$$|\mathcal{M}\chi(\alpha+i\beta)| \le \frac{C_{p,V}+C_{0,V}}{\varepsilon}\frac{1}{1+|\beta|^p} \qquad \text{for } |\alpha+i\beta| > \varepsilon, \quad \alpha \in V$$

and thus (31) follows (see Exercise 11(d)).

Remark 2. Note that

$$(\mathcal{M}\chi)(z) = \begin{cases} (\mathcal{L}\kappa)(z) & \text{for } \operatorname{Re} z < 0 \\ (\mathcal{L}(\kappa-1))(z) & \text{for } \operatorname{Re} z > 0. \end{cases}$$

In fact,

$$\begin{aligned} \left(\mathcal{L}\left(\frac{d\kappa}{dy}\right)\right)(z) &= \mathcal{L}\left(\frac{d(\kappa-1)}{dy}\right)(z) = \int_{-\infty}^{+\infty} e^{zy}\frac{d(\kappa-1)}{dy}\,dy \\ &= -z\int_{-\infty}^{+\infty} e^{zy}(\kappa-1)\,dy = -z(\mathcal{L}(\kappa-1))(z) \qquad \text{for} \quad \operatorname{Re} z > 0, \end{aligned}$$

hence by (32) we get: $-z(\mathcal{M}\chi)(z) = -z(\mathcal{L}(\kappa-1))(z)$ for $\operatorname{Re} z > 0$. The formula for $\operatorname{Re} z < 0$, follows from Corollary 2 since $\chi \in M_{-1}$.

Remark 3. Let χ be a fixed function as in Proposition 4. The function $G(z) = (\mathcal{M}\chi)(z)$ for $z \in \mathbf{C} \setminus \{0\}$ is called a *modified Cauchy kernel* and will be applied in Section 12.

Corollary 3. *Let χ be as in Proposition 4. Then for any $j \in \mathbf{N}_0$*

(i) $\mathcal{M}((\ln x)^j \cdot \chi(x)) = (-1)^{j+1}\dfrac{j!}{z^{j+1}} + \widetilde{G}(z)$ *with $\widetilde{G}$ entire on* $\mathbf{C}$.

(ii) *For any $p \in \mathbf{N}$, $\varepsilon > 0$ and any bounded set $V \subset \mathbf{R}$ there exists a constant $C_{\varepsilon,p,V}$ such that*

$$|\mathcal{M}((\ln x)^j \cdot \chi(x))(\alpha+i\beta)| \le \frac{C_{\varepsilon,p,V}}{\langle\beta\rangle^p} \qquad \text{for } |\alpha+i\beta| \ge \varepsilon, \quad \alpha \in V.$$

Proof: We have for $j \in \mathbf{N}_0$

$$\mathcal{M}((\ln x)^j \cdot \chi(x))(z) = (-1)^j \frac{d^j}{dz^j} \mathcal{M}\chi(z)$$

hence (i) follows by assertion (i) of Proposition 4 and by (32) we get

$$\begin{aligned}\mathcal{M}((\ln x)^j \chi(x))(z) &= (-1)^{j+1} \frac{d^j}{dz^j} \left(\frac{1}{z} \mathcal{L} \left(\frac{d\kappa}{dy} \right)(z) \right) \\ &= \sum_{k=0}^{j} (-1)^k \frac{j!}{(j-k)!} \frac{1}{z^{k+1}} \mathcal{L} \left(y^{j-k} \frac{d\kappa}{dy} \right)(z).\end{aligned}$$

Suppose now that $|z| \geq \varepsilon$. Since $y^{j-k} \frac{d\kappa}{dy} \in C_0^\infty(\mathbf{R})$ we derive, as in the proof of Proposition 4, assertion (ii) of the corollary.

We shall use this corollary in Subsection 13.1.

In Subsection 13.4 we shall need information on the behaviour of the Mellin transform of cut-off functions as $\operatorname{Re} z \to \infty$. It is provided by Proposition 5 below.

Let $0 < \tilde{r} = r\delta$ for some fixed δ, $0 < \delta < 1$ and take a function $\chi \in C^\infty(\mathbf{R}_+)$ such that $\chi(x) = 1$ for $0 < x \leq \delta$, $\chi(x) = 0$ for $x \geq 1$, $0 \leq \chi \leq 1$. Define

$$\chi_{\tilde{r},r}(x) = \chi\left(\frac{x}{r}\right) \qquad \text{for } x > 0.$$

It is clear that $\chi_{\tilde{r},r} \in C^\infty(\mathbf{R}_+)$, $0 \leq \chi_{\tilde{r},r} \leq 1$ and that $\chi_{\tilde{r},r}(x) = 1$ for $0 < x \leq \tilde{r}$, $\chi_{\tilde{r},r}(x) = 0$ for $x \geq r$.

Proposition 5. *The Mellin transform of the cut-off function $\chi_{\tilde{r},r}$ defined above has the following properties:*

(i) $\mathcal{M}\chi_{\tilde{r},r}(z) = -\frac{1}{z} + \widetilde{G}(z)$ *with $\widetilde{G}$ entire on* $\mathbf{C}$,

(ii) *For every $\varepsilon > 0$, $p \in \mathbf{N}$ there exists a constant $C_{\varepsilon,p} < \infty$, (depending on δ and independent of r) such that with $z = \alpha + i\beta$*

$$(34) \qquad |\mathcal{M}\chi_{\tilde{r},r}(z)| \leq \begin{cases} C_{\varepsilon,p} \dfrac{r^{-\alpha}}{\langle\beta\rangle^p} & \text{for } \alpha \leq -1 \\ C_{\varepsilon,p} \dfrac{\tilde{r}^{-\alpha}}{\langle\beta\rangle^p} & \text{for } \alpha > -1,\ |z| \geq \varepsilon. \end{cases}$$

Proof: Since $\chi_{\tilde{r},r}$ verifies the assumptions of Proposition 4 we get the assertion (i). For the proof of (ii) note that for $p \in \mathbf{N}$ and $z \in \mathbf{C}$:

$$z^p \mathcal{M}\chi_{\tilde{r},r}(z) = \mathcal{M}\Big(\Big(x\frac{d}{dx}\Big)^p \chi_{\tilde{r},r}\Big)(z) = \int_{\tilde{r}}^{r} x^{-z-1}\Big(x\frac{d}{dx}\Big)^p \chi_{\tilde{r},r}(x)\,dx. \tag{35}$$

Let $m_j = \sup\left|\chi^{(j)}(x)\right|$ $(j = 1, 2, \ldots)$. It is easy to see that

$$\sup_{\tilde{r}\le x\le r} x^{-\alpha-1} = \begin{cases} r^{-\alpha-1} & \text{if } \alpha \le -1 \\ \tilde{r}^{-\alpha-1} & \text{if } \alpha \ge -1, \end{cases}$$

$$\sup_{\tilde{r}\le x\le r}\left|\Big(x\frac{d}{dx}\Big)^p \chi_{\tilde{r},r}(x)\right| \le \sum_{j=1}^{p} m_j = C_p < \infty.$$

Hence from (35) we get for $p \in \mathbf{N}$ the estimation

$$|z|^p\,|\mathcal{M}\chi_{\tilde{r},r}(z)| \le \begin{cases} C_p(r-\tilde{r})r^{-\alpha-1} = C_p(1-\delta)r^{-\alpha} & \text{for } \alpha \le -1 \\ C_p(r-\tilde{r})\tilde{r}^{-\alpha-1} = C_p\Big(\frac{1}{\delta}-1\Big)\tilde{r}^{-\alpha} & \text{for } \alpha > -1. \end{cases} \tag{36}$$

Observe that

$$\frac{1}{|z|} \le \frac{\sqrt{2}}{\langle\beta\rangle} \quad \text{for } \alpha \le -1, \qquad \frac{1}{|z|} \le \Big(1+\frac{1}{\varepsilon}\Big)\frac{1}{\langle\beta\rangle} \quad \text{for } |z| \ge \varepsilon,$$

hence by (36) we get

$$|\mathcal{M}\chi_{\tilde{r},r}(z)| \le \begin{cases} C_p(1-\delta)\dfrac{2^{p/2}}{\langle\beta\rangle^p} r^{-\alpha} & \text{for } \alpha \le -1 \\ C_p\Big(\dfrac{1}{\delta}-1\Big)\Big(1+\dfrac{1}{\varepsilon}\Big)^p \dfrac{1}{\langle\beta\rangle^p}\tilde{r}^{-\alpha} & \text{for } \alpha > -1,\ |z| \ge \varepsilon. \end{cases}$$

Taking $C_{\varepsilon,p} = C_p \max\left((1-\delta)2^{p/2}, \left(\frac{1}{\delta}-1\right)\left(1+\frac{1}{\varepsilon}\right)^p\right)$ we obtain inequalities (34).

3.2. n-Dimensional Smooth Cut-Off Functions with a Parameter

Let $\varphi \in C_0^\infty(\mathbf{R}^n)$, $\varphi \equiv 1$ in a neighbourhood of zero, $\operatorname{supp}\varphi \subset B(1)$, $|\varphi| \le 1$. Let

$$\chi_r(x) = \varphi\Big(\frac{1}{r}x\Big) \qquad \text{for } x \in \mathbf{R}^n \qquad (r \in \mathbf{R}_+). \tag{37}$$

Then χ_r is smooth on $\mathbf{R}^n$, $\chi_r \equiv 1$ in a neighbourhood of zero, $\operatorname{supp}\chi_r \subset B(r)$, $|\chi_r| \le 1$.

Proposition 6. *Let a be a smooth function on $B(r_0)$, $r_0 \in \mathbf{R}_+$. Then for an arbitrary $h \in \mathbf{N}_0^n$ there exists a constant C_h (independent of $r \le r_0$) such that*

$$\sup_x \left| \left(x \frac{\partial}{\partial x} \right)^h (a\chi_r)(x) \right| < C_h \qquad \text{for } r \le r_0 \tag{38}$$

and for every $m \in \mathbf{N}$ there exists a constant C_m (independent of $r \le r_0$) such that

$$|\mathcal{M}(a\chi_r)(z)| \le C_m \frac{1}{\langle z \rangle^m} \frac{\mathbf{r}^{-\alpha}}{(-\alpha)^1} \qquad \text{for } \alpha = \operatorname{Re} z < 0,\ r \le r_0 \tag{39}$$

(here $\mathbf{r} = (r, \ldots, r)$).

Proof: We shall prove (38) for $h = (1, 0, \ldots, 0)$. For other h the proof goes by symmetry and induction. Let C_1, c, c_1 be such that $\sup_{x \in \mathbf{R}^n} \left| \frac{\partial \varphi}{\partial x_1} \right| < C_1$, $|a| \le c$, $\left| \frac{\partial a}{\partial x_1} \right| \le c_1$ on $B(r)$. Then

$$\left| x_1 \frac{\partial}{\partial x_1} (a\chi_r)(x) \right| \le \left| x_1 \frac{\partial a}{\partial x_1} \chi_r(x) \right| + \left| a x_1 \frac{1}{r} \frac{\partial \varphi}{\partial x_1} \left(\frac{1}{r} x \right) \right| \le c_1 r_0 + c C_1.$$

For the proof of (39) observe that $a\chi_r \in M'_{(0)}(I)$, $I = (0, \mathbf{r}]$, $\mathbf{r} = (r, \ldots, r)$, $\mathcal{M}(a\chi_r)(z)$ is holomorphic for $\operatorname{Re} z < 0$ and that by (38) there exists $C_{j,m}$ such that

$$\begin{aligned} \left| \mathcal{M}\left(\left(x_j \frac{\partial}{\partial x_j} \right)^m (a\chi_r) \right)(z) \right| &\le \sup \left| \left(x_j \frac{\partial}{\partial x_j} \right)^m (a\chi_r)(x) \right| \int_I x^{-\operatorname{Re} z - 1}\, dx \\ &\le C_{j,m} \frac{\mathbf{r}^{-\alpha}}{(-\alpha)^1} \\ &\qquad \text{for } \operatorname{Re} z < 0 \quad (j = 1, \ldots, n;\ m = 1, 2, \ldots). \end{aligned}$$

By (11)

$$\mathcal{M}\left(\left(x_j \frac{\partial}{\partial x_j} \right)^m (a\chi_r) \right)(z) = z_j^m \mathcal{M}(a\chi_r)(z) \qquad \text{for } \operatorname{Re} z < 0.$$

Putting $\widetilde{C}_m = c + \sum_{j=1}^n C_{j,m}$ we get the estimation

$$\left(1 + \sum_{j=1}^n |z_j|^m \right) |\mathcal{M}(a\chi_r)(z)| \le \widetilde{C}_m \frac{\mathbf{r}^{-\alpha}}{(-\alpha)^1} \qquad \text{for } \operatorname{Re} z < 0.$$

Hence by Exercise 11(d) we get (39).

Proposition 6 will be useful in Subsection 13.1 in Chapter III.

In the next proposition we shall consider yet another family of functions.[1] For their definition we choose a function λ defined on the real line: $\lambda \in C^\infty(\mathbf{R})$, $0 \le \lambda \le 1$, $\lambda(s) = 0$ for $s \le 1$, $\lambda = 1$ for $s \ge 2$. Let $r \in \mathbf{R}_+$, $t \in \mathbf{R}^n_+$, $t_i > 2r$ $(i = 1, \dots, n)$. We put

$$\eta_r(x) = \lambda\Big(\frac{x_1}{r}\Big) \cdot \ldots \cdot \lambda\Big(\frac{x_n}{r}\Big) \qquad \text{for } x \in I = (0, t]. \tag{40}$$

It is easy to see that $\eta_r \in C^\infty(I)$, $0 \le \eta_r \le 1$, $\eta_r(x) = 1$ for $x_1 \ge 2r, \dots, x_n \ge 2r$, $\operatorname{supp} \eta_r \subset \{x \in \mathbf{R}^n \colon x_j \ge r \text{ for some } j = 1, \dots, n\}$.

Proposition 7. *Let η_r be the functions defined by (40). Then $\eta_r \in M_{[-1]}(I)$ and for every $\alpha \in \mathbf{N}^n_0$ there exists a constant $C_\alpha < \infty$ (independent of r) such that*

$$\sup_{x \in I} \Big| \Big(x \frac{\partial}{\partial x}\Big)^\alpha \eta_r(x) \Big| \le C_\alpha \tag{41}$$

and

$$\eta_r \underset{r \to 0}{\longrightarrow} 1 \quad in \quad M_{[-1]}(I).$$

Proof: The estimation (41) can be proved in the same way as (38). For the proof of the convergence of the sequence $\{\eta_r\}$ in the projective limit topology in $M_{[-1]}$ (see Subsection 5.2) take arbitrarily $\delta \in \mathbf{R}^n_+$. It suffices to show that $\eta_r - 1 \underset{r \to 0}{\longrightarrow} 0$ in $M_{-1+\delta}(I)$. Let

$$J_r = I \setminus \{x \in \mathbf{R}^n \colon x_j > 2r \ (j = 1, \dots, n)\}.$$

It is easy to see that $\lim_{r \to 0} \sup_{x \in J_r} |x^\delta| = 0$ and that by (41) $\tilde{\varrho}_{-1+\delta,\alpha}(\eta_r - 1) \le C_\alpha \sup_{x \in J_r} |x^\delta|$.

Exercises

1. Prove the continuity of $\mathcal{M}u$ in the set Ω_u directly without referring to Theorem 1.

2. Let $f \in M_b((0,t])$, $t \in \mathbf{R}_+$. Then $\mathcal{M}(\frac{d}{dx} f)(z) = (z+1)(\mathcal{M}f)(z+1)$ for $\operatorname{Re} z < -b-2$ if $\frac{d}{dx} f$ is understood as the distributional derivative of f regarded as a Mellin distribution $f \in M'_{(-b-1)}$. Note that if $f' \in M_{b+1}$ denotes the point-wise derivative of f on I then

$$(Mf')(z) = f(t)t^{-z-1} + (z+1)(Mf)(z+1) \qquad \text{for} \quad \operatorname{Re} z < -b-2.$$

[1] Such functions have already been used in the proof of Proposition 5.3 with the parameter $\nu = \frac{1}{r}$.

3. Suppose $u \in M'_{(\omega)}(I)$ $(\omega \in \mathbf{R}^n)$, $\psi \in S(\mathbf{R}^n)$ and let for some $a < \omega$ $\mathbf{R}^n \ni \beta \mapsto g(\cdot,\beta) \in M_a(I)$ be a bounded continuous family of functions in M_a. Show that

$$u\left[\int_{\mathbf{R}^n} g(\cdot,\beta)\psi(\beta)\,d\beta\right] = \int_{\mathbf{R}^n} u[g(\cdot,\beta)]\psi(\beta)\,d\beta.$$

HINT. Consider the Riemann sums for the integral $\int_{\mathbf{R}^n} g(\cdot,\beta)\psi(\beta)\,d\beta$ and show that they converge to the integral in the space M_a. Make use of the linearity and continuity of u (cf. Exercise 4.4).

4. Suppose that $v \in S'(\mathbf{R}^n)$ is supported by a set $\{y \in \mathbf{R}^n\colon \mathbf{r} < y\}$ for some $\mathbf{r} \in \mathbf{R}^n$. Show that the (right-sided) Laplace transform $\mathcal{L}v(z) = v[e^{zy}]$ is holomorphic for $\operatorname{Re} z < 0$.

5. Let $u \in D'_K(\mathbf{R}^n_+)$, K a compact subset of $\mathbf{R}^n_+$. Show that for any I such that $K \subset I$ and any $\omega \in \mathbf{R}^n$, $u \in M'_{(\omega)}(I)$ and consequently $\mathcal{M}u$ is an entire function on $\mathbf{C}^n$.

6. Let $\alpha,\beta \in \mathbf{C}$, $u(x,y) = x^\alpha y^\beta$ for $0 < x \le t$, $0 < y \le t$. Show that $\mathcal{M}u(z) = \frac{1}{(z_1-\alpha)(z_2-\beta)} t^{\alpha+\beta-z_1-z_2}$ for $z \in \mathbf{C}^2 \setminus \{z \in \mathbf{C}^2\colon z_1 = \alpha \vee z_2 = \beta\}$ is holomorphic in this set.

7. Suppose g is differentiable at zero of any order α, $|\alpha| < k$ (but not necessarily of class C^k in a neighbourhood of zero). Prove that for a sufficiently small polyinterval $I = (0,t]$

$$\mathcal{M}(\chi_I g)(z) = \sum_{|\alpha| \le k-1} C_\alpha \frac{t^{\alpha-z}}{(\alpha-z)^{\mathbf{1}}} + \sum_{j=1}^{n} N_j(z),$$

where N_j is holomorphic on a set $\{z \in \mathbf{C}^n\colon \operatorname{Re} z_l < 0 \text{ for } l \ne j,\ \operatorname{Re} z_j < k\}$.

8. Let $\alpha \in \mathbf{R}_+$, $B = \{(x_1,x_2)\colon 0 < x_1 \le 1,\ 0 < x_2 \le x_1^\alpha\}$, $I = (0,\mathbf{1}]$ and let χ_B be the characteristic function of the set B. Show that:

$$\mathcal{M}\chi_B(z) = \frac{1}{z_2}\frac{1}{z_1+\alpha z_2} \quad \text{for } z \in \mathbf{C}^2 \setminus \{z \in \mathbf{C}^2\colon z_2 = 0 \vee z_1 + \alpha z_2 = 0\},$$

$$\mathcal{M}\chi_I(z) = \frac{1}{z_1}\frac{1}{z_2} \quad \text{for } z \in \mathbf{C}^2 \setminus \{z \in \mathbf{C}^2\colon z_1 = 0 \vee z_2 = 0\}$$

and are holomorphic there. However if f were an arbitrary bounded measurable function on I we could only assert that $\mathcal{M}(\chi_I f)(z)$ is holomorphic for $\operatorname{Re} z < 0$.

9. Show that the function

$$\mathbf{C}^2 \ni (z_1,z_2) \mapsto (z_1+z_2-\theta)\mathcal{M}((x_1+x_2)^\theta \chi_{a,b})(z)$$

($\chi_{a,b}$ as in Example 9) is holomorphic on $\mathbf{C}^2$ for $\theta \in \mathbf{C}$ and $0 < a < b$, $b \in \mathbf{R}$.

OUTLINE OF THE PROOF. Denote $S_{a,b} = \Delta_{a,b} \cap \{x\colon \|x\| \le 1\}$. Since $\overline{\Delta_{a,b} \setminus S_{a,b}}$ is a compact subset of $\mathbf{R}^2_+$ by Exercise 5 it is enough to consider only $S_{a,b}$. Introducing polar coordinates in $\mathbf{R}^2$ find that

$$\int_{S_{a,b}} (x_1 + x_2)^\theta x_1^{-z_1-1} x_2^{-z_2-1}\, dx_1\, dx_2 = \int_0^1 r^{\theta - z_2 - z_1 - 1} h(z_1, z_2)\, dr$$

where

$$h(z_1, z_2) = \int_{\varphi_1}^{\varphi_2} (\cos\varphi + \sin\varphi)^\theta (\cos\varphi)^{-z_1-1} (\sin\varphi)^{-z_2-1}\, d\varphi$$

and $\varphi_1 = \operatorname{arctg} a$, $\varphi_2 = \operatorname{arctg} b$. Since $\varphi_1 > 0$, $\varphi_2 < \frac{\pi}{2}$ show by differentiating under integral sign that the function h is holomorphic in $\mathbf{C}^2$. To end the proof note that

$$\int_0^1 r^{\theta - z_2 - z_1 - 1}\, dr = \frac{1}{\theta - z_1 - z_2} \qquad \text{for } \operatorname{Re}(z_1 + z_2 - \theta) < 0.$$

10. Let χ be the characteristic function of the cube $\{0 < x \le 1\}$ and let $m \in \mathbf{N}_0$. Derive the following formula from Example 11

$$\begin{aligned} \mathcal{M}(\chi \cdot (x_1 + x_2)^m)(z) &= \frac{1}{z_1 + z_2 - m} \left\{ \sum_{j=0}^m \binom{m}{j} \left(\frac{1}{z_2 - j} + \frac{1}{z_1 - m + j} \right) \right\} \\ &= \sum_{j=0}^m \binom{m}{j} \frac{1}{(z_2 - j)(z_1 - m + j)} \end{aligned}$$

HINT. Observe that $\binom{m}{j} = \binom{m}{m-j}$.

11. Let $\langle z \rangle = 1 + |z_1| + \dots |z_n|$. Show the following estimations with suitable constant, $C < \infty$:

(a) $\langle z \rangle^m \le C \sum_{j=1}^n (1 + |z_j + 1| \cdot \ldots \cdot |z_j + m|)$,

(b) $\langle z_1 \rangle^2 \cdot \ldots \cdot \langle z_n \rangle^2 \le C \langle z \rangle^m$ if $m \ge 2n$,

(c) $|p_\alpha(z)| \le C \langle z \rangle^{|\alpha|}$ for $p_\alpha(z) = \prod_{j=1}^n \prod_{k=1}^{\alpha_j} (z_j + k)$, $\alpha \in \mathbf{N}_0^n$,

(d) $\langle z \rangle^m \le C(1 + \sum_{j=1}^n |z_j|^m) \le C \langle z \rangle^m$.

12. Show that for every $h \in \mathbf{N}_0^n$ there exist $m \in \mathbf{N}_0$ and $C < \infty$ such that

$$(1+|z_1|^{h_1})\cdot\ldots\cdot(1+|z_n|^{h_n}) \le C\Big(1+\sum_{j=1}^{n}|z_j|^m\Big).$$

13. Let $\chi_{\bar r,r}$ be the cut-off function considered in Proposition 5, and let $j \in \mathbf{N}_0$. Prove that $\mathcal{M}((\ln x)^j\chi_{\bar r,r})(z)$ fulfills the assertion (ii) of Proposition 5 with the constants $C_{\varepsilon,p}$ depending on r and j.
HINT. Make use of the estimate (34) and the formulas

$$\mathcal{M}((\ln x)^j\chi_{\bar r,r})(z) = (-1)^j\frac{d^j}{dz^j}\mathcal{M}\chi_{\bar r,r}(z),$$

$$F(\mathring{z}) = \frac{1}{2\pi i}\int_{|\zeta-\mathring{z}|=\mathring{\varepsilon}}\frac{F(\zeta)}{\zeta-\mathring{z}}\,d\zeta$$

valid for F holomorphic in a neighbourhood of $\{\zeta\colon |\zeta-\mathring{z}| \le \mathring{\varepsilon}\}$.
14. Show that the formula of Example 7 also holds for $t = e^c$.

§8. THE STRUCTURE OF MELLIN DISTRIBUTIONS

1. Characterizations of Mellin Distributions

Theorem 1.(1) *The space $M'(I)$ of Mellin distributions coincides with the space of restrictions to $\mathbf{R}_+^n$ of distributions on $\mathbf{R}^n$ with support in $\bar I$.*

Proof: If $w \in D'_{\bar I}(\mathbf{R}^n)$ i.e. w is a distribution on $\mathbf{R}^n$ with support in $\bar I$, then by Theorem 3.1 there exist constants $C < +\infty$ and $m \in \mathbf{N}_0$ such that

$$(1)\qquad |w[\psi]| \le C\sum_{|\alpha|\le m}\sup_{x\in I}\Big|\Big(\frac{\partial}{\partial x}\Big)^{\alpha}\psi(x)\Big| \qquad \text{for } \psi \in C^m(\mathbf{R}^n).$$

Let $a = -\mathbf{m}-\mathbf{1}$ and choose arbitrarily a sequence $\{\varphi_\nu\}$ such that

$$(2)\qquad C^\infty_{(0)}(I) \ni \varphi_\nu \to 0 \quad \text{in } M_{(a)}.$$

Hence there exists an $\varepsilon > 0$ such that

$$\varrho_{a-\varepsilon,\alpha}(\varphi_\nu) \to 0 \qquad \text{for every } \alpha \in \mathbf{N}_0^n.$$

(1) See [Sz2] theorem 5. Theorem 1 is a modification of proposition 5 from [Ziel], see Exercise 1.

Let $\tilde{u} = w|_{\mathbf{R}^n_+}$, hence $\tilde{u} \in D'_I(\mathbf{R}^n_+)$. By Proposition 3.1 $D'_I(\mathbf{R}^n_+)$ and $D'(I)$ are isomorphic under a canonical isomorphism L, $L\tilde{u} = u$, $u[\varphi_\nu] = \tilde{u}[\tilde{\varphi}_\nu]$ where $\tilde{\varphi}_\nu$ is an arbitrary smooth extension of φ_ν from I to $\mathbf{R}^n_+$ ($\nu = 1, 2, \ldots$). Thus by (1) we get

$$|L\tilde{u}[\varphi_\nu]| \leq C \sum_{|\alpha|\leq m} \sup_{x\in I} \left| \left(\frac{\partial}{\partial x}\right)^\alpha \varphi_\nu(x) \right| \leq C \sum_{|\alpha|\leq m} \varrho_{a-\varepsilon,\alpha}(\varphi_\nu) \sup_{x\in I} \left| x^{-a-\alpha-1+\varepsilon} \right|$$

and therefore $u[\varphi_\nu] \to 0$. By the arbitrariness of sequence (2) $u \in M'_{(a)}$. In view of the isomorphism L we identify $\tilde{u}$ and $L\tilde{u}$ and write $\tilde{u} \in M'_{(a)}$.

We begin the proof of the second part of Theorem 1 with the following lemma

Lemma 1. *Let $\Omega = \mathbf{R}^n \setminus (\bar{I} \setminus I)$ and Ω_0 be a bounded open set $\Omega_0 \subset \Omega$. Suppose $C_0^\infty(\Omega_0) \ni \psi_\nu \to 0$ in $D(\mathbf{R}^n)$ and let $\varphi_\nu = \psi_\nu|_I$ ($\nu = 1, 2, \ldots$). Then for every $a \in \mathbf{R}^n$ $M_a(I) \ni \varphi_\nu \to 0$ in $M_a(I)$.*

Proof: Take $C_0^\infty(\Omega_0) \ni \psi_\nu \to 0$ in $D(\mathbf{R}^n)$. Thus $\operatorname{supp} \psi_\nu \subset \Omega_0$, $\sup_{x\in\mathbf{R}^n} \left|\left(\frac{\partial}{\partial x}\right)^\beta \psi_\nu(x)\right|$ $\to 0$ for every $\beta \in \mathbf{N}_0^n$. Fix arbitrarily $a \in \mathbf{R}^n$, $\alpha \in \mathbf{N}_0^n$ and take $\mathbf{p} = (p, \ldots, p) \in \mathbf{N}^n$ such that $\mathbf{p} + a + \alpha + \mathbf{1} > 0$. Write $g_\nu(x_n) = \left(\frac{\partial}{\partial x}\right)^\alpha \psi_\nu(x_1, \ldots, x_{n-1}, x_n)$ for $x \in I$ ($\nu = 1, 2, \ldots$). By Taylor formula there exists θ_n, $0 < \theta_n < 1$ such that

$$g_\nu(x_n) = \frac{x_n^p}{p!} \frac{d^p}{dx_n^p} g(\theta_n x_n).$$

Proceeding in the same way with respect to the variables $x_1, \ldots, x_{n-1}$ we arrive at the formula

$$\left(\frac{\partial}{\partial x}\right)^\alpha \psi_\nu(x) = \frac{x^{\mathbf{p}}}{(p!)^n} \left(\frac{\partial}{\partial x}\right)^{\alpha+\mathbf{p}} \psi_\nu(\theta_1 x_1, \ldots, \theta_n x_n),$$

where $0 < \theta_j < 1$ ($j = 1, \ldots, n$).

From the inequality

$$\varrho_{a,\alpha}(\varphi_\nu) \leq \frac{1}{(p!)^n} \sup_{x\in I} \left| x^{a+\alpha+1+\mathbf{p}} \right| \sup_{x\in I} \left| \left(\frac{\partial}{\partial x}\right)^{\alpha+\mathbf{p}} \psi_\nu(x) \right|$$

it follows at once that $\varrho_{a,\alpha}(\varphi_\nu) \to 0$.

Returning to the proof of Theorem 1 suppose that $u \in M'(I)$ it is $u \in M'_a(I)$ for some $a \in \mathbf{R}^n$. Hence $u \in D'(I)$ (more precisely $u|_{C^\infty_{(0)}(I)} \in D'(I)$) and by Proposition 3.1 there exists a unique $v \in D'_I(\mathbf{R}^n_+)$ such that $Lv = u$. By Lemma 1 the formal definition

$$(3) \qquad T[\psi] = u[\psi|_I] \qquad \text{for } \psi \in C_0^\infty(\Omega)$$

defines correctly $T \in D'(\Omega)$. In fact, if $C_0^\infty(\Omega) \ni \psi_\nu \to 0$ in $D(\Omega)$ then there exists a bounded open set $\Omega_0 \subset \Omega$ such that $\psi_\nu \in C_0^\infty(\Omega_0)$ and if $\varphi_\nu = \psi_\nu|_I$ then $\varphi_\nu \to 0$ in M_a. Hence $u[\varphi_\nu] \to 0$ and therefore $T[\psi_\nu] \to 0$.

For the proof that T extends to a distribution on $\mathbf{R}^n$ take an arbitrary bounded open set $\Omega_0 \subset \Omega$. By Proposition 3.4 it suffices to prove that $T|_{\Omega_0}$ extends to $\mathbf{R}^n$. To this aim we apply condition (iii) of Proposition 3.3. Let $\{\psi_\nu\}$ be an arbitrary sequence such that $C_0^\infty(\Omega_0) \ni \psi_\nu \to 0$ in $D(\mathbf{R}^n)$. By Lemma 1 $\psi_\nu|_I \to 0$ in M_a and hence $T[\psi_\nu] \to 0$.

Let w be an extension of T:

$$w \in D'(\mathbf{R}^n), \qquad w = T \text{ on } \Omega. \tag{4}$$

Take $\tilde{\varphi} \in C_0^\infty(\mathbf{R}_+^n) \subset C_0^\infty(\Omega)$ and put $\varphi = \tilde{\varphi}|_I$. From (3) and from Proposition 3.1 we get $T[\tilde{\varphi}] = u[\varphi] = v[\tilde{\varphi}]$, thus $T = v$ on $\mathbf{R}_+^n$ and by (4) $w = v$ on $\mathbf{R}_+^n$. To see that $\operatorname{supp} w \subset \bar{I}$ take $\psi \in C_0^\infty(\Omega)$, $\operatorname{supp} \psi \subset \mathbf{R}^n \setminus \bar{I}$. By (4) and (3) $w[\psi] = T[\psi] = u[\psi|_I] = u[0] = 0$.

Remark 1. It follows from Theorem 1 that every Mellin distribution is a distribution of finite order (see Subsections 3.3, 3.4, also see Remark 3).

Theorem 2 ([Ły1]). *Let $\omega \in (\mathbf{R} \cup \{+\infty\})^n$. In order that u belong to $M'_{(\omega)}(I)$ it is necessary and sufficient that for every $a < \omega$ there exist $m \in \mathbf{N}_0$ and functions $h_\alpha \in L^\infty(I)$ for $\alpha \in \mathbf{N}_0^n$ $|\alpha| \le m$ such that*

$$u = \sum_{|\alpha| \le m} \left(x\frac{\partial}{\partial x}\right)^\alpha (x^a h_\alpha) \quad \text{in } M'_{(a)}. \tag{5}$$

Proof: The condition is sufficient. Indeed, by Exercise 5.11 $h_\alpha \in M'_{(0)}$, hence (see Subsection 6.1) the right hand side of (5) is in $M'_{(a)}$. For the proof of the second part of Theorem 2 observe that since $M'_{(\omega)} = \bigcap_{a<\omega} M'_a$ (see Exercise 5.6) we need only prove the following

Proposition 1. *Let $a \in \mathbf{R}^n$, $u \in M'_a$. Then there exist $m \in \mathbf{N}_0$ and bounded functions h_α for $\alpha \in \mathbf{N}_0^n$, $|\alpha| \le m$ such that (5) holds.*

We start the proof of Proposition 1 with

Lemma 2. *Let $\psi \in C^1_{(0)}(I)$, $I = (0,t]$, $t \in \mathbf{R}_+^n$. Then*

$$\sup_{x \in I} |x^1 \psi(x)| \le \sum_{\gamma \in \{0,1\}^n} \left\| x^\gamma \left(\frac{\partial}{\partial x}\right)^\gamma \psi(x) \right\|_{L^1(I)}. \tag{6}$$

Proof: For $n = 1$ we have the formula

$$x\psi(x) = \int_0^x y\psi'(y)\,dy + \int_0^x \psi(y)\,dy.$$

Hence

$$|x\psi(x)| \le \int_0^t \left|y\frac{d}{dy}\psi(y)\right| dy + \int_0^t |\psi(y)|\,dy \qquad \text{for } x \in I,$$

which proves (6) for $n = 1$. For $n > 1$ the proof goes by induction with respect to n.

Proof of Proposition 1: Let $a \in \mathbf{R}^n$, $u \in M'_a$. Then there exist constants $C < \infty$, $p \in \mathbf{N}_0$ such that

$$|u[\varphi]| \le C \sum_{|\alpha| \le p} \varrho_{a,\alpha}(\varphi) \qquad \text{for } \varphi \in C^\infty_{(0)}(I). \tag{7}$$

Take any $\varphi \in C^\infty_{(0)}(I)$ and let $\psi(x) = x^{a+\alpha}\left(\frac{\partial}{\partial x}\right)^\alpha \varphi(x)$. By Exercise 5.1 there exist constants $C_{\alpha\beta}$ such that $\psi(x) = \sum_{\beta \le \alpha} C_{\alpha\beta}\left(x\frac{\partial}{\partial x}\right)^\beta (x^a\varphi(x))$. Thus applying Lemma 2 to the function ψ we derive

$$\begin{aligned}\varrho_{a,\alpha}(\varphi) = \sup_{x \in I} |x^1\psi(x)| &\le \sum_{\gamma \in \{0,1\}^n} \left\| x^\gamma \left(\frac{\partial}{\partial x}\right)^\gamma \sum_{\beta \le \alpha} C_{\alpha\beta}\left(x\frac{\partial}{\partial x}\right)^\beta (x^a\varphi(x))\right\|_{L^1(I)} \\ &\le \sum_{\lambda \le \alpha+1} \widetilde{C}_{\alpha\lambda}\left\|\left(x\frac{\partial}{\partial x}\right)^\lambda (x^a\varphi(x))\right\|_{L^1(I)}.\end{aligned}$$

Hence by (7) we get with some constant $C_1 = C_1(C, p)$

$$|u[\varphi]| \le C_1 \max_{|\lambda| \le p+n} \left\|\left(x\frac{\partial}{\partial x}\right)^\lambda (x^a\varphi(x))\right\|_{L^1(I)} \qquad \text{for } \varphi \in C^\infty_{(0)}(I). \tag{8}$$

Let $m = p + n$, $N = \operatorname{card}\{\lambda \in \mathbf{N}_0^n \colon |\lambda| \le m\} = \binom{m+n}{n}$. We define[(1)] the operator

$$C^\infty_{(0)}(I) \ni \varphi \mapsto A\varphi = \left\{\left(x\frac{\partial}{\partial x}\right)^\lambda (x^a\varphi)\right\}_{|\lambda| \le m}.$$

Note that $A\colon C^\infty_{(0)} \to AC^\infty_{(0)}$ is one to one and that $A(C^\infty_{(0)}(I)) \subset (L^1(I))^N$. We also define the functional

$$A(C^\infty_{(0)}(I)) \ni \psi \mapsto T[\psi] = u[A^{-1}\psi].$$

(1) This part of the proof is similar to the proof of the Theorem 3.4 (See Exercises 3.10–3.12).

Due to (8) T is continuous on $A(C^\infty_{(0)})$ in the norm of $(L^1)^N$. By the Hahn–Banach theorem there exists a continuous extension $\tilde{T}$ of T to the whole of $(L^1)^N$. So by the Riesz theorem there exist functions $\tilde{h}_\lambda \in L^\infty(I)$ for $|\lambda| \le m$ such that

$$\tilde{T}[\{\sigma_\lambda\}_{|\lambda|\le m}] = \sum_{|\lambda|\le m} \tilde{h}_\lambda[\sigma_\lambda] \qquad \text{for } \sigma_\lambda \in L^1.$$

Hence for $\varphi \in C^\infty_{(0)}(I)$

$$\begin{aligned} u[\varphi] = T[A\varphi] &= \tilde{T}\Big[\Big\{\Big(x\frac{\partial}{\partial x}\Big)^\lambda (x^a\varphi)\Big\}_{|\lambda|\le m}\Big] = \sum_{|\lambda|\le m} \tilde{h}_\lambda\Big[\Big(x\frac{\partial}{\partial x}\Big)^\lambda (x^a\varphi)\Big] \\ &= \sum_{|\lambda|\le m} (-1)^{|\lambda|} x^a \Big(\frac{\partial}{\partial x}x\Big)^\lambda \tilde{h}_\lambda[\varphi] \end{aligned}$$

and by Exercise 5.1 this is equal to

$$\sum_{|\alpha|\le m} \Big(x\frac{\partial}{\partial x}\Big)^\alpha (x^a h_\alpha)[\varphi]$$

where for every α, $|\alpha| \le m$ the function h_α is a linear combination of $\tilde{h}_\lambda$, $|\lambda| \le m$. Since $C^\infty_{(0)}(I)$ is dense in $M_{(a)}$ we get (5) with $h_\alpha \in L^\infty(I)$.

Now we proceed to characterizations of Mellin distributions supported by smaller cubes.

Denote by $\mathring{M}_a((0,t])$ the space of functions $\varphi \in M_a((0,t])$ vanishing with all derivatives $\big(\frac{\partial}{\partial x}\big)^\alpha \varphi$ on the set $(0,t]\setminus(0,t)$ with the topology induced by the topology of $M_a((0,t])$. It is easy to see that the space $\mathring{M}_a((0,t])$ is complete. We define $\mathring{M}_{(\omega)}((0,t]) = \varprojlim_{a<\omega} \mathring{M}_a((0,t])$. Here as usual $\omega \in (\mathbf{R}\cup\{\infty\})^n$, $t \in \mathbf{R}^n_+$, $a \in \mathbf{R}^n$.

By Theorem 5.1 we get

Corollary 1. *Let $0 < t < \tilde{t} \in \mathbf{R}^n_+$. Then to every $0 < \varepsilon < \tilde{t} - t$, $\varepsilon < t$ there exists a linear extension mapping $\mathcal{E}_\varepsilon$*

$$\mathcal{E}_\varepsilon\colon M_{(\omega)}((0,t]) \to \mathring{M}_{(\omega)}((0,\tilde{t}])$$

continuous in the respective topologies and such that $\mathcal{E}_\varepsilon\varphi(x) = 0$ if $\varphi \in M_{(\omega)}((0,t])$ and $t_j + \varepsilon_j < x_j < \tilde{t}_j$ for some $1 \le j \le n$.

Theorem 3. *Let $u \in M'_{(\omega)}((0,\tilde{t}])$ and $\operatorname{supp} u \subset (0,t]$ for some $t < \tilde{t}$. Then for any $b < \omega$ there exist constants $C = C(b)$ and $k = k(b) \in \mathbf{N}_0$ such that*

$$(9) \qquad |u[\varphi]| \le C \sum_{|\alpha| \le k} \sup_{x \in (0,t]} \left| x^{b+1} \left(x \frac{\partial}{\partial x} \right)^{\alpha} \varphi(x) \right| \qquad \text{for } \varphi \in M_b((0,\tilde{t}]).$$

Hence the restriction mapping

$$M_{(\omega)}((0,\tilde{t}]) \ni \varphi \mapsto \varphi|_{(0,t]} \in M_{(\omega)}((0,t])$$

induces a linear isomorphism of the spaces

$$(10) \qquad \{u \in M'_{(\omega)}((0,\tilde{t}]) \colon \operatorname{supp} u \subset (0,t]\} \simeq M'_{(\omega)}((0,t]).$$

Proof: Let η_r be the functions defined by (7.40). Let $\widetilde{I} = (0,\tilde{t}]$. Then by Propositions 7.7 and 6.7

$$(11) \qquad \lim_{r \to 0} u_r = u \ \text{ in } \ M'_{(\omega)}(\widetilde{I}), \quad \text{where } \ u_r = \eta_r u \ \text{ for } \ r > 0.$$

Observe that $\operatorname{supp} u_r$ is a compact set $K_r \subset (0,t] \cap \{x \in \mathbf{R}^n \colon x \ge \mathbf{r}\}$. Hence by Theorem 3.1 there exist constants $C = C_r < \infty$, $k = k_r \in \mathbf{N}_0$ such that for every φ of class C^k in a neighbourhood of K_r

$$(12) \qquad |u_r[\varphi]| \le C \sum_{|\alpha| \le k} \sup_{x \in K_r} \left| \left(\frac{\partial}{\partial x} \right)^{\alpha} \varphi(x) \right|.$$

Take arbitrarily $b < \omega$, a function $\gamma \in M_b((0,t])$ and its extension $\tilde{\gamma} \in M_b((0,\tilde{t}])$ (e.g. $\tilde{\gamma} = \mathcal{E}_\varepsilon \gamma$ see Th. 3). From (12) we derive that

$$(13) \qquad |u_r[\tilde{\gamma}]| \le C_{rb} \sum_{|\alpha| \le k} \sup_{x \in K_r} \left| x^{b+1} \left(x \frac{\partial}{\partial x} \right)^{\alpha} \gamma(x) \right|.$$

Define

$$v_r[\gamma] = u_r[\tilde{\gamma}] \qquad \text{for } \gamma \in M_b((0,t]), \ \ r > 0.$$

The definition of the functional v_r is correct (i.e. does not depend on the choice of the extension $\tilde{\gamma}$ in view of (12)) and by (13) $v_r \in M'_b((0,t])$ for $r > 0$ since $K_r \subset (0,t]$. We can also write $v_r[\varphi|_{(0,t]}] = u_r[\varphi]$ for $\varphi \in M_{(\omega)}((0,\tilde{t}])$ and putting $\gamma = \varphi|_{(0,t]}$ we get by (11)

$$v_r[\gamma] \underset{r \to 0}{\longrightarrow} u[\varphi] \qquad \text{for } \varphi \in M_{(\omega)}((0,\tilde{t}]).$$

Thus in view of Corollary 2.1 for every $b < \omega$ there exist constants $C_b < \infty$ and $k \in \mathbf{N}_0$ such that

$$|v_r[\gamma]|, |u[\varphi]| \le C_b \sum_{|\alpha| \le k} \sup_{x \in (0,t]} \left| x^{b+1} \left(x \frac{\partial}{\partial x} \right)^{\alpha} \gamma(x) \right| \qquad \text{for } \varphi \in M_b((0,\tilde{t}])$$

and hence (9) holds since $\gamma(x) = \varphi(x)$ for $x \in (0,t]$.

To prove the isomorphism (10) take u satisfying the assumptions of the theorem. Define

$$\tilde{u}[\varphi] = u[\mathcal{E}_\varepsilon \varphi] \qquad \text{for } \varphi \in M_{(\omega)}((0,t]).$$

It follows from Corollary 1 that $\tilde{u} \in M'_{(\omega)}((0,t])$ and from (9) we see that the definition of $\tilde{u}$ is independent of the choice of the extension $\mathcal{E}_\varepsilon \varphi$. Conversely, given a $\tilde{u} \in M'_{(\omega)}((0,t])$ the formula

$$u[\varphi] = \tilde{u}[\varphi|_{(0,t]}] \qquad \text{for } \varphi \in M_{(\omega)}((0,\tilde{t}])$$

defines $u \in M'_{(\omega)}((0,\tilde{t}])$ with support in $(0,t]$. This ends the proof.

2. Substitution in a Mellin Distribution

Let $u \in M'_{(\omega)}(I)$. By (5.16) $u \in D'(I)$ and we may define *substitution* $u \circ f$ by formula (3.4) if f^{-1} is a one-to-one mapping of $\Omega = \mathbf{R}^n_+$ onto $\Omega_1 \subset \mathbf{R}^n$ of class C^∞ with a non vanishing Jacobian. Then $u \circ f \in D'(H)$, where $H = f^{-1}(I)$ is relatively closed in $f^{-1}(\mathbf{R}^n_+)$ but need not be bounded. We describe below such a situation, important because of its applications to the study of singular operators in Section 15. By imposing some restrictions on the support of the Mellin distribution u we obtain that $u \circ f$ is itself a Mellin distribution.

Let $\mathbf{A} \in \mathrm{GL}(n;\mathbf{R})$ (the space of invertible $n \times n$ matrices with real entries) have non-negative entries. Define S: $\mathbf{R}^n_+ \to \mathbf{R}^n_+$, $S(y) = \exp(\mathbf{A}^{\mathrm{tr}} \ln y)$ for $y \in \mathbf{R}^n_+$. Since the transpose $\mathbf{A}^{\mathrm{tr}}$ of $\mathbf{A}$ has also non-negative entries it follows that the set (see Exercise 5)

$$Z_t^{\mathbf{A}} = S((0,t]) \subset \mathbf{R}^n_+, \qquad (t \in \mathbf{R}^n_+)$$

is bounded. Actually $S((0,t]) \subset (0, S(t)]$. Note that

$$S^{-1}(x) = \exp((\mathbf{A}^{\mathrm{tr}})^{-1} \ln x) \qquad \text{for } x \in \mathbf{R}^n_+$$

and the Jacobian of S^{-1} equals

$$JS^{-1}(x) = \frac{1}{\det \mathbf{A}} (S^{-1}(x))^{\mathbf{1}} x^{-\mathbf{1}}.$$

Denote $h = S(t)$, $I = (0,h]$. Let $u \in M'_{(\omega)}(I)$ for some $\omega \in (\mathbf{R} \cup \{+\infty\})^n$. As observed in the beginning, formula (3.3) for $\psi \in C^\infty_{(0)}(S^{-1}(I))$:

$$(14) \qquad u \circ S[\psi] = \frac{1}{|\det \mathbf{A}|} u[\psi \circ S^{-1}(x) \cdot (S^{-1}(x))^{\mathbf{1}} x^{-\mathbf{1}}]$$

defines correctly $u \circ S \in D'(\mathbf{R}_+^n)$ with $\operatorname{supp} u \circ S \subset S^{-1}(I)$ in general not bounded. Suppose that $\operatorname{supp} u \subset Z_t^{\mathbf{A}}$. Then $S^{-1}(\operatorname{supp} u) \subset (0,t]$ hence it is to be expected that $u \circ S$ extends to a Mellin distribution on $(0,t]$ and in fact such an extension belongs to $M'_{(\mathbf{A}\omega)}((0,t])$. However it is not clear that (14) makes sense for all $\psi \in M_{(\mathbf{A}\omega)}((0,t])$. To this end we need more information about Mellin distributions supported by the set $Z_t^{\mathbf{A}}$. We start by introducing suitable spaces:

Let $a \in \mathbf{R}^n$; define

$$M_a(Z_t^{\mathbf{A}}) = \{\varphi \in C^\infty(Z_t^{\mathbf{A}})\colon \varrho_{a,\alpha}^{\mathbf{A}}(\varphi) < \infty \quad \text{for} \quad \alpha \in \mathbf{N}_0^n\},$$

where

$$\varrho_{a,\alpha}^{\mathbf{A}}(\varphi) = \sup_{x \in Z_t^{\mathbf{A}}} \left| x^{a+\alpha+1} \left(\frac{\partial}{\partial x}\right)^\alpha \varphi(x) \right| \qquad \text{for} \quad \alpha \in \mathbf{N}_0^n$$

and, as in the case of $C^\infty(I)$, $C^\infty(Z_t^{\mathbf{A}})$ denotes the space of restrictions to $Z_t^{\mathbf{A}}$ of functions $C^\infty(\mathbf{R}_+^n)$. We also define for $\omega \in (\mathbf{R} \cup \{\infty\})^n$

$$M_{(\omega)}(Z_t^{\mathbf{A}}) = \varinjlim_{a<\omega} M_a(Z_t^{\mathbf{A}}).$$

Let $t < \tilde{t} \in \mathbf{R}_+^n$ and $\mathcal{E}_\varepsilon$ be the linear extension mapping of Corollary 1:

$$\mathcal{E}_\varepsilon\colon M_{(\mathbf{A}\omega)}((0,t]) \to \mathring{M}_{(\mathbf{A}\omega)}((0,\tilde{t}]).$$

Define

$$\mathcal{E}^{\mathbf{A}}(\varphi) = JS^{-1} \cdot \mathcal{E}_\varepsilon(JS \cdot \varphi \circ S) \circ S^{-1} \qquad \text{for} \quad \varphi \in M_{(\omega)}(Z_t^{\mathbf{A}}).$$

Proposition 2. *Let* $h = S(t)$, $\tilde{h} = S(\tilde{t})$. $\mathcal{E}^{\mathbf{A}}$ *is a continuous linear extension mapping*

$$\mathcal{E}^{\mathbf{A}}\colon M_{(\omega)}(Z_t^{\mathbf{A}}) \to M_{(\omega)}((0,\tilde{h}]).$$

In the proof we shall need assertions (15) and (16) of the following

Lemma 3. *Let* $\omega \in (\mathbf{R}\cup\{\infty\})^n$, $t < \tilde{t}$, $h = S(t)$, $\tilde{h} = S(\tilde{t})$. *The following mappings are continuous in the respective topologies.*

$$M_{(\omega)}(Z_t^{\mathbf{A}}) \ni \varphi \mapsto JS \cdot (\varphi \circ S) \in M_{(\mathbf{A}\omega)}((0,t]), \tag{15}$$

$$\mathring{M}_{(\mathbf{A}\omega)}((0,\tilde{t}]) \ni \psi \mapsto JS^{-1} \cdot (\psi \circ S^{-1}) \in \mathring{M}_{(\omega)}((0,\tilde{h}]), \tag{16}$$

$$M_{(\mathbf{A}\omega)}((0,t]) \ni \psi \mapsto JS^{-1} \cdot (\psi \circ S^{-1}) \in M_{(\omega)}(Z_t^{\mathbf{A}}). \tag{17}$$

Proof: We consider first the mapping (16). Take arbitrarily $\psi \in \mathring{M}_{(\mathbf{A}\omega)}((0,\tilde{t}])$, choose $\delta \in \mathbf{R}^n_+$ such that $\psi \in \mathring{M}_{\mathbf{A}\omega-\delta}((0,\tilde{t}])$ and extend ψ by zero to $\mathbf{R}^n_+$. Hence $\psi \circ S^{-1}(x)$ is smooth on $\mathbf{R}^n_+$ and equal to zero on $\mathbf{R}^n_+ \setminus \operatorname{Int} S((0,\tilde{t}])$. Select $\tilde{\delta} \in \mathbf{R}^n_+$ such that $\mathbf{A}\tilde{\delta} < \delta$. We shall prove that $(S^{-1}(x))^{\mathbf{1}}x^{-\mathbf{1}} \cdot \psi \circ S^{-1}(x)$ is in $M_{\omega-\tilde{\delta}}((0,\tilde{h}])$ and thus in $M_{(\omega)}((0,\tilde{h}])$. By Exercise 4 $(S^{-1}(x))^{\mathbf{1}}x^{-\mathbf{1}} = x^{\mathbf{A}^{-1}\mathbf{1}-\mathbf{1}}$ and by (6.1) $x^{\mathbf{A}^{-1}\mathbf{1}-\mathbf{1}}\psi \circ S^{-1}(x) \in M_{\omega-\tilde{\delta}}((0,\tilde{h}])$ if and only if $\psi\circ S^{-1}(x) \in M_{\omega-\tilde{\delta}+\mathbf{A}^{-1}\mathbf{1}-\mathbf{1}}((0,\tilde{h}\,])$. Hence it suffices to prove that $\psi \circ S^{-1}(x) \in M_{\omega-\tilde{\delta}+\mathbf{A}^{-1}\mathbf{1}-\mathbf{1}}((0,\tilde{h}\,])$. Observe that under the notation $x\frac{\partial}{\partial x} = \left(x_1\frac{\partial}{\partial x_1}, \dots, x_n\frac{\partial}{\partial x_n}\right)$ we have the following vector formula

$$x\frac{\partial}{\partial x}(\psi(S^{-1}(x))) = \left(\mathbf{A}^{-1}\left(y\frac{\partial}{\partial y}\right)\psi\right)(S^{-1}(x)).$$

Then for any $\alpha \in \mathbf{N}^n_0$ using once more Exercise 4 we get

$$\begin{aligned}(18)\qquad &\sup_{x\in(0,\tilde{h}\,]}\left|x^{\omega-\tilde{\delta}+\mathbf{A}^{-1}\mathbf{1}}\left(x\frac{\partial}{\partial x}\right)^{\alpha}(\psi(S^{-1}(x)))\right| \\ &\le \sup_{y\in(0,\tilde{t}\,]}\left|S(y)^{\omega-\tilde{\delta}+\mathbf{A}^{-1}\mathbf{1}}\left(\mathbf{A}^{-1}\left(y\frac{\partial}{\partial y}\right)\right)^{\alpha}\psi(y)\right| \\ &= \sup_{y\in(0,\tilde{t}\,]}\left|y^{\mathbf{A}\omega+\mathbf{1}-\mathbf{A}\tilde{\delta}}\left(\mathbf{A}^{-1}\left(y\frac{\partial}{\partial y}\right)\right)^{\alpha}\psi(y)\right|.\end{aligned}$$

The last expression is finite (cf. Def. 5.1′) since $M_{\mathbf{A}\omega-\delta}((0,\tilde{t}\,]) \subset M_{\mathbf{A}\omega-\mathbf{A}\tilde{\delta}}((0,\tilde{t}\,])$ by the choice of $\tilde{\delta}$. Thus

$$(\psi \circ S^{-1})(x)(S^{-1}(x))^{\mathbf{1}}x^{-\mathbf{1}} \in M_{(\omega)}((0,\tilde{h}\,]).$$

For the proof of (17) take arbitrarily $\psi \in M_{(\mathbf{A}\omega)}((0,t])$ and observe that by (16) and Corollary 1 $JS^{-1} \cdot (\mathcal{E}_\varepsilon\psi \circ S^{-1}) \in \mathring{M}_{(\omega)}((0,\tilde{h}])$. Since $\psi \circ S^{-1}$ (defined on $Z^{\mathbf{A}}_t$) is a restriction of $\mathcal{E}_\varepsilon\psi \circ S^{-1}$ to $Z^{\mathbf{A}}_t$ the assertion (17) follows.

To prove (15) let $\varphi \in M_b(Z^{\mathbf{A}}_t)$ for some $b < \omega$. Denote

$$\psi(y) = \det \mathbf{A} \cdot \varphi(S(y))(S(y))^{\mathbf{1}}y^{-\mathbf{1}} \qquad \text{for } y \in (0,t].$$

Proceeding in an analogous way as in the proof of (16) we prove for any $k \in \mathbf{N}_0$ the estimates

$$\begin{aligned}&\sum_{|\alpha|\le k}\sup_{y\in(0,t]}\left|y^{\mathbf{A}b+\mathbf{1}}\left(y\frac{\partial}{\partial y}\right)^{\alpha}\psi(y)\right| \\ &\le \tilde{C}\sum_{|\alpha|\le k}\sup_{x\in Z^{\mathbf{A}}_t}\left|(S^{-1}(x))^{\mathbf{A}b+\mathbf{1}}x^{\mathbf{1}}(S^{-1}(x))^{-\mathbf{1}}\left(\mathbf{A}\left(x\frac{\partial}{\partial x}\right)\right)^{\alpha}\varphi(x)\right| \\ &= C\sum_{|\alpha|\le k}\sup_{x\in Z^{\mathbf{A}}_t}\left|x^{b+\mathbf{1}}\left(x\frac{\partial}{\partial x}\right)^{\alpha}\varphi(x)\right|,\end{aligned}$$

which ends the proof of the lemma.

Proof of Proposition 2: Let $\varphi \in M_{(\omega)}(Z_t^{\mathbf{A}})$. Then by Lemma 3, (15) $\psi = JS \cdot (\varphi \circ S) \in M_{(\mathbf{A}\omega)}((0,t])$ hence by Corollary 1 $\mathcal{E}_\varepsilon\psi \in \overset{\circ}{M}_{(\mathbf{A}\omega)}((0,\tilde{t}])$ and again by Lemma 3, (16) $JS^{-1} \cdot ((\mathcal{E}_\varepsilon\psi) \circ S^{-1})$ is in $M_{(\omega)}((0,\tilde{h}])$. Clearly $\mathcal{E}^{\mathbf{A}}$ is linear and continuous.

Now in the same way as in the proof of Proposition 5.1, by using the extension mapping $\mathcal{E}^{\mathbf{A}}$, we prove

Proposition 3. *The space $M_a(Z_t^{\mathbf{A}})$ with the topology given by the seminorms $\varrho^{\mathbf{A}}_{a,\alpha}$, $\alpha \in \mathbf{N}_0^n$ is complete.*

Proceeding in an analogous way as in the proof of Theorem 3 we get the following generalization of that theorem:

Theorem 3′. *Let $u \in M'_{(\omega)}((0,\tilde{h}])$ and $\operatorname{supp} u \subset Z_t^{\mathbf{A}}$ for some $t < \tilde{t} = S^{-1}(\tilde{h})$. Then for any $b < \omega$ there exist constants $C = C(b)$ and $k = k(b) \in \mathbf{N}_0$ such that*

$$|u[\varphi]| \le C \sum_{|\alpha|\le k} \sup_{x\in Z_t^{\mathbf{A}}} \left| x^{b+1}\left(x\frac{\partial}{\partial x}\right)^\alpha \varphi(x)\right| \qquad \text{for } \varphi \in M_b((0,\tilde{h}]).$$

Hence the restriction mapping

$$M_{(\omega)}((0,\tilde{h}]) \ni \varphi \mapsto \varphi|_{Z_t^{\mathbf{A}}} \in M_{(\omega)}(Z_t^{\mathbf{A}})$$

induces a linear isomorphism of the spaces

$$\{u \in M'_{(\omega)}((0,\tilde{h}]) \colon \operatorname{supp} u \subset Z_t^{\mathbf{A}}\} \simeq M'_{(\omega)}(Z_t^{\mathbf{A}}).$$

Now, we are in a position to state the final result:

Theorem 4. *Let $t \in \mathbf{R}^n_+$, $S(y) = \exp(\mathbf{A}^{\mathrm{tr}} \ln y)$ for $y \in \mathbf{R}^n_+$ where $\mathbf{A} \in \mathrm{GL}(n;\mathbf{R})$ has non-negative entries. Let $u \in M'_{(\omega)}((0,h])$, $h = S(t)$ and $\operatorname{supp} u \subset Z_t^{\mathbf{A}} = S((0,t])$. Define for $\psi \in M_{(\mathbf{A}\omega)}((0,t])$*

$$u \circ S[\psi] = u[\varphi], \tag{19}$$

where

$$\varphi(x) = \frac{1}{|\det \mathbf{A}|}\psi \circ S^{-1}(x) \cdot (S^{-1}(x))^{1} x^{-1} \qquad \text{for } x \in Z_t^{\mathbf{A}}$$

(here u is regarded as an element of $M'_{(\omega)}(Z_t^{\mathbf{A}})$). Then $u \circ S \in M'_{(\mathbf{A}\omega)}((0,t])$ and

$$\mathcal{M}(u \circ S)(\zeta) = \frac{1}{|\det \mathbf{A}|}(\mathcal{M}u) \circ \mathbf{A}^{-1}(\zeta) \qquad \text{for } \operatorname{Re}\zeta < \mathbf{A}\omega.$$

Proof: Let $\tilde{t} > t$, $\tilde{h} = S(\tilde{t})$ and observe that u can be considered as a functional in $M'_{(\omega)}((0,\tilde{h}])$ (namely $\tilde{u}[\psi] = u[\psi|_{(0,h]}]$ for $\psi \in M_{(\omega)}((0,\tilde{h}])$. Hence by Theorem 3′ u can be also considered as an element of $M'_{(\omega)}(Z_t^{\mathbf{A}})$ and, by Lemma 3 (17), formula (21) defines correctly the functional $u \circ S \in M'_{(\mathbf{A}\omega)}((0,t])$. From (19) we get

$$\mathcal{M}(u \circ S)(\zeta) = u \circ S[y^{-\zeta-1}] = u\left[(S^{-1}(x))^{-\zeta-\mathbf{1}}\frac{1}{|\det \mathbf{A}|}(S^{-1}(x))^{\mathbf{1}}x^{-\mathbf{1}}\right]$$
$$= \frac{1}{|\det \mathbf{A}|}u[x^{-\mathbf{A}^{-1}\zeta-\mathbf{1}}] = \frac{1}{|\det \mathbf{A}|}(\mathcal{M}u) \circ \mathbf{A}^{-1}\zeta,$$

which ends the proof.

Remark 2. Proposition 2 generalizes Corollary 1 ($\mathcal{E}_\varepsilon = \mathcal{E}^{\mathrm{Id}}$) and hence also the Seeley extension theorem.

3. Mellin Order of a Mellin Distribution

Recall that $u \in M'_a(I)$ was said to be of Mellin order $\leq m$, $m \in \mathbb{N}_0$ (see formula (5.9)) if

$$(20) \qquad |u[\varphi]| \leq C(a) \sum_{|\alpha|\leq m} \sup_{x \in I}\left|x^{a+\alpha+1}\left(\frac{\partial}{\partial x}\right)^{\alpha}\varphi(x)\right| \qquad \text{for } \varphi \in M_a(I)$$

for some constant $C(a) < \infty$.

Now, if $u \in M'_{(\omega)}(I)$ is a Mellin distribution we say that the *Mellin order* of u is $\leq m$ if (20) holds for all $a < \omega$. If such m does not exist u is said to be of *infinite Mellin order.*

Remark 3. Note that the notion of Mellin order is a concept essentially different from the classical order of a distribution (see Definition 3.1) which for Mellin distributions is always finite (see Remark 1). Indeed, there exist Mellin distributions of infinite Mellin order. Such an example is provided by the function

$$u(x) = \sum_{k=1}^{\infty}\left(x\frac{d}{dx}\right)^k \sin x^{a_k} \qquad \text{for } x \in (0,1],$$

where $a_k = -\frac{1}{k!}$ for $k \in \mathbb{N}$. See [Ly1] for the proof that $u \in M'_{(0)}$ has infinite Mellin order.

Exercises

1. Show that the space $M'(I)$ of Mellin distributions restricted to $(0,r)$ coincides with the space of resrictions to $(0,r)$ of distributions from $D'(\mathbf{R}^n)$ ([Ziel], proposition 5 and [Sz2], theorem 6).

2. Give a characterization theorem for the space $\mathfrak{M}'_\alpha$. The theorem can be of the following form: $u \in \mathfrak{M}'_\alpha(\mathbf{R}^n_+)$ if and only if there exist $\lambda \in \mathbf{N}_0^n$, $C < \infty$, $k \in \mathbf{N}_0$ and a function $f \in C^0(\mathbf{R}^n_+)$ such that

$$u = x^\alpha \left(x\frac{\partial}{\partial x}\right)^\lambda f, \qquad |f(x)| < C(1 + \|\ln x\|^2)^k \qquad \text{for } x > 0.$$

3. Prove that $\mathring{M}'_{(\omega)}((0,t]) \subset D'(\mathbf{R}^n)$, where $\mathring{M}'_{(\omega)}((0,t])$ is the dual space of the space $\mathring{M}_{(\omega)}((0,t])$ defined in Subsection 2.

HINT. See Theorem 1 and Proposition 3 (iii).

4. Let $\mathbf{A} \in \mathrm{GL}(n,\mathbf{R})$, $b \in \mathbf{C}^n$. Show the following identity

$$(\exp(\mathbf{A}^{\mathrm{tr}}(\ln y)))^b = y^{\mathbf{A}b} \qquad \text{for } y \in \mathbf{R}^n_+.$$

5. Let $\mathbf{A} \in \mathrm{GL}(n,\mathbf{R})$, $S(y) = \exp(\mathbf{A}^{\mathrm{tr}} \ln y)$ for $y \in \mathbf{R}^n_+$, $Z_t^{\mathbf{A}} = S((0,t])$, $t \in \mathbf{R}^n_+$. Show that the set $Z_t^{\mathbf{A}}$ is bounded if and only if all the entries of $\mathbf{A}$ are non-negative.

6. Let $\mathbf{A} = (a_{ij})^n_{i,j=1}$, $a_{1j} = 1$ for $j = 1,\dots,n$, $a_{ij} = 0$ for $j \neq i$, $a_{ii} = 1$ $(i = 2,\dots,n)$. Show that the transformation S: $\mathbf{R}^n_+ \to \mathbf{R}^n_+$ defined by $S(y) = \exp(\mathbf{A}^{\mathrm{tr}} \ln y)$ has the form: $x_1 = y_1$, $x_j = y_1 y_j$ for $j = 2,\dots,n$.

7. Let $\mathbf{A}$, S be defined as in Exercise 5. Assume that the set $Z^{\mathbf{A}} = S((0,\mathbf{1}])$ is bounded and denote by $\chi^{\mathbf{A}}$ its characteristic function. Show that

$$(\mathcal{M}\chi^{\mathbf{A}})(z) = (-1)^n\, |\det \mathbf{A}|\, (\mathbf{A}z)^{-\mathbf{1}}.$$

HINT. See Theorem 4 and the beginning of Subsection 7.3.

8. Prove Theorem 3.1 in the case of a closed polyinterval K.

HINT. Apply Seeley extension theorem (see Exercise 5.14).

§9. PALEY–WIENER TYPE THEOREMS FOR THE MELLIN TRANSFORMATION

By *Paley–Wiener* type *theorems for the Mellin transformation* we mean (on analogy with the Fourier–Laplace transformation) theorems which relate the regularity and support properties of a Mellin distribution to estimates of the growth order of its Mellin transform.

Theorem 1. *Let $u \in M'_{(\omega)}(I)$ $(I = (0,t], \omega \in (\mathbf{R} \cup \{\infty\})^n)$ and let $F(z) = (\mathcal{M}u)(z)$ for $z \in \Omega = \{z \in \mathbf{C}^n\colon \operatorname{Re} z < \omega\}$.*

Then F is a holomorphic function on Ω and for every $b < \omega$ there exist $s(b) \in \mathbf{N}_0$ and $C(b) < \infty$ such that

$$(1) \qquad |F(z)| \le C(b)\langle z\rangle^{s(b)} t^{-\operatorname{Re} z} \qquad \text{for } \operatorname{Re} z \le b$$

where $\langle z\rangle = 1 + |z_1| + \cdots + |z_n|$. If u is of Mellin order $\le s$, $s \in \mathbf{N}_0$, then for every $b < \omega$ the estimation (1) may be written in the form

$$(2) \qquad |F(z)| \le C(b)\langle z\rangle^{s} t^{-\operatorname{Re} z} \qquad \text{for } \operatorname{Re} z \le b$$

with some constant $C(b)$.

Proof: By Theorem 7.1, F is holomorphic on Ω. If $b < \omega$, $u \in M'_b$ and $x^{-z-1} \in M_b$ for $\operatorname{Re} z \le b$. Hence there exist $C_1(b)$ and $s(b) \in \mathbf{N}_0$ such that

$$\begin{aligned} |F(z)| = \left|u[x^{-z-1}]\right| &\le C_1(b) \sum_{|\alpha| \le s(b)} \varrho_{b,\alpha}(x^{-z-1}) \\ &= C_1(b) \sum_{|\alpha| \le s(b)} \sup_{x \in I} \left|p_\alpha(z) x^{b-z}\right| \\ &= C_2(b) \sum_{|\alpha| \le s(b)} |p_\alpha(z)|\, t^{-\operatorname{Re} z} \qquad \text{for } \operatorname{Re} z \le b \end{aligned}$$

where

$$(3) \qquad p_\alpha(z) = \prod_{j=1}^{n} \prod_{k=1}^{\alpha_j} (z_j + k) \qquad \text{for } |\alpha| \le s(b) \text{ and } C_2(b) = t^b C_1(b).$$

Thus we get (1) (see Exercise 7.11(c)). If u is of Mellin order $\le s$ then $s(b) = s$ for every $b < \omega$ and we get (2).

Theorem 2 (uniqueness). *Let $\omega_1 \le \omega_2 \in (\mathbf{R} \cup \{\infty\})^n$, $u \in M'_{(\omega_1)}(I)$, $v \in M'_{(\omega_2)}(I)$ and*

$$(4) \qquad \mathcal{M}u(z) = \mathcal{M}v(z) \qquad \text{for } \operatorname{Re} z < \omega_1.$$

Then $u = v$ in $M'_{(\omega_1)}$.

Proof: We have to prove that $u[\varphi] = v[\varphi]$ for $\varphi \in M_{(\omega_1)} = \bigcup_{a < \omega_1} M_a$. Choose arbitrarily $b < \omega_1$, $\psi \in M_b$ and take a so that $b < a < \omega_1$. By Lemma 5.2 ψ belongs to the closure of X_a in M_a $(X_a = \operatorname{span}\{x^{-c-1}\}_{c \le a})$. Thus it suffices to prove that

$$u[x^{-c-1}] = v[x^{-c-1}] \qquad \text{for } c \le a$$

which is clear from (4).

Another proof of Theorem 2 is proposed in Exercise 2.

Now we intend to prove the converse of Theorem 1. We start with two lemmas:

Lemma 1. *Let $b \in \mathbf{R}^n$. Suppose that G is holomorphic in the set $\{z \in \mathbf{C}^n\colon \operatorname{Re} z \leq b\}$ and satisfies there the estimation*

$$|G(z)| \leq \frac{K}{\langle z_1 \rangle^2 \cdots \langle z_n \rangle^2} t^{-\operatorname{Re} z} \tag{5}$$

where $K < \infty$, $t \in \mathbf{R}^n_+$.

Then the formal definition

$$g(x) = \frac{1}{(2\pi i)^n} \int_{c+i\mathbf{R}^n} G(z) x^z \, dz \qquad \text{for } x > 0 \quad (c \leq b) \tag{6}$$

defines correctly a function g (continuous for $x > 0$) which does not depend on the choice of $c \leq b$. Moreover:

(a) $\operatorname{supp} g \subset I = (0, t]$,

(b) *$x^{-b} g(x)$ is bounded and $g \in M'_{(b)}(I)$ is a distribution of Mellin order ≤ 0,*

(c) *$G(z) = (\mathcal{M} g)(z)$ for $\operatorname{Re} z \leq b$.*

Proof: Let $c \leq b$ and observe that by (5):

$$\left| G(c + i\eta) x^{c+i\eta} \right| \leq \frac{K t^{-c} x^c}{\langle c_1 + i\eta_1 \rangle^2 \ldots \langle c_n + i\eta_n \rangle^2}. \tag{7}$$

Hence

$$g_c(x) \overset{\text{def}}{=} \frac{1}{(2\pi)^n} \int_{\mathbf{R}^n} G(c + i\eta) x^{c+i\eta} \, d\eta \qquad \text{for } x > 0 \tag{8}$$

defines a continuous function because the integral on the right hand side is uniformly and absolutely convergent. The proof that $g_c = g_{\tilde{c}}$ for $c \leq b$, $\tilde{c} \leq b$ can be performed successively for c, $\tilde{c}$ differing in one component. For simplicity take $c = (c_1, c')$, $\tilde{c} = (\tilde{c}_1, c')$, $x = (x_1, x')$, $z = (z_1, z')$ with $c' = (c_2, \ldots, c_n)$, $z' = (z_2, \ldots, z_n)$, $x' = (x_2, \ldots, x_n)$, $b' = (b_2, \ldots, b_n)$. We have

$$g_c(x) - g_{\tilde{c}}(x) = \left(\frac{1}{2\pi i}\right)^n \int_{c'+i\mathbf{R}^{n-1}} (x')^{z'} \left\{ \int_{c_1+i\mathbf{R}} G(z_1, z') x_1^{z_1} \, dz_1 - \right.$$
$$\left. - \int_{\tilde{c}_1+i\mathbf{R}} G(z_1, z') x_1^{z_1} \, dz_1 \right\} dz'.$$

Now it suffices to prove that for every fixed $z' \in \mathbf{C}^{n-1}$, $\operatorname{Re} z' \leq b'$

$$\int_{c_1+i\mathbf{R}} G(z_1, z') x_1^{z_1} \, dz_1 = \int_{\tilde{c}_1+i\mathbf{R}} G(z_1, z') x_1^{z_1} \, dz_1.$$

For the proof take (for a fixed z') a rectangle in the complex z_1-plane with the vertices $A_r = c_1 - ir$, $B_r = c_1 + ir$, $C_r = \tilde{c}_1 + ir$, $D_r = \tilde{c}_1 - ir$. By the Cauchy theorem we have

$$\int_{[A_r,B_r]} + \int_{[B_r,C_r]} + \int_{[C_r,D_r]} + \int_{[D_r,A_r]} G(z_1,z')x_1^{z_1}\,dz_1 = 0.$$

Thus it remains to prove that

$$\lim_{r\to\infty}\int_{[B_r,C_r]} G(z_1,z')x_1^{z_1}\,dz_1 = 0 = \lim_{r\to\infty}\int_{[D_r,A_r]} G(z_1,z')x_1^{z_1}\,dz_1. \tag{9}$$

Write $t = (t_1, t')$, $t' = (t_2, \dots, t_n)$. By assumption (5)

$$\left|\int_{[B_r,C_r]} G(z_1,z')x_1^{z_1}\,dz_1\right| \le \frac{K(t')^{-\operatorname{Re} z'}}{\langle z_2\rangle^2\cdots\langle z_n\rangle^2}\,\frac{|\tilde{c}_1 - c_1|}{r^2}\cdot\max_{\xi_1\in[c_1,\tilde{c}_1]}\left(\frac{x_1}{t_1}\right)^{\xi_1}.$$

Hence follows the first of the relations (9). The second is proved in the same way.

To prove the assertion (a) observe that for every $c \le b$ we have by (6) and (7) the estimations

$$\begin{aligned}
|g(x)| &\le \frac{Kt^{-c}x^c}{(2\pi)^n}\int_{\mathbf{R}^n}\frac{d\eta_1\cdots d\eta_n}{\langle c_1+i\eta_1\rangle^2\cdot\ \dots\ \cdot\langle c_n+i\eta_n\rangle^2}\\
&\le \frac{Kt^{-c}x^c}{(2\pi)^n}\,\frac{\pi^n}{\sqrt{1+c_1^2}\cdot\ \dots\ \cdot\sqrt{1+c_n^2}}\\
&\le K\,2^{-n/2}\left(\frac{x}{t}\right)^c\frac{1}{\langle c_1\rangle\cdot\ \dots\ \cdot\langle c_n\rangle}.
\end{aligned}\tag{10}$$

Choose arbitrarily $x > 0$, $x \notin I$ and $1 \le j \le n$ such that $x_j > t_j$. Let $c_j \to -\infty$. Then $(x_j/t_j)^{c_j} \to 0$ and the right hand side of (10) tends to 0. Since $g(x)$ does not depend on c we have $g(x) = 0$.

Note that from (10) we have

$$|g(x)| \le K2^{-n/2}\left(\frac{x}{t}\right)^b\frac{1}{\langle b_1\rangle\cdot\ \dots\ \cdot\langle b_n\rangle} \tag{11}$$

hence

$$\left|x^{-b}g(x)\right| \le K2^{-n/2}t^{-b}\frac{1}{\langle b_1\rangle\cdot\ \dots\ \cdot\langle b_n\rangle} < \infty$$

and by (6.2) and Exercise 5.11 we get assertion (b). Now we shall prove that $G(z) = (\mathcal{M}g)(z)$ for $\operatorname{Re} z < b$. Let $c < b$. Putting $x = e^{-\tau}$ in (8) we find from (8) and (7)

$$g(e^{-\tau}) = \frac{1}{(2\pi)^n}\int_{\mathbf{R}^n} G(c+i\eta)e^{-\tau c - i\tau\eta}\,d\eta = (2\pi)^{-n/2}e^{-\tau c}\mathcal{F}(G(c+i\,\cdot))(\tau). \tag{12}$$

Let

$$h(\tau) = (2\pi)^{n/2} g(e^{-\tau}) e^{\tau c} \qquad \text{for } \tau \in \mathbf{R}^n.$$

We shall show that $h \in L^1(\mathbf{R}^n)$. Put $\varepsilon = b - c$. Observe that $\operatorname{supp} h \subset \{\tau \in \mathbf{R}^n : \tau \geq - \ln t\}$ and that by (11)

$$\begin{aligned} |h(\tau)| &\leq (2\pi)^{n/2} e^{\tau(b-\varepsilon)} K 2^{-n/2} \left(\frac{e^{-\tau}}{t}\right)^b \frac{1}{\langle b_1\rangle \cdot \ldots \cdot \langle b_n\rangle} \\ &\leq \pi^{n/2} \cdot K t^{-b} e^{-\varepsilon\tau} \frac{1}{\langle b_1\rangle \cdot \ldots \cdot \langle b_n\rangle} = \widetilde{K} e^{-\varepsilon\tau} \end{aligned}$$

i.e. h is dominated on its support by an integrable function. Thus $h \in L^1(\mathbf{R}^n)$ and so by (12) we get

$$G(c + i\eta) = (\mathcal{F}^{-1}h)(\eta) = \int_{\mathbf{R}^n} g(e^{-\tau}) e^{\tau(c+i\eta)}\, d\tau$$

hence

$$G(z) = \int_{\mathbf{R}^n} e^{\tau z} g(e^{-\tau})\, d\tau = \int_I g(x) x^{-z-1}\, dx = (\mathcal{M}g)(z) \qquad \text{for } \operatorname{Re} z < b$$

and by hypothesis that G is holomorphic on $\{z \in \mathbf{C}^n : \operatorname{Re} z \leq b\}$ we get assertion (c).

Lemma 2. *Let $\alpha \in \mathbf{N}_0$, $P(z) = \sum_{|\alpha| \leq s} a_\alpha |z_1|^{\alpha_1} \cdot \ldots \cdot |z_n|^{\alpha_n}$ and let $\tilde{\omega} \in \mathbf{R}^n$. Define (cf. Exercise 5)*

$$Q(z) = (z_1 - \tilde{\omega}_1 - 1)^{s+2} \cdot \ldots \cdot (z_n - \tilde{\omega}_n - 1)^{s+2}.$$

Then there exists $K < \infty$ such that

$$(13) \qquad |Q(z)| \geq 1, \qquad \frac{|P(z)|}{|Q(z)|} \leq \frac{K}{\langle z_1\rangle^2 \cdot \ldots \cdot \langle z_n\rangle^2} \qquad \text{for } \operatorname{Re} z \leq \tilde{\omega}.$$

Proof: It is easy to see that

$$\frac{|z_j|}{|z_j - \tilde{\omega}_j - 1|} \leq \sqrt{1 + \tilde{\omega}_j^2}, \qquad |z_j - \tilde{\omega}_j - 1| \geq 1, \qquad \frac{|z_j| + 1}{|z_j - \tilde{\omega}_j - 1|} \leq 2 + |\tilde{\omega}_j + 1|$$

for $\operatorname{Re} z_j < \tilde{\omega}_j$, $j = 1, 2, \ldots, n$. Hence for $\operatorname{Re} z \leq \tilde{\omega}$ we have $|Q(z)| \geq 1$ and

$$\frac{|P(z)|}{|Q(z)|} \leq \sum_{|\alpha| \leq s} |a_\alpha| \frac{C_\alpha}{\langle z_1\rangle^2 \cdot \ldots \cdot \langle z_n\rangle^2} \prod_{j=1}^n \frac{(|z_j| + 1)^2}{|z_j - \tilde{\omega}_j - 1|^2} \leq \frac{K}{\langle z_1\rangle^2 \cdot \ldots \cdot \langle z_n\rangle^2},$$

where

$$C_\alpha = (1 + \tilde{\omega}_1^2)^{\frac{\alpha_1}{2}} \cdot \ldots \cdot (1 + \tilde{\omega}_n^2)^{\frac{\alpha_n}{2}},$$

$$K = (2 + |\tilde{\omega}_1 + 1|)^2 \cdots (2 + |\tilde{\omega}_n + 1|)^2 \sum_{|\alpha| \leq s} |a_\alpha| C_\alpha.$$

Theorem 3 (inversion theorem for the Mellin transformation). *Let $\omega \in (\mathbf{R} \cup \{+\infty\})^n$ and let F be a holomorphic function on the set $\{z \in \mathbf{C}^n: \operatorname{Re} z < \omega\}$ such that for every $b < \omega$ there exist $C(b) < \infty$ and $s(b) \in \mathbf{N}_0$ such that*

$$|F(z)| \leq C(b)\langle z\rangle^{s(b)} t^{-\operatorname{Re} z} \quad \text{for } \operatorname{Re} z \leq b. \tag{14}$$

Then there exists a unique distribution $u \in M'_{(\omega)}((0,t])$ such that

$$F(z) = (\mathcal{M}u)(z) \quad \text{for } \operatorname{Re} z < \omega. \tag{15}$$

Proof: Take $b < \omega$ and suppose (14) holds. Choose $\tilde{\omega} \in \mathbf{R}^n$ such that $b < \tilde{\omega} \leq \omega$ (for example we take $\tilde{\omega}_j = \omega_j$ if $\omega_j \in \mathbf{R}$ and $\tilde{\omega}_j = b_j + 1$ if $\omega_j = \infty$). Put

$$Q(z) = (z_1 - \tilde{\omega}_1 - 1)^{s(b)+2} \cdot \ldots \cdot (z_n - \tilde{\omega}_n - 1)^{s(b)+2}.$$

By Lemma 2 with $P(z) = \langle z\rangle^{s(b)}$ we have the inequalities (13) and by (14) the function

$$G(z) = \frac{F(z)}{Q(z)} \quad \text{for } \operatorname{Re} z \leq \tilde{\omega}$$

satisfies all the assumptions of Lemma 1. Thus the function g (defined independently of the choice of $c \leq b$)

$$g(x) = \frac{1}{(2\pi i)^n} \int_{c+i\mathbf{R}^n} \frac{F(z)}{Q(z)} x^z \, dz \quad \text{for } x > 0, \ c \leq b$$

is continuous. Moreover $g \in M'_{(b)}(I)$ is a distribution of Mellin order ≤ 0 and $(\mathcal{M}g)(z) = \frac{F(z)}{Q(z)}$ for $\operatorname{Re} z \leq b$. Define

$$u = Q\left(x_1 \frac{\partial}{\partial x_1}, \ldots, x_n \frac{\partial}{\partial x_n}\right) g.$$

By Corollary 6.1 $u \in M'_{(b)}(I)$ and by (7.11)

$$\mathcal{M}u(z) = Q(z)(\mathcal{M}g)(z) = F(z) \quad \text{for } \operatorname{Re} z \leq b.$$

Since Q is a polynomial of degree $n(s(b)+2)$, the Mellin order of u is $\leq n(s(b)+2)$.

Now take any nondecreasing sequence $b_\nu \in \mathbf{R}^n$, $b_\nu \underset{\nu\to\infty}{\longrightarrow} \omega$. We have just proved that there exist $u_\nu \in M'_{(b_\nu)}(I)$ such that $(\mathcal{M}u_\nu)(z) = F(z)$ for $\operatorname{Re} z \leq b_\nu$ ($\nu = 1, 2, \ldots$). The Mellin order of u_ν is $\leq n(s(b_\nu) + 2)$. By Theorem 2

$$u_\nu = u_{\nu'} \text{ in } M'_{(b_\nu)}, \text{ if } \nu \leq \nu' \quad (\text{i.e. } b_\nu \leq b_{\nu'}).$$

Thus the definition

$$u = u_\nu \text{ on } M_{(b_\nu)}, \qquad \nu = 1, 2, \ldots \tag{16}$$

defines correctly a functional u on $M_{(\omega)}$. To show that $u \in M'_{(\omega)}$ take $\varphi_k \to 0$ in $M_{(\omega)}$. Hence $\varphi_k \to 0$ in M_b for some $b < \omega$, and since $b_\nu \to \omega$ there exists b_ν such that $b < b_\nu < \omega$ and $\varphi_k \to 0$ in M_{b_ν}. By (16) $u[\varphi_k] = u_\nu[\varphi_k] \to 0$ as $k \to \infty$.

For the proof of (15) take arbitrarily a point $\mathring{z}$, $\operatorname{Re} \mathring{z} < \omega$ and find $\mathring{\nu}$ such that $\operatorname{Re} \mathring{z} \leq b_{\mathring{\nu}} < \infty$. Thus $x^{-\mathring{z}-1} \in M_{(b_{\mathring{\nu}})}$ and $(\mathcal{M}u)(\mathring{z}) = u[x^{-\mathring{z}-1}] = u_{\mathring{\nu}}[x^{-\mathring{z}-1}] = \mathcal{M}u_{\mathring{\nu}}(\mathring{z}) = F(\mathring{z})$. This ends the proof since the uniqueness of u follows from Theorem 2.

Remark 1. If in addition to the assumptions of Theorem 3 we assume that there exists $\mathring{s}$ such that $s(b) \leq \mathring{s}$ for $b < \omega$ then it follows from the proof of the theorem that u is a Mellin distribution in $M'_{(\omega)}(I)$ of Mellin order not greater than $n(\mathring{s} + 2)$. This estimation is far from the best and could be much improved by introducing suitable weighted L^2 spaces.

Combining Theorems 1 and 3 we obtain

Theorem 4 (Paley–Wiener type theorem). *In order that a function* $F(z) = F(z_1, \ldots, z_n)$ *be the Mellin transform of a unique Mellin distribution* $u \in M'_{(\omega)}((0,t])$ *it is necessary and sufficient that* F *be holomorphic in the set* $\{z \in \mathbf{C}^n\colon \operatorname{Re} z < \omega\}$ *and for every* $b < \omega$ *there exist* $C(b)$ *and* $s(b)$ *such that* (14) *holds.*

This theorem corresponds to the Paley–Wiener–Schwartz theorem for the Fourier–Laplace transforms of distributions with compact support (theorem 7.3.1 [Hö2]).

Now, in the case of the Mellin transformation we give a theorem analogous to the Paley–Wiener theorem for the Fourier–Laplace transform of $C_0^\infty(\mathbf{R}^n)$ functions.

In our case the set which corresponds to $C_0^\infty(\mathbf{R}^n)$ is the subset $\mathring{M}_{[-1]}(I)$ of the set $M_{[-1]}(I)$ consisting of the functions $\varphi \in M_{[-1]}(I)$ satisfying the boundary conditions (cf. Exercise 5)

$$\left.\frac{\partial^p \varphi(x)}{\partial x_j^p}\right|_{x_j = t_j} = 0 \qquad \text{for } x \in I \qquad (j = 1, \ldots n, \quad p = 0, 1, \ldots). \tag{17}$$

Recall that the space $M_{[-1]}(I)$ defined in Subsection 5.2 as $M_{[-1]} = \varprojlim_{a > -1} M_a$ coincides, as shown in Subsection 6.2, with the space of Mellin multipliers.

Denote by

$$\varphi_{jp}(x) = \left(x_j \frac{\partial}{\partial x_j}\right)^p \varphi(x) \qquad \text{for } x \in I \qquad (j = 1, \ldots, n, \quad p = 0, 1, \ldots)$$

the classical derivatives of a $\varphi \in C^\infty_{(0)}(I)$.

Lemma 3. *Let $\varphi \in \mathring{M}_{[-1]}(I)$. Then $\varphi \in M'_{(0)}$, $\left(x_j \frac{\partial}{\partial x_j}\right)^m \varphi \in M'_{(0)}$ are Mellin distributions of Mellin order ≤ 0 and*

$$\text{(18)} \qquad \left(x_j \frac{\partial}{\partial x_j}\right)^m \varphi = \varphi_{jm} \qquad \text{in } M'_{(0)} \qquad (j = 1, \ldots, n, \quad m = 1, 2, \ldots).$$

Proof: Let $\varphi \in \mathring{M}_{[-1]}$. By (5.19) and operational rules from Subsection 6.1, $\varphi_{jm} \in M_{-1+\varepsilon}$ for $\varepsilon > 0$, $\varphi \in \varinjlim_{\varepsilon>0} M'_{-\varepsilon} = M'_{(0)}$ and $\left(x_j \frac{\partial}{\partial x_j}\right)^m \varphi \in M'_{(0)}$ $(j = 1, \ldots, n,$ $m = 1, 2, \ldots)$. By Exercise 6.3 we get (18). We shall now prove that $\left(x_j \frac{\partial}{\partial x_j}\right)^m \varphi$ are distributions of Mellin order ≤ 0. To this aim we choose arbitrarily $\eta > 0$ and $\lambda \in M_{-\eta}$. By (18) we obtain

$$\begin{aligned}\left|\left(\left(x_j \frac{\partial}{\partial x_j}\right)^m \varphi\right)[\lambda]\right| &= \left|\int_I \varphi_{jm}(x)\lambda(x)\, dx\right| \\ &\leq \sup_{x\in I}\left|x^{-\eta+1}\lambda(x)\right| \sup_{x\in I}\left|x^{\eta/2}\varphi_{jm}(x)\right| \int_I x^{\eta/2-1}\, dx \\ &\leq C_\eta \varrho_{-\eta,0}(\lambda)\end{aligned}$$

where $C_\eta = \varrho_{-1+\frac{\eta}{2},0}(\varphi_{jm}) \int_I x^{\frac{\eta}{2}-1}\, dx < \infty$.

Theorem 5 (Paley–Wiener type theorem for $\mathring{M}_{[-1]}$). *In order that a function G be the Mellin transform of a unique function $\varphi \in \mathring{M}_{[-1]}((0, t])$ it is necessary and sufficient that G be holomorphic in the set $\{z \in \mathbf{C}^n\colon \operatorname{Re} z < 0\}$ and that for every $m \in \mathbf{N}_0$ and $b < 0$ there exists constant $C_{mb} < \infty$ such that*

$$\text{(19)} \qquad |G(z)| \leq C_{mb} \frac{1}{\langle z\rangle^m} t^{-\operatorname{Re} z} \qquad \text{for } \operatorname{Re} z \leq b.$$

Proof: 1^0. Suppose first that $\varphi \in \mathring{M}_{[-1]}(I)$, $I = (0, t]$. By Lemma 3, $\varphi \in M'_{(0)}$ and $\left(x_j \frac{\partial}{\partial x_j}\right)^m \varphi \in M'_{(0)}$ $(j = 1, 2, \ldots, n;\ m = 1, 2, \ldots,)$ are Mellin distributions of Mellin order ≤ 0. Therefore $G(z) = (M\varphi)(z)$ is holomorphic for $\operatorname{Re} z < 0$ and by (7.11)

$$\text{(20)} \quad \mathcal{M}\left(\left(x_j \frac{\partial}{\partial x_j}\right)^m \varphi\right)(z) = z_j^m (\mathcal{M}\varphi)(z) \qquad \text{for } \operatorname{Re} z < 0 \qquad (j = 1, 2, \ldots, n).$$

By Theorem 1 for every $b < 0$ there exist constants C_b, C_{jmb} such that

$$(21)\qquad \begin{aligned} |(\mathcal{M}\varphi)(z)| &\le C_b\, t^{-\operatorname{Re} z}, \\ \Big|\mathcal{M}\Big(\Big(x_j\frac{\partial}{\partial x_j}\Big)^m \varphi\Big)(z)\Big| &\le C_{jmb}\, t^{-\operatorname{Re} z} \end{aligned} \qquad \text{for } \operatorname{Re} z \le b.$$

Putting

$$\widetilde{C}_{mb} = C_b + \sum_{j=1}^{n} C_{jmb}$$

we get from (20) and (21) the estimation

$$\Big(1 + \sum_{j=1}^{n} |z_j|^m\Big)\, |(\mathcal{M}\varphi)(z)| \le \widetilde{C}_{mb}\, t^{-\operatorname{Re} z} \qquad \text{for } \operatorname{Re} z \le b.$$

Hence by Exercise 7.11(d)

$$|(\mathcal{M}\varphi)(z)| \le C_{mb} \frac{1}{\langle z\rangle^m}\, t^{-\operatorname{Re} z} \qquad \text{for } \operatorname{Re} z \le b.$$

2^0. Suppose now, that G is a holomorphic function on the set $\{z \in \mathbf{C}^n : \operatorname{Re} z < 0\}$ such that for every $m \in \mathbf{N}_0$ and $b < 0$ there exists a constant $C_{mb} < \infty$ satisfying (19). Take $m \ge 2n$. Then there exists a constant C such that (cf. Exercise 7.11 (b))

$$\langle z_1\rangle^2 \cdot \ldots \cdot \langle z_n\rangle^2 \le C\langle z\rangle^m.$$

By (19) we get the estimation

$$|G(z)| \le CC_{mb} \frac{1}{\langle z_1\rangle^2 \cdot \ldots \cdot \langle z_n\rangle^2}\, t^{-\operatorname{Re} z} \qquad \text{for } \operatorname{Re} z \le b.$$

By Lemma 1 the function g

$$(22)\qquad g(x) = (2\pi)^{-n} \int_{\mathbf{R}^n} G(c+iy) x^{c+iy}\, dy \qquad \text{for } x > 0 \qquad (c \le b)$$

is continuous for $x > 0$, its definition does not depend on the choice of $c \le b$, $\operatorname{supp} g \subset I$, $g \in M'_{(b)}(I)$, $G(z) = Mg(z)$ for $\operatorname{Re} z \le b$. We shall show that $g \in M_{[-1]}$. For this end it is enough to show that $g \in M_{-1-b}$ that is

$$\sup_{x\in I}\Big| x^{-b}\Big(x\frac{\partial}{\partial x}\Big)^\alpha g(x)\Big| < \infty \qquad \text{for } \alpha \in \mathbf{N}_0^n.$$

Fix arbitrarily $\alpha \in \mathbf{N}_0^n$ and take m such that $m > \max(2n, |\alpha| + n)$. By (22), (19) and Exercise 7.11(d) we get for $c = b$ the estimation

$$\left|\left(x\frac{\partial}{\partial x}\right)^\alpha g(x)\right| \le (2\pi)^{-n} C_{mb}\, t^{-b} x^b \int_{\mathbf{R}^n} \frac{|(b+iy)^\alpha|}{1 + \sum_{j=1}^n |b_j + iy_j|^m}\, dy \qquad \text{for } x > 0.$$

Hence it follows from the choice of m that $x^{-b}\left(x\frac{\partial}{\partial x}\right)^\alpha g$ is bounded on I. Since $\alpha \in \mathbf{N}_0^n$ was arbitrary $g \in M_{-1-b}$. Thus we have proved that $g \in M_{[-1]}$. Since $g \in C^\infty(\mathbf{R}_+^n)$ and $\operatorname{supp} g \subset I$, we have $g \in \mathring{M}_{[-1]}(I)$.

Remark 2. Condition (19) in Theorem 5 can be equivalently replaced by one of the following:

for every $h \in \mathbf{N}_0^n$ and every $b < 0$ there exists a constant C_{hb} such that

$$(19') \qquad |G(z)| \le C_{hb} \frac{1}{(1+|z_1|)^{h_1} \cdot \ldots \cdot (1+|z_n|)^{h_n}}\, t^{-\operatorname{Re} z} \qquad \text{for } \operatorname{Re} z \le b,$$

or

for every $m \in \mathbf{N}_0$ and every $b < 0$ there exists a constant C_{mb} such that

$$(19'') \qquad |G(z)| \le C_{mb} \frac{1}{1 + \sum_{j=1}^n |z_j|^m}\, t^{-\operatorname{Re} z} \qquad \text{for } \operatorname{Re} z \le b.$$

(See Exercise 7.12 and Exercise 7.11 (d)).

The following modifications of Theorems 4 and 5 will turn out useful in Sections 11 and 13.

Theorem 4*. *In order that a function $F(z) = F(z_1, \ldots, z_n)$ be the Mellin transform of a unique Mellin distribution $u \in M'_{(\omega)}((0,t])$ it is necessary and sufficient that F be holomorphic in the set $\{z \in \mathbf{C}^n\colon \operatorname{Re} z < \omega\}$ and that for every $b < \omega$ and every $\tilde t > t$ there exist $C(b, \tilde t)$ and $s(b) \in \mathbf{N}_0$ such that*

$$(14^*) \qquad |F(\alpha + i\beta)| \le C(b, \tilde t)\langle\beta\rangle^{s(b)} \tilde t^{-\alpha} \qquad \text{for } \alpha \le b.$$

Proof: i) The condition is sufficient: since $s(b) \ge 0$ then from (14*): $|F(\alpha + i\beta)| \le C(b, \tilde t)\langle z\rangle^{s(b)} \tilde t^{-\alpha}$ for $\alpha \le b$. Thus by Theorem 3 there exists a unique distribution $u \in M'_{(\omega)}((0, \tilde t])$ such that $F(z) = (\mathcal{M}u)(z)$ for $\operatorname{Re} z < \omega$. Since $u \in M'_{(\omega)}((0, \tilde t])$ for all $t < \tilde t$ it follows that $\operatorname{supp} u \subset (0, t]$ hence by Theorem 8.3 $u \in M'_{(\omega)}((0, t])$.

ii) The condition is necessary: Let $u \in M'_{(\omega)}((0,t])$ and let $F(z) = (\mathcal{M}u)(z)$ for $\{z \in \mathbf{C}^n\colon \operatorname{Re} z < \omega\}$. Then by Theorem 1 there exist $s(b) \in \mathbf{N}_0$ and $C(b) < \infty$ such that (14) holds. If $s(b) = 0$ we clearly see that (14) implies (14*). For $s(b) > 0$ we use the inequality $\langle z\rangle \le \langle\alpha\rangle \cdot \langle\beta\rangle$ and the fact that for any fixed $\mathbf{1} < h \in \mathbf{R}^n$ $h^\alpha \langle\alpha\rangle^{s(b)} \le C^*(b)$ for $\alpha \le b$.

Theorem 5*. *In order that a function $G(z)$ be the Mellin transform of a unique function $\varphi \in \overset{\circ}{M}_{[-1]}((0,t])$ it is necessary and sufficient that G be holomorphic in the set $\{z \in \mathbf{C}^n\colon \operatorname{Re} z < 0\}$ and that for every $m \in \mathbf{N}_0$, $b < 0$ and $t < \tilde{t}$ there exist a constant $C(m,b,\tilde{t}\,)$ such that*

$$|G(z)| \le C(m,b,\tilde{t}\,)\frac{1}{\langle\beta\rangle^m}\tilde{t}^{-\alpha} \qquad \text{for } \alpha \le b \qquad (z = \alpha + i\beta). \tag{19*}$$

Proof: Assume (19*) and take $\tilde{\tilde{t}} = h\tilde{t}$, $h > 1$. Since $\frac{1}{\langle\beta\rangle} \le \frac{\langle\alpha\rangle}{\langle z\rangle}$ and $\langle\alpha\rangle^m h^\alpha$ is bounded for $\alpha \le b$ we get

$$|G(z)| \le \frac{1}{\langle z\rangle^m} C(m,b,\tilde{t}\,)\langle\alpha\rangle^m h^\alpha \tilde{\tilde{t}}^{-\alpha} \le C(m,b,\tilde{t},\tilde{\tilde{t}}\,)\frac{1}{\langle z\rangle^m}\tilde{\tilde{t}}^{-\alpha}.$$

By Theorem 5 there exists a function $\varphi \in \overset{\circ}{M}_{[-1]}((0,\tilde{\tilde{t}}\,])$ such that $\mathcal{M}\varphi(z) = G(z)$ for $\operatorname{Re} z < 0$. Since $\tilde{\tilde{t}} > \tilde{t} > t$ were arbitrary it follows that $\varphi \in \overset{\circ}{M}_{[-1]}((0,t])$.

To prove the necessity of the condition (19*) suppose that $\varphi \in \overset{\circ}{M}_{[-1]}((0,t])$, $G(z) = (\mathcal{M}\varphi)(z)$. Then by Theorem 5 for every $m \in \mathbf{N}_0$ and $b < 0$ there exists a constant $C(m,b) < \infty$ such that (19) holds. Thus

$$|G(\alpha + i\beta)| \le \frac{1}{\langle\beta\rangle^m}\tilde{t}^{-\alpha}C(m,b)\left(\sup_{\alpha \le b}\frac{\tilde{t}^\alpha}{t^\alpha}\right) \le C(m,b,\tilde{t}\,)\frac{1}{\langle\beta\rangle^m}\tilde{t}^{-\alpha}$$

for $\alpha \le b$ since $\tilde{t} > t$.

Corollary 1. *In order that a function $F(z)$ be the Mellin transform of a compactly supported distribution $u \in D'([t^*,t])$ for $t^* < t < \infty$, $(t^*, t \in \mathbf{R}^n_+)$, it is necessary and sufficient that F be an entire function on $\mathbf{C}^n$ and to every $b \in \mathbf{R}^n$, $\varrho > 0$ there exists $0 < C = C(b,\varrho) < \infty$, $s = s(b) \in \mathbf{N}_0$, such that*

$$|F(\alpha + i\beta)| \le \begin{cases} C\langle\beta\rangle^s (e^\varrho t)^{-\alpha} & \text{for } \alpha \le b, \\ C\langle\beta\rangle^s (e^{-\varrho} t^*)^{-\alpha} & \text{for } \alpha > b. \end{cases} \tag{23}$$

Proof: Suppose first $u \in D'([t^*,t])$. By Proposition 3.1 and Theorem 3.1 $u \in M'_a$ for every $a \in \mathbf{R}^n$ and hence by Theorem 9.4* $\mathcal{M}u$ is entire on $\mathbf{C}^n$ and to every $b \in \mathbf{R}^n$ and $\varrho > 0$

$$|\mathcal{M}u(\alpha + i\beta)| \le C\,|\beta|^s\,(e^\varrho t)^{-\alpha} \qquad \text{for } \alpha \le b.$$

Let λ be the mapping $\mathbf{R}^n_+ \ni x \mapsto \frac{1}{x} \overset{\text{def}}{=} \left(\frac{1}{x_1},\dots,\frac{1}{x_n}\right) \in \mathbf{R}^n_+$. Note that

$$\mathcal{M}(u \circ \lambda)(z) = \mathcal{M}u(-z), \qquad \operatorname{supp}(u \circ \lambda) \subset \left[\frac{1}{t},\frac{1}{t^*}\right].$$

Hence as before we get

$$(24)\qquad |\mathcal{M}u(\alpha+i\beta)| \le C\langle\beta\rangle^s (e^{-\varrho}t^*)^{-\alpha} \qquad \text{for } \alpha > b.$$

To prove the second part of Corollary 1, assume that F is entire on $\mathbf{C}^n$ and that (23) holds. By the first inequality (23) and Theorem 9.4* there exists a unique $u \in M'_{(\infty)}((0,t])$, such that $\mathcal{M}u(z) = F(z)$ for $z \in \mathbf{C}^n$. By second inequality (23) after changing z to $-z$, b to $-b$, again by Theorem 9.4* there exists a unique $\tilde{u} \in M'_{(\infty)}((0,1/t^*])$ such that $\mathcal{M}\tilde{u}(z) = F(-z)$ for $z \in \mathbf{C}^n$.

By (4.6) and (7.12) with $\alpha = 0$ we get

$$\mathcal{F}^{-1}(\tilde{u}\circ\mu)(-\beta) = \mathcal{F}^{-1}(u\circ\mu)(\beta) \qquad \text{for } \beta \in \mathbf{R}^n$$

and hence $(u\circ\mu)^{\vee} = \tilde{u}\circ\mu$ in S', where $^{\vee}$ denotes the reflection $y \to -y$ and $\mu(y) = e^{-y}$ for $y \in \mathbf{R}^n$. Since $(u\circ\mu)^{\vee} = (u\circ\lambda)\circ\mu$ by Exercise 5.5 we infer that $u\circ\lambda = \tilde{u}$.

Hence $u\circ\lambda \in M'((0,\frac{1}{t^*}])$ and by definition of the function λ, $\operatorname{supp} u \subset \{x\colon x \ge t^*\}$ which together with $\operatorname{supp} u \subset [0,t]$ give the desired assertion: $\operatorname{supp} u \subset [t^*,t]$.

Exercises

1. Let $u \in M'_{(\omega)}(I)$ and $F(z) = (\mathcal{M}u)(z)$ for $z \in \Omega = \{z \in \mathbf{C}^n,\ \operatorname{Re} z < \omega\}$. Let $r \in \mathbf{R}^n_+$, $c \in \mathbf{R}^n$, $c < \omega$ and let

$$u_{rc}(x) = (2\pi i)^{-n}\int_{c-ir}^{c+ir} F(z)x^z\,dz \qquad \text{for } x \in \mathbf{R}^n_+.$$

show that independently of the choice of $c < \omega$

$$u = \lim_{r\to\infty} u_{rc} \qquad \text{in } D'(\mathbf{R}^n_+)$$

($r \to \infty$ means $r_j \to \infty$, $j = 1,\dots,n$).

OUTLINE OF THE PROOF. Let $Q_r = \{\xi \in \mathbf{R}^n\colon -r < \xi_j < r,\ j = 1,\dots,n\}$ and observe that $u_{rc}(x) = (2\pi)^{-n}\int_{Q_r} F(c+i\xi)x^{c+i\xi}\,d\xi$ is continuous on $\mathbf{R}^n_+$. Then, proceeding analogously as in Exercise 7.3 we get for arbitrary $\varphi \in D(\mathbf{R}^n_+)$

$$u_{rc}[\varphi] = (2\pi)^{-\frac{n}{2}}\int_{Q_r} F(c+i\xi)\psi(\xi)\,d\xi = (2\pi)^{-\frac{n}{2}} u\left[\int_{Q_r} \tau^{-c-1-i\xi}\psi(\xi)\,d\xi\right]$$

where $\psi(\xi) = (2\pi)^{-n/2}\int_{\mathbf{R}^n_+} \varphi(x)x^{c+i\xi}\,dx = \mathcal{F}(\varphi(e^{-y})e^{-y(c+1)})(\xi)$. Observe that $\varphi(\tau) = (2\pi)^{-n/2}\int_{\mathbf{R}^n}\psi(\xi)\tau^{-c-1-i\xi}\,d\xi$. Hence $u[\varphi] - u_{rc}[\varphi] = (2\pi)^{-n/2}u[\lambda_r]$ where $\lambda_r(\tau) = \int_{\mathbf{R}^n\setminus Q_r}\psi(\xi)\tau^{-c-1-i\xi}\,d\xi$. Take $b \in \mathbf{R}^n$ such that $c < b < \omega$ and prove that $\lambda_r \to 0$ in M_b.

2. Give another proof of Theorem 2 using Exercise 1 and Proposition 5.3.

3. Let $\sigma \in \overset{\circ}{M}_a$ and define (cf. Definition 4.3)

$$(\mathcal{M}^{\mathrm{tr}})^{-1}\sigma(z) = (2\pi)^n \int_{\mathbf{R}^n_+} x^z \sigma(x)\,dx \qquad \text{for} \quad \operatorname{Re} z > a.$$

Prove that $(\mathcal{M}^{\mathrm{tr}})^{-1}\sigma$ is holomorphic for $\operatorname{Re} z > a$ and for any $c > a$ and any $m \in \mathbf{N}_0$

$$\left|(\mathcal{M}^{\mathrm{tr}})^{-1}\sigma(c+i\beta)\right| \le \frac{C_c}{\langle\beta\rangle^m} \qquad \text{for } \beta \in \mathbf{R}^n$$

for some constant C_c.

4. DISTRIBUTIONAL INVERSION FORMULA FOR THE MELLIN TRANSFORMATION. Let $u \in M'_{(\omega)}(I)$, $I = (0,t]$, $F(z) = \mathcal{M}u(z)$ for $\operatorname{Re} z < \omega$. Let $\sigma \in M_{(\omega)}(I)$ and denote by $\tilde{\sigma}$ an extension of σ to a function in $\overset{\circ}{M}_{(\omega)}(\tilde{I})$, $\tilde{I} = (0,\tilde{t}\,]$, $t < \tilde{t}$. Choose $a < c < \omega$ so that $\tilde{\sigma} \in \overset{\circ}{M}_a$. Prove that

$$u[\sigma] = \frac{1}{(2\pi i)^n} \int_{c+i\mathbf{R}^n} F(z)\left(\int_{\mathbf{R}^n_+} x^z \tilde{\sigma}(x)\,dx\right) dz.$$

HINT. Apply Propositions 4.2 and 7.1, Exercise 3 and Theorem 3.

5. Let $P(z_1,z_2) = (|z_1|^2+1)(|z_2|^2+1)$, $Q(z_1,z_2) = (z_1+z_2-\omega_1-\omega_2-2)^4$. Show that P/Q is not bounded in the set $\{z \in \mathbf{C}^2 \colon \operatorname{Re} z_1 \le \omega_1,\ \operatorname{Re} z_2 \le \omega_2\}$ but is bounded in the set $\{z \in \mathbf{R}^2 \colon z_1 \le \omega_1, z_2 \le \omega_2\}$.

6. Denote

$$\overset{\circ}{M}_a((0,t]) = \left\{\varphi \in M_a((0,t]) \colon \left.\frac{\partial^p \varphi(x)}{\partial x_j^p}\right|_{x_j=t_j} = 0 \quad (j=1,\dots,n,\ p \in \mathbf{N}_0)\right\}.$$

Show that

$$\overset{\circ}{M}_a((0,t]) = \left\{\varphi \in M_a((0,t]) \colon \left.\left(\frac{\partial}{\partial x}\right)^\alpha \varphi(x)\right|_{x_j=0} = 0 \quad (j=1,\dots,n,\ \alpha \in \mathbf{N}_0^n)\right\}.$$

and hence (see Subsection 8.1)

$$\overset{\circ}{M}_{(\omega)}((0,t]) = \varinjlim_{a<\omega} \overset{\circ}{M}_a((0,t]), \qquad \overset{\circ}{M}_{[-1]}((0,t]) = \varprojlim_{-1<a} \overset{\circ}{M}_a((0,t]).$$

7. Let $u \in M'_{(\omega)}(I)$, $\omega \in (\mathbf{R} \cup \{\infty\})^n$, $I = (0,t]$ be of Mellin order $\le m$, $m \in \mathbf{N}_0$. Show that for every $a < \omega$ and $\tilde{t} > t$ there exists $C = C(a,\tilde{t}) < \infty$ such that

$$|\mathcal{M}u(\alpha+i\beta)| \le C\langle\beta\rangle^m \tilde{t}^{-\alpha} \qquad \text{for} \quad \alpha \le a,\ \beta \in \mathbf{R}^n.$$

HINT. See Theorem 4* and estimation (2).

8. Show that for an entire function F the estimations (23) for arbitrary $b \in \mathbf{R}^n$ are equivalent to those for a fixed $\mathring{b} \in \mathbf{R}^n$ (e.g. $\mathring{b} = -\mathbf{1}$; see Subsection 13.3 for an application).

§10. MELLIN TRANSFORMS OF CUT-OFF FUNCTIONS (continued)

Owing to its importance in the local Mellin analysis, the investigation of the Mellin transforms of different classes of cut-off functions started in Subsection 7.3 is continued in this section. It is worth while to note that conical cut-off functions though singular at the origin behave like smooth functions from the point of view of the Mellin analysis. The reader interested in the important applications of conical cut-off functions to the study of non-linear propagation of singularities is referred to [B] and [Zie–K].

Subsections: 1–3 are developed for the use in Sections 11, 14 and Subsection 4 will be useful in Section 15.

1. Conical Cut-Off Functions

Consider the transformation $S\colon \mathbf{R}_+ \times \mathbf{R}_+^{n-1} \to \mathbf{R}_+^n$ defined by (cf. Exercise 8.6)

$$x_1 = y_1, \qquad x_j = y_1 y_j \qquad \text{for } j = 2, \ldots, n. \tag{1}$$

Let Ω, Ω_1 be bounded open sets[(1)] $\Omega \subset\subset \Omega_1 \subset\subset \mathbf{R}_+^{n-1}$. Denote $K = \overline{\Omega}$, $K_1 = \overline{\Omega}_1$ and consider the cones $\Gamma_\Omega = S(\mathbf{R}_+ \times \Omega)$, $\Gamma_K = S(\mathbf{R}_+ \times K)$ (and analogously Γ_{Ω_1} and Γ_{K_1}). Let $y' = (y_2, \ldots, y_n)$ and take a function $\kappa' \in C^\infty(\mathbf{R}_+^{n-1})$ such that $\kappa' = 1$ on $\overline{\Omega}$, $0 \le \kappa' \le 1$, $\operatorname{supp} \kappa' \subset \overline{\Omega}_1$. As in Subsection 7.3.2 take $\varphi \in C_0^\infty(\mathbf{R}^n)$, $0 \le \varphi \le 1$, $\varphi \equiv 1$ in a neighbourhood of zero, $\operatorname{supp} \varphi \subset B(1)$. Let $r \in \mathbf{R}_+$ and define a conical cut-off function κ_r by

$$\kappa_r(x) = \kappa'\left(\frac{x_2}{x_1}, \ldots, \frac{x_n}{x_1}\right) \varphi\left(\frac{1}{r} x\right) \qquad \text{for } x \in \mathbf{R}_+^n. \tag{2}$$

It is clear that κ_r is smooth on $\mathbf{R}_+^n$, $0 \le \kappa_r \le 1$, $\operatorname{supp} \kappa_r \subset \overline{B(r)} \cap \Gamma_{K_1}$, $\kappa_r = 1$ on the intersection of Γ_K with a neighbourhood of zero.

---

(1) For two subsets $\Omega \subset \Omega_1 \subset \mathbf{R}^n$ we write $\Omega \subset\subset \Omega_1$ if $\overline{\Omega} \subset \Omega_1$ where the closure of Ω is taken in $\mathbf{R}^n$.

Remark 1. Observe that κ_r can be written in the following equivalent form

$$\kappa_r(x) = \lambda\left(\frac{x}{\|x\|}\right)\varphi\left(\frac{1}{r}x\right) \qquad \text{for } x \in \mathbf{R}^n_+, \tag{3}$$

where $\lambda(\omega) = \kappa'(\omega'/\omega_1)$ is a smooth, compactly supported function on the unit quarter sphere $S^{n-1}_+ = S^{n-1} \cap \mathbf{R}^n_+$. In other words

$$\kappa_r(x) = \tilde{\kappa}(x)\varphi\left(\frac{1}{r}x\right),$$

where $\tilde{\kappa} \in C^\infty(\mathbf{R}^n_+)$ is homogeneous of order zero and $\overline{\operatorname{supp}\tilde{\kappa}} \subset \mathbf{R}^n_+ \cup \{0\}$.

In the sequel a will denote a smooth function in a neighbourhood U of zero and r a positive parameter not bigger than $r_0 < 1$ where r_0 is so small that $B(r_0) \subset U$.

Proposition 1. *Let $\tau \in \mathbf{N}^n_0$, $\eta \in \mathbf{R}^n$, $\sum \eta_j > 0$ and κ_r be given by (2). Then for an arbitrary $\nu \in \mathbf{N}^n_0$ there exists a constant $C = C(a, \eta, \tau, \nu)$ (independent of r for $0 < r \le r_0 < 1$) such that*

$$\sup_{x \in \operatorname{supp}\kappa_r} \left|\left(x\frac{\partial}{\partial x}\right)^\nu x^\eta (\ln x)^\tau (a\kappa_r)(x)\right| < C. \tag{4}$$

If $\tau = 0$ we can put $\sum \eta_j \ge 0$. If $\sum \eta_j > 0$, $\tau \in \mathbf{N}^n_0$, the function

$$\operatorname{supp}\kappa_r \ni x \mapsto x^{-1}\left(x\frac{\partial}{\partial x}\right)^\nu (x^\eta(\ln x)^\tau(a\kappa_r)(x))$$

is integrable.

Proof: We begin with the case $\tau = 0$, $\eta = 0$. We shall prove (4) for $|\nu| = \nu_1 + \cdots + \nu_n = 1$. For other ν the proof goes by induction. Let c_j, μ, μ_j, p_i be such that

$$\left|\frac{\partial\varphi}{\partial x_j}\right| < c_j, \qquad |a| < \mu,$$

$$\left|\frac{\partial a}{\partial x_j}\right| < \mu_j \quad (j = 1, \ldots, n), \qquad \left|\frac{\partial\kappa'}{\partial y_i}\right| < p_i \quad (i = 2, \ldots, n)$$

and let $\varrho_i = \max_{y \in \overline{\Omega}_1} y_i$ $(i = 2, \ldots, n)$. Note that on the support of κ_r

$$|\varphi| \le 1, \quad |\kappa'| \le 1, \quad |\kappa_r| \le 1, \quad |x_j| < r < r_0 < 1 \qquad (j = 1, \ldots, n).$$

Hence

$$\left|x_1\frac{\partial}{\partial x_1}(a\kappa_r)(x)\right| \le \left|x_1\frac{\partial a}{\partial x_1}\kappa_r\right| + \left|a\varphi\left(\frac{1}{r}x\right)\sum_{i=2}^n \frac{\partial\kappa'}{\partial y_i}\frac{x_i}{x_1}\right| + \left|x_1 a\kappa'\frac{1}{r}\frac{\partial\varphi}{\partial x_1}\right|$$

$$\le r_0\mu_1 + \mu\sum_{i=2}^n \varrho_i p_i + \mu c_1 < \infty,$$

$$\left|x_i\frac{\partial}{\partial x_i}(a\kappa_r)\right| \le \left|x_i\frac{\partial a}{\partial x_i}\kappa_r\right| + \left|a\varphi\left(\frac{1}{r}x\right)\frac{\partial\kappa'}{\partial y_i}\frac{x_i}{x_1}\right| + \left|x_i a\kappa'\frac{1}{r}\frac{\partial\varphi}{\partial x_i}\right|$$
$$\le r_0\mu_i + \mu p_i\varrho_i + \mu c_i < \infty \qquad (i = 2,\dots,n).$$

Suppose now that $\eta_1 + \dots + \eta_n > 0$, $\tau \in \mathbf{N}_0^n$. Then[(1)]

$$x^\eta(\ln x)^\tau\Big|_{x=S(y)} = y_1^{\sum\eta_j}(\ln y_1)^{\tau_1}y_2^{\eta_2}\cdot\ldots\cdot y_n^{\eta_n}\prod_{j=2}^n(\ln y_1 + \ln y_j)^{\tau_j}$$
$$= \sum_{k_2=0}^{\tau_2}\cdots\sum_{k_n=0}^{\tau_n}\binom{\tau_2}{k_2}\cdot\ldots\cdot\binom{\tau_n}{k_n}\times$$
$$\times\, y_1^{\sum\eta_j}(\ln y_1)^{\tau_1+k_2+\dots+k_n}y_2^{\eta_2}\cdot\ldots\cdot y_n^{\eta_n}(\ln y_2)^{\tau_2-k_2}\cdot\ldots\cdot(\ln y_n)^{\tau_n-k_n}$$

is bounded for $0 < y_1 < r$, $y' = (y_2,\dots,y_n) \in K_1$ and hence

$$\sup_{x\in\operatorname{supp}\kappa_r}|x^\eta(\ln x)^\tau| \quad\text{is bounded.}$$

Clearly (4) results for $\nu = 0$. The case of an arbitrary $\nu \in \mathbf{N}_0^n$ follows by the Leibniz rule and the estimations of $x_j\frac{\partial}{\partial x_j}(a\kappa_r)$ $(j = 1,2,\dots,n)$ given above. To prove the last assertion of Proposition 1 it is enough to note that

$$\int_{\operatorname{supp}\kappa_r}\left|(\ln x)^\tau x^{\eta-1}\right|dx$$
$$\le \int_{K_1}\Big\{\int_0^r\sum_{k_2=0}^{\tau_2}\cdots\sum_{k_n=0}^{\tau_n}\binom{\tau_2}{k_2}\cdot\ldots\cdot\binom{\tau_n}{k_n}y_1^{\sum\eta_j-1}|\ln y_1|^{\tau_1+k_2+\dots+k_n}\times$$
$$\times y_2^{\eta_2-1}\cdot\ldots\cdot y_n^{\eta_n-1}|\ln y_2|^{\tau_2-k_2}\cdot\ldots\cdot|\ln y_n|^{\tau_n-k_n}\,dy_1\Big\}\,dy'$$
$$\le \int_0^r y_1^{\sum\eta_j-1}(|\ln y_1|^{|\tau|}+1)\,dy_1\times$$
$$\times\sum_{k_2=0}^{\tau_2}\cdots\sum_{k_n=0}^{\tau_n}\binom{\tau_2}{k_2}\cdot\ldots\cdot\binom{\tau_n}{k_n}\int_{K_1}y_2^{\eta_2-1}\cdot\ldots\cdot y_n^{\eta_n-1}|\ln y_2|^{\tau_2-k_2}\cdot\ldots\cdot|\ln y_n|^{\tau_n-k_n}\,dy'$$
$$< \infty.$$

Proposition 2. *Let b be a measurable bounded function* $\operatorname{supp} b \subset \operatorname{supp}\kappa_r \subset$ *$B(r)\cap\Gamma_{K_1}$, $0 < r \le r_0 < 1$ and denote $m = \sup|b(x)|$ (e.g. $b = a\kappa_r$ with a smooth). Then the Mellin transform $\mathcal{M}b$ is holomorphic in the set*

$$A_0 = \Big\{z \in \mathbf{C}^n\colon \sum_{i=1}^n \operatorname{Re} z_i < 0\Big\}$$

---

(1) If no range is given under summation sign it is understood to be the set $\{1,\dots,n\}$.

and

$$(5) \qquad |\mathcal{M}b(\alpha+i\beta)| \le mG(r,\alpha) \quad \text{on } A_0,$$

where

$$(6) \qquad G(r,\alpha) = \frac{r^{-\sum \alpha_j}}{-\sum \alpha_j} \int_K (y')^{-\alpha'-1}\, dy'$$

$$\alpha' = (\alpha_2, \ldots, \alpha_n), \qquad y' = (y_2, \ldots, y_n).$$

Moreover for $\delta \in \mathbf{N}_0^n$ *and* $\sum \alpha_j < 0$ *there exists a constant* $C_{\delta\alpha} < \infty$ *such that*

$$(7) \qquad \left| \left(\frac{\partial}{\partial \beta}\right)^{\delta} (\mathcal{M}b)(\alpha+i\beta) \right| \le C_{\delta\alpha}.$$

Proof: Let $I = (0, \mathbf{r}_0]$. Then

$$(8) \qquad \begin{aligned}(\mathcal{M}b)(z) &= \int_I x^{-z-1} b(x)\, dx \\ &= \int_{K_1} \left\{ \int_0^r y_1^{-z_1-1} (y_1y_2)^{-z_2-1} \cdot \ldots \cdot (y_1y_n)^{-z_n-1} \times \right. \\ &\quad \left. \times\, b(y_1, y_1y_2, \ldots, y_1y_n) y_1^{n-1}\, dy_1 \right\} dy'.\end{aligned}$$

Hence

$$\begin{aligned}|(\mathcal{M}b)(z)| &\le m \int_{K_1} \left\{ \int_0^r y_1^{-\sum \alpha_j - 1} y_2^{-\alpha_2-1} \cdot \ldots \cdot y_n^{-\alpha_n-1}\, dy_1 \right\} dy' \\ &= m \frac{r^{-\sum \alpha_j}}{-\sum \alpha_j} \int_{K_1} (y')^{-\alpha'-1}\, dy' \qquad \text{for } \textstyle\sum \alpha_j < 0\end{aligned}$$

and thus (5) holds. To prove that $(\mathcal{M}b)(z)$ is holomorphic on A_0 we verify that

$$(9) \qquad \left.\begin{aligned}\frac{\partial}{\partial \alpha_j}(\mathcal{M}b)(\alpha+i\beta) &= -\mathcal{M}(b \ln x_j)(\alpha+i\beta) \\ \frac{\partial}{\partial \beta_j}(\mathcal{M}b)(\alpha+i\beta) &= -i\mathcal{M}(b \ln x_j)(\alpha+i\beta)\end{aligned}\right\} \quad \text{for } j = 1, \ldots, n,$$

which shows that $\mathcal{M}b$ satisfies the Cauchy–Riemann equations and therefore is holomorphic (as a function of several variables). For the proof of (9) fix $\mathring{z} = \mathring{\alpha} + i\mathring{\beta}$ in A_0 and choose $\varepsilon \in \mathbf{R}_+$ such that $\sum \mathring{\alpha}_j + 2\varepsilon < 0$. Proceeding in an analogous way as in the proof of (5) we deduce that $\int_I |\ln x_j|\, x_j^{-\varepsilon} x^{-\mathring{\alpha}-1}\, dx < \infty$ and hence the limit of the difference quotient for the derivative $\frac{\partial}{\partial \alpha_j}$ equals

$$\lim_{h \to 0} \int_I \frac{x_j^{-h} - 1}{h} x^{-\mathring{z}-1} b(x) = -\int_I \ln x_j x^{-\mathring{z}-1} b(x)\, dx = -\mathcal{M}(b \ln x_j)(\mathring{z})$$

$(j = 1, \ldots, n)$ i.e. the first formulas (9) hold. The proof of the remaining formulas (9) is analogous.

To prove (7) observe that by repeating (9) we have

$$\left|\left(\frac{\partial}{\partial\beta}\right)^{\delta}(\mathcal{M}b)(\alpha+i\beta)\right| = \left|\mathcal{M}((\ln x)^{\delta} b)(\alpha+i\beta)\right| \qquad \text{for } \delta\in \mathbf{N}_0^n$$

and proceeding as in the proof of (5) we get

$$\begin{aligned}&\left|\mathcal{M}((\ln x)^{\delta} b)(\alpha+i\beta)\right| \\ &\quad \le m\int_{K_1}\left\{\int_0^r y_1^{-(\sum\alpha_j)-1}\,|\ln y_1|^{\delta_1}\, y_2^{-\alpha_2-1}\cdot\ldots\cdot y_n^{-\alpha_n-1}\times\right.\\ &\qquad\qquad \left.\times\,|\ln(y_1y_2)|^{\delta_2}\cdot\ldots\cdot|\ln(y_1y_n)|^{\delta_n}\,dy_1\right\}dy'.\end{aligned}$$

Observe that

$$|\ln(y_1y_j)|^{\delta_j} \le \sum_{\nu_j=0}^{\delta_j}\binom{\delta_j}{\nu_j}|\ln y_1|^{\nu_j}\cdot|\ln y_j|^{\delta_j-\nu_j} \qquad (j=2,\ldots,n)$$

and since $\sum\alpha_j<0$,

$$\int_0^r y_1^{(\sum\alpha_j)-1}\,|\ln y_1|^{\delta_1+\nu_2+\cdots+\nu_n}\,dy_1 \le C_{\nu_2\ldots\nu_n}<\infty.$$

Therefore

$$\begin{aligned}\left|\mathcal{M}((\ln x)^{\delta} b)(\alpha+i\beta)\right| &\le m\sum_{\nu_2=0}^{\delta_2}\cdots\sum_{\nu_n=0}^{\delta_n} C_{\nu_2\ldots\nu_n}\tbinom{\delta_2}{\nu_2}\cdot\ldots\cdot\tbinom{\delta_n}{\nu_n}\times\\ &\times\int_{K_1}(y')^{-\alpha'-1}\,|\ln y_2|^{\delta_2-\nu_2}\cdot\ldots\cdot|\ln y_n|^{\delta_n-\nu_n}\,dy' \le C_{\delta\alpha}<\infty\end{aligned}$$

since the last integral over a compact set $K_1\subset\mathbf{R}_+^{n-1}$ is finite.

Remark 2. From definition (6) of $G(r,\alpha)$ we deduce at once that for every $\sum\varepsilon_j>0$ there exists a constant $C^*(\varepsilon)$ such that $G(r,-\varepsilon)\le C^*(\varepsilon)$ for every $0<r\le\frac{1}{2}$. If $\alpha,\ \tilde\alpha\in\mathbf{R}^n$, $\sum\alpha_j<0$, $\sum\alpha_j+\tilde\alpha_j<0$ then there exists a constant $C=C(\alpha,\tilde\alpha)$, independent of r, $0<r\le\frac{1}{2}$ such that

$$G(r,\alpha+\tilde\alpha)\le CG(r,\alpha)r^{-\sum\tilde\alpha_j}.$$

Corollary 1. *Take $\tau \in \mathbf{N}_0^n$, $\sum \varepsilon_j > 0$, $\sum \eta_j > 0$ ($\sum \eta_j \geq 0$ if $\tau = 0$). Then $\mathcal{M}(x^{\varepsilon+\eta}(\ln x)^\tau a\kappa_r)$ is holomorphic on $\{z \in \mathbf{C}^n \colon \sum \operatorname{Re} z_j - \varepsilon_j < 0\}$ and for every $h \in \mathbf{N}_0^n$ there exists a constant $C = C(a, \eta, \tau, h)$ (independent of $r \leq \frac{1}{2}$) such that*

$$\left|\mathcal{M}(x^{\varepsilon+\eta}(\ln x)^\tau a\kappa_r)(z)\right| \leq \frac{CG(r, \alpha - \varepsilon)}{\langle z_1 - \varepsilon_1\rangle^{h_1} \cdot \ldots \cdot \langle z_n - \varepsilon_n\rangle^{h_n}} \tag{10}$$

for $\sum \alpha_j - \varepsilon_j < 0$, $\alpha = \operatorname{Re} z$ with G defined by (6). Moreover the function

$$g(\beta) = \mathcal{M}(x^{\varepsilon+\eta}a\kappa_r)(i\beta) \qquad \text{for } \beta \in \mathbf{R}^n \tag{11}$$

belongs to the Schwartz space $S(\mathbf{R}^n)$.

Proof: Let $b(x) = x^\eta(\ln x)^\tau(a\kappa_r)(x)$ for $x \in \mathbf{R}_+^n$. By Proposition 1 b is bounded, hence by Proposition 2 $\mathcal{M}b$ is holomorphic for $\sum \alpha_j < 0$ and therefore the function $\mathcal{M}(x^{\varepsilon+\eta}(\ln x)^\tau a\kappa_r) = \mathcal{M}(x^\varepsilon b)$ is holomorphic for $\sum \alpha_j - \varepsilon_j < 0$. Let $h \in \mathbf{N}_0^n$. From the obvious inequality $\langle z_1\rangle^{h_1} \cdot \ldots \cdot \langle z_n\rangle^{h_n} < C(h) \sum_{0 \leq \nu \leq h} |z^\nu|$ we deduce from (7.11), (4) and (5) that

$$\langle z_1\rangle^{h_1} \cdot \ldots \cdot \langle z_n\rangle^{h_n} \, |\mathcal{M}b(z)| \leq C(h) \sum_{0 \leq \nu \leq h} \left|\mathcal{M}\Big(\Big(x\frac{\partial}{\partial x}\Big)^\nu b\Big)(z)\right| \leq CG(r, \alpha)$$
$$\text{for } \textstyle\sum \alpha_j < 0,$$

where $C = C(a, \eta, \tau, h)$ is independent of r, $0 < r \leq \frac{1}{2}$. This in view of (7.9) gives (10).

Before proving the last assertion we first note that the function g is well-defined because $\sum -\varepsilon_j < 0$. Next we observe that for arbitrary τ, $\nu \in \mathbf{N}_0^n$

$$\beta^\nu \Big(\frac{\partial}{\partial \beta}\Big)^\tau g(\beta) = (-i)^{\tau+\nu} \mathcal{M}\Big(\Big(x\frac{\partial}{\partial x}\Big)^\nu (x^{\varepsilon+\eta}(\ln x)^\tau a\kappa_r)\Big)(i\beta)$$

and by the last assertion of Proposition 1 we conclude that $\beta^\nu \big(\frac{\partial}{\partial \beta}\big)^\tau g(\beta)$ is bounded on $\mathbf{R}^n$.

2. The K-Inequalities

Let K be a compact set in $\mathbf{R}_+^{n-1}$.

For $\alpha \in \mathbf{R}^n$ define

$$l(\alpha, \varrho) = \alpha_1 + \varrho_1\alpha_2 + \cdots + \varrho_{n-1}\alpha_n \qquad \text{for } \varrho \in K. \tag{12}$$

We shall write $\alpha <_K 0$ ($\alpha \leq_K 0$, resp.) for $\max_{\varrho\in K} l(\alpha,\varrho) < 0$ ($\max_{\varrho\in K} l(\alpha,\varrho) \leq 0$, resp.). If $\alpha, \delta \in \mathbf{R}^n$ by $\alpha <_K \delta$ ($\alpha \leq_K \delta$) we mean $\alpha - \delta <_K 0$ ($\alpha - \delta \leq_K 0$). Note that

if $\alpha <_K \delta$ and $\delta \leq_K \gamma$ then $\alpha <_K \gamma$;
if $K \subset \widetilde{K}$ and $\alpha <_{\widetilde{K}} 0$ ($\alpha \leq_{\widetilde{K}} 0$) then $\alpha <_K 0$ ($\alpha \leq_K 0$).

Take $u \in M'_{(\omega)}$ for $\omega <_K \alpha$ ($\omega \leq_K \alpha$, resp.). Then for every $\varepsilon \in \mathbf{R}^n$, $x^\varepsilon u \in M'_{(\gamma)}$ for $\gamma <_K \alpha + \varepsilon$ ($\gamma \leq_K \alpha + \varepsilon$, resp.). Hence (see notation (1.1))

$$(13) \qquad x_1 u + \cdots + x_n u \in M'_{(\gamma)} \qquad \text{for } \gamma <_K \alpha + ce^1 \ \ (\gamma \leq_K \alpha + ce^1, \text{ resp.}),$$

where

$$(14) \qquad c = \min_{\varrho\in K}(1, \varrho_1, \ldots, \varrho_{n-1}) > 0 \qquad (\text{since } K \subset \mathbf{R}^{n-1}_+ \text{ is compact}).$$

Further for every $\nu \in \mathbf{N}_0^n$ we have

$$\left(\frac{\partial}{\partial x}\right)^\nu u \in M'_{(\gamma)} \qquad \text{for } \gamma <_K \alpha - \nu \qquad (\gamma \leq_K \alpha - \nu, \ \text{resp.}).$$

Remark 3. If $\varepsilon \in \mathbf{R}^n$, $0 <_K \varepsilon$, then for $\delta \in \mathbf{R}^n_+$ sufficiently small we have $0 < \delta <_K \varepsilon$.

Corollary 2. *Fix arbitrarily $\alpha \in \mathbf{R}^n$ and $\varepsilon \in \mathbf{R}^n$ such that $0 <_K \varepsilon$. Then in the set $\{\gamma \in \mathbf{R}^n\colon \gamma \leq_K \alpha + \varepsilon\}$ there exist points γ, $\alpha < \gamma$.*

Proof: Take $\gamma = \alpha + \delta$ with δ given by Remark 3.

3. The "Tangent Cones" Ξ_K and Related Cut-Off Functions

Let Ω be (as in Subsection 2) a bounded open set in $\mathbf{R}^{n-1}_+$ such that $K = \overline{\Omega} \subset \mathbf{R}^{n-1}_+$. Denote

$$(15) \qquad \Xi_\Omega = \Psi(\mathbf{R}_+ \times \Omega), \qquad \Xi_K = \Psi(\mathbf{R}_+ \times K),$$

where

$$\Psi\colon \mathbf{R}_+ \times K \ni (t,\varrho) \mapsto x \in \mathbf{R}^n_+$$

is defined by

$$(16) \qquad x_1 = t, \quad x_2 = t^{\varrho_1}, \quad \ldots, \quad x_n = t^{\varrho_{n-1}}.$$

The following examples motivate the terminology "*tangent cones*" for the sets Ξ_Ω, Ξ_K.

Example 1. Let $n = 2$, $\Omega = (\varrho^1, \varrho^2)$, and suppose that $0 < \varrho^1 < 1 < \varrho^2$. Then the set Ξ_Ω is tangent at zero of order $\frac{1}{\varrho^1} - 1$ to the line $x_1 = 0$ and of order $\varrho^2 - 1$ to $x_2 = 0$. If $1 \le \varrho^1 < \varrho^2$ there is no tangency to the line $x_1 = 0$.

Example 2. Let $n = 3$, $\Omega = (\varrho_1^1, \varrho_1^2) \times (\varrho_2^1, \varrho_2^2)$. Suppose $0 < \varrho_1^1 < 1 < \varrho_1^2$, $0 < \varrho_2^1 < 1 < \varrho_2^2$. Then Ξ_Ω is tangent to the planes $x_i = 0$ $(i = 1, 2, 3)$.

Take a function κ' smooth on $\mathbf{R}_+^{n-1}$, bounded by 1 and equal to zero outside Ω and a function $\varphi \in C_0^\infty(\mathbf{R}^n)$, $|\varphi| \le 1$, $\varphi \equiv 1$ in a neighbourhood of zero, $\operatorname{supp}\varphi \subset B(1)$. Let $r \in \mathbf{R}_+$, $r \le \frac{1}{2}$ and define

$$\kappa_r^K(x) = \kappa'\left(\frac{\ln x_2}{\ln x_1}, \dots, \frac{\ln x_n}{\ln x_1}\right)\varphi\left(\frac{1}{r}x\right) \qquad \text{for } x \in \mathbf{R}_+^n. \tag{17}$$

We see that κ_r^K is a smooth function on $\mathbf{R}_+^n$ and $\operatorname{supp}\kappa_r^K \subset \{x\colon \|x\| \le r\} \cap \Xi_K$.

In the sequel a will denote a smooth function in a neighbourhood U of zero and r a positive parameter $\le \frac{1}{2}$ so small that $B(r) \subset U$.

Proposition 3. *Let $\tau \in \mathbf{N}_0^n$, $0 <_K \eta$ and κ_r^K be given by (17). Then for an arbitrary $\nu \in \mathbf{N}_0^n$ there exists a constant $C = C(a, \eta, \tau, \nu)$ (independent of r for $0 < r \le \frac{1}{2}$) such that*

$$\sup_{x \in \operatorname{supp}\kappa_r^K} \left| \left(x\frac{\partial}{\partial x}\right)^\nu x^\eta (\ln x)^\tau (a\kappa_r^K)(x) \right| < C. \tag{18}$$

If $\tau = 0$ we can put $\eta = 0$. If $0 <_K \eta$, $\tau \in \mathbf{N}_0^n$ the function $\operatorname{supp}\kappa_r^K \ni x \mapsto x^{-1}\left(x\frac{\partial}{\partial x}\right)^\nu (x^\eta (\ln x)^\tau (a\kappa_r^K)(x))$ is integrable.

We get formally Proposition 3 by replacing in Proposition 1 κ_r by κ_r^K and $\eta \in \mathbf{R}_+^n$ by $0 <_K \eta$. The proof follows by the same method (see Exercises 2 and 3).

The following proposition is a reformulation of Proposition 2 in the case considered in this subsection.

Proposition 4. *Let b be a measurable bounded function, $\operatorname{supp} b \subset \operatorname{supp}\kappa_r^K$, and denote $m = \sup|b(x)|$ (e.g. $b = a\kappa_r^K$ with a smooth). Then the Mellin transform $\mathcal{M}b$ is holomorphic in the set*

$$B_0 = \{z \in \mathbf{C}^n\colon \operatorname{Re} z <_K 0\} = \{\alpha + i\beta\colon \max_{\varrho \in K} l(\alpha, \varrho) < 0\}$$

and

$$|\mathcal{M}b(\alpha + i\beta)| \le mH(r, \alpha) \qquad \text{on } B_0 \tag{19}$$

where

$$H(r,\alpha)=\sum_{j=0}^{n-1}\frac{(n-1)!}{j!}|\ln r|^j\int_K\frac{r^{-l(\alpha,\varrho)}}{(-l(\alpha,\varrho))^{n-j}}\,d\varrho. \tag{20}$$

Moreover for every $\delta\in\mathbf{N}_0^n$ and $\alpha<_K 0$ there exists a constant $C_{\delta\alpha}<\infty$ such that

$$\left|\left(\frac{\partial}{\partial\beta}\right)^{\delta}(\mathcal{M}b)(\alpha+i\beta)\right|<C_{\delta\alpha}. \tag{21}$$

Proof: The proof proceeds along the same lines as that of Proposition 2. From the definition of the set Ξ_K we see that for $z=(z_1,\ldots,z_n)\in B_0$

$$\mathcal{M}b(z)=\int_K\int_0^r b(t,t^{\varrho_1},\ldots,t^{\varrho_{n-1}})t^{-(z_1+\varrho_1z_2+\cdots+\varrho_{n-1}z_n)-1}(\ln t)^{n-1}\,d\varrho_1\ldots d\varrho_{n-1}\,dt.$$

Hence

$$|\mathcal{M}b(z)|\le m\int_K\left(\int_0^r t^{-l(\alpha,\varrho)-1}|\ln t|^{n-1}\,dt\right)d\varrho=mH(r,\alpha).$$

Similarly from

$$\left|\left(\frac{\partial}{\partial\beta}\right)^{\delta}(\mathcal{M}b)(\alpha+i\beta)\right|=|\mathcal{M}((\ln x)^{\delta}b)(\alpha+i\beta)|\qquad\text{for }\delta\in\mathbf{N}_0^n$$

we deduce (21) with

$$C_{\delta\alpha}=m\int_K\varrho_1^{\delta_2}\cdot\ldots\cdot\varrho_{n-1}^{\delta_n}\,d\varrho_1\ldots d\varrho_n\cdot\int_0^r|\ln t|^{|\delta|+n-1}\,t^{\min(-l(\alpha,\varrho)-1)}\,dt.$$

Propositions 3 and 4 lead to the following

Corollary 3. *Take $\tau\in\mathbf{N}_0^n$, $0<_K\varepsilon$, $0<_K\eta$ ($\eta=0$ if $\tau=0$). Then $\mathcal{M}(x^{\varepsilon+\eta}(\ln x)^{\tau}a\kappa_r^K)$ is holomorphic on $\{z\in\mathbf{C}^n\colon \operatorname{Re}z<_K\varepsilon\}$ and for every $h\in\mathbf{N}_0^n$ there exists a constant $C=C(a,\eta,\tau,h)$ (independent of r) such that*

$$\left|\mathcal{M}(x^{\varepsilon+\eta}(\ln x)^{\tau}a\kappa_r^K)(z)\right|\le\frac{CH(r,\alpha-\varepsilon)}{\langle z_1-\varepsilon_1\rangle^{h_1}\cdots\langle z_n-\varepsilon_n\rangle^{h_n}} \tag{22}$$

for $\alpha=\operatorname{Re}z<_K\varepsilon$ with H defined by (20). Moreover the function

$$g(\beta)=\mathcal{M}(x^{\varepsilon+\eta}a\kappa_r^K)(i\beta)\qquad\text{for }\beta\in\mathbf{R}^n \tag{23}$$

belongs to the Schwartz space $S(\mathbf{R}^n)$.

The proof runs as that of Corollary 1 based on Propositions 1 and 2. To Remark 2 corresponds now the following

Remark 4. The function $H(r,\alpha)$ defined by (20) has the properties:

i) for every $0 <_K \varepsilon$ there exists a constant $C^*(\varepsilon)$ such that

$$H(r,-\varepsilon) \le C^*(\varepsilon) \qquad \text{for every } \ 0 < r \le \tfrac{1}{2}.$$

ii) If $\alpha, \tilde{\alpha} \in \mathbf{R}^n$, $\alpha <_K 0$, $\alpha + \tilde{\alpha} <_K 0$ then there exists a constant $C = C(\alpha, \tilde{\alpha})$ independent of r, $0 < r \le \frac{1}{2}$ such that

$$H(r, \alpha + \tilde{\alpha}) \le C H(r,\alpha) r^{-\max l(\tilde{\alpha},\varrho)},$$

where the maximum is taken over $\varrho \in K$.

Proof: The first estimation follows easily from definition (20) since

$$\min_{\varrho\in K}(-l(-\varepsilon,\varrho)) = \min_{\varrho\in K} l(\varepsilon,\varrho) > 0.$$

For the proof of ii) observe that by (12) and (20) we get for $0 < r < \frac{1}{2}$

$$H(r,\alpha+\tilde{\alpha}) \le C_1 r^{-\max l(\tilde{\alpha},\varrho)} G(r),$$

$$H(r,\alpha) \ge C_2 G(r),$$

where

$$G(r) = \sum_{j=0}^{n-1} \frac{(n-1)!}{j!} |\ln r|^j \int_K r^{-l(\alpha,\varrho)}\, d\varrho,$$

$$C_1 = C_1(\alpha,\tilde{\alpha}) = \max_{0\le j\le n-1} \frac{1}{(-\max_{\varrho\in K} l(\alpha+\tilde{\alpha},\varrho))^{n-j}},$$

$$C_2 = C_2(\alpha,\tilde{\alpha}) = \min_{0\le j\le n-1} \frac{1}{(-\min_{\varrho\in K} l(\alpha,\varrho))^{n-j}}$$

and $0 < C_1 < \infty$, $0 < C_2 < \infty$. The assertion now follows with $C = C_1/C_2$.

Remark 5. Let $\sum \eta_j > 0$, $\tau \in \mathbf{N}_0^n$ ($\sum \eta_j \ge 0$ if $\tau = 0$) and a be a smooth function in a neighbourhood of zero. Let κ_r be given by (2) and κ_r^K by (17). Then the cut-off functions

$$\mathbf{R}_+^n \ni x \mapsto x^\eta (\ln x)^\tau (a\kappa_r)(x), \qquad \mathbf{R}_+^n \ni x \mapsto x^\eta (\ln x)^\tau (a\kappa_r^K)(x)$$

belong to M_{-1} hence are Mellin multipliers and belong to $M'_{(0)}$ (see (6.4) and (5.19)).

The Remark follows from (4) and (18). The same holds true for the cut-off functions considered in Section 7 (see (7.38) and (7.41)).

Proposition 5. *Suppose $u \in \mathfrak{M}'_\alpha(\mathbf{R}^n_+)$ and a is a smooth function in a neighbourhood of zero. Then $a\kappa^K_r u \in M'_{(\omega)}((0,\mathbf{r}])$ for $\omega \leq_K \alpha$ (r small enough) and the multiplication by $a\kappa^K_r$ is continuous in the respective topologies.*

Proof: Let $\varphi_j \to 0$ in $M_{(\omega)}$. This means that there is a $0 < \delta$ such that $\varphi_j \to 0$ in $M_{\omega-\delta}$. Put $\varepsilon = \alpha - \omega + \delta$. Since $\omega \leq_K \alpha$ and $0 <_K \delta$ we have $0 <_K \varepsilon$. To see that $a\kappa^K_r u \in M'_{(\omega)}$ we need only to show that for every $0 <_K \varepsilon$ if $\varphi_j \to 0$ in $M_{\alpha-\varepsilon}$ i.e. if

$$\sup_{x\in(0,\mathbf{r}]}\left|x^{\alpha-\varepsilon+1}\left(x\frac{\partial}{\partial x}\right)^\nu \varphi_j\right| \to 0 \qquad \text{for every } \nu \in \mathbf{N}^n_0 \tag{24}$$

then $a\kappa^K_r \varphi_j \to 0$ in $\mathfrak{M}_\alpha$, i.e. $(x^{\alpha+1}a\kappa^K_r\varphi_j)\circ\mu \to 0$ in $S(\mathbf{R}^n)$.

Assume (24), choose arbitrarily $\kappa, \nu \in \mathbf{N}^n_0$ and put

$$\psi_j(y) = y^\kappa\left(\frac{\partial}{\partial y}\right)^\nu (x^{\alpha+1}a\kappa^K_r\varphi_j)\circ\mu.$$

It suffices to show that $\sup_{y\in\mathbf{R}^n}|\psi_j(y)| \to 0$.

To this aim observe that by (24) and Proposition 3

$$\sup_{(0,\mathbf{r}]}\left|\left(x\frac{\partial}{\partial x}\right)^\nu (x^{\alpha-\varepsilon+1}a\kappa^K_r\varphi_j)\right| \to 0 \tag{25}$$

and that by (4.7)

$$\sup_{y\in\mathbf{R}^n}|\psi_j(y)| = \sup_{x\in\operatorname{supp}\kappa^K_r}\left|(\ln x)^\kappa\left(x\frac{\partial}{\partial x}\right)^\nu (x^{\alpha+1}a\kappa^K_r\varphi_j)\right|.$$

The desired assertion now follows from (25) and Exercise 2.

The easy proof of the continuity of the multiplication $\kappa^K_r a\colon \mathfrak{M}'_\alpha \to M'_{(\omega)}$ is left to the reader.

Proposition 6. *Let $\psi = ax^\varepsilon\kappa^K_r$ with $0 <_K \varepsilon$ and a smooth in a neighbourhood of zero. If $u \in \mathfrak{M}'_\alpha$ then*

$$(2\pi)^n\mathcal{M}(\psi u)(\alpha + i\beta) = (\mathcal{M}\psi(i\,\cdot\,) * \mathcal{M}_\alpha u)(\beta), \qquad \beta \in \mathbf{R}^n, \tag{26}$$

where the right hand side is the convolution of a function $\mathcal{M}\psi(i\,\cdot\,) \in S$ with a tempered distribution $\mathcal{M}_\alpha u$ (i.e. can be written as $\mathcal{M}_\alpha u[\mathcal{M}\psi(i(\beta-\gamma))]$).

Proof: By Exercise 4 and Proposition 4.4 we get

$$(2\pi)^n \mathcal{M}_\alpha(\psi u) = \mathcal{M}_0\psi * \mathcal{M}_\alpha u. \tag{27}$$

By Proposition 5 $\psi u \in M'_{(\gamma)}$ for $\gamma \leq_K \alpha + \varepsilon$. Since by Corollary 2 there exists $\tilde{\gamma}$ such that $\alpha < \tilde{\gamma} \leq_K \alpha + \varepsilon$ Proposition 7.1 yields $\psi u \in \mathfrak{M}'_\alpha$. Taking $u \equiv 1 \in \mathfrak{M}'_0$ we get $\psi \in M'_{(\gamma)}$ for $\gamma \leq_K \varepsilon$. Similarly $\psi \in \mathfrak{M}'_0$ and

$$\mathcal{M}_\alpha(\psi u)(\beta) = \mathcal{M}(\psi u)(\alpha + i\beta), \qquad \mathcal{M}_0\psi(\beta) = \mathcal{M}\psi(i\beta),$$

so (26) follows from (27).

4. Further Investigation of the Mellin Transform of a Conical Cut-Off Function

We shall now investigate the properties of the Mellin transform of a conical cut-off function in the spirit of Proposition 7.4.

Definition 1. Let $\delta\mathring{x} \in \mathbf{R}^n_+$. By a *conical cut-off function* at $(0; \delta\mathring{x})$ we understand any function $\kappa \in C^\infty(\mathbf{R}^n_+)$ of the form $\kappa = \varphi \cdot \tilde{\kappa}$ where $\varphi \in C_0^\infty(\mathbf{R}^n)$, $\varphi \equiv 1$ in a neighbourhood of zero and $\tilde{\kappa} \in C^\infty(\mathbf{R}^n_+)$ is homogeneous of order zero, $\overline{\operatorname{supp}\tilde{\kappa}} \subset \mathbf{R}^n_+ \cup \{0\}$ and $\tilde{\kappa}(\delta\mathring{x}) \neq 0$.

By Remark 1

$$\kappa(x) = \varphi(x)\lambda\left(\frac{x}{\|x\|}\right) \qquad \text{for } x \in \mathbf{R}^n_+, \tag{28}$$

where λ is a smooth compactly supported function on the unit quarter-sphere S^{n-1}_+ and $\lambda\left(\frac{\delta\mathring{x}}{\|\delta\mathring{x}\|}\right) \neq 0$.

Since the singularities of the Mellin transform of κ are related to the variable $z_1 + \cdots + z_n$, we shall use the change of variables $\zeta = \mathbf{A}z$ where

$$\zeta_1 = z_1 + \cdots + z_n, \qquad \zeta_j = z_j \qquad \text{for } j = 2, \ldots, n.$$

The change of variables $\zeta = \mathbf{A}z$ in the Mellin transform corresponds to the change $x = S(y)$ of the initial variables $S : \mathbf{R}^n_+ \to \mathbf{R}^n_+$ given by (1) (see Theorem 8.4 and Exercise 8.6).

Denote $k(y) = \kappa \circ S(y)$, $k'(y') = \lim_{y_1 \to 0_+} k(y_1, y')$ where $y = (y_1, y')$. We have

$$k'(y') = k(0, y') = \lambda\left(\frac{(1, y')}{\|(1, y')\|}\right).$$

Proposition 7. *The Mellin transform $\mathcal{M}\kappa$ of the conical cut-off function (28) has the following properties*

(i) $((\mathcal{M}\kappa)\circ \mathbf{A}^{-1})(\zeta) = -\dfrac{\mathcal{M}'k'(\zeta')}{\zeta_1} + \widetilde{K}(\zeta)$ *with* $\widetilde{K} \in \mathcal{O}(\mathbf{C}^n)$, *where* $\mathcal{M}'$ *denotes the Mellin transformation in variables* $y' = (y_2, \dots, y_n)$.

(i′) $k' \in C_0^\infty(\mathbf{R}_+^{n-1})$ *and* $\mathcal{M}'k' \in \mathcal{O}(\mathbf{C}^{n-1})$.

(ii) *For every* $a \in \mathbf{R}^n$ *the function*

$$\mathbf{R}^n \ni b \mapsto (a_1 + ib_1)\mathcal{M}\kappa(\mathbf{A}^{-1}(a+ib))$$

is in $S(\mathbf{R}^n)$ *locally uniformly with respect to* a.

(ii′) *For every* $a' \in \mathbf{R}^{n-1}$ *the function*

$$\mathbf{R}^{n-1} \ni b' \mapsto \mathcal{M}'k'(a' + ib')$$

is in $S(\mathbf{R}^{n-1})$ *locally uniformly with respect to* a'.

Proof: Denote $\omega(s) = k(e^{-s})$, $\omega'(s') = k'(e^{-s'})$ and observe that

$$\omega(s) = \varphi\circ S(e^{-s})\lambda\Big(\frac{(1,e^{-s'})}{\|(1,e^{-s'})\|}\Big), \qquad \omega'(s') = \lambda\Big(\frac{(1,e^{-s'})}{\|(1,e^{-s'})\|}\Big), \tag{29}$$

$$\omega(s_1, s') = \begin{cases} \omega'(s') & \text{for } s_1 \text{ big enough,} \\ 0 & \text{for } (-s_1) \text{ big enough.}\end{cases}$$

Since λ is compactly supported on S_+^{n-1} we deduce from the above relations that $\omega'(s') \in C_0^\infty(\mathbf{R}^{n-1})$ and that $\frac{\partial\omega}{\partial s_1} \in C_0^\infty(\mathbf{R}^n)$. Hence $\mathcal{L}'\omega'(\zeta') \in \mathcal{O}(\mathbf{C}^{n-1})$ and $\mathcal{L}(\frac{\partial\omega}{\partial s_1})(\zeta) \in \mathcal{O}(\mathbf{C}^n)$. Observe that by (4.5) $\frac{\partial\omega}{\partial s_1} = -(y_1\frac{\partial}{\partial y_1}k)\circ\mu$ with $\mu(s) = e^{-s}$. Thus

$$\mathcal{L}\Big(\frac{\partial\omega}{\partial s_1}\Big)(\zeta) = \int_{\mathbf{R}^n} e^{\zeta s}\frac{\partial\omega}{\partial s_1}\,ds = -\int_{\mathbf{R}_+^n} y^{-\zeta-1}y_1\frac{\partial k}{\partial y_1}\,dy = -\zeta_1(\mathcal{M}k)(\zeta)$$

and therefore

$$\zeta_1\mathcal{M}(\kappa\circ S)(\zeta) = -\mathcal{L}\Big(\frac{\partial\omega}{\partial s_1}\Big)(\zeta) \in \mathcal{O}(\mathbf{C}^n). \tag{30}$$

Note that

$$\begin{aligned}(\zeta_1\mathcal{M}(\kappa\circ S)(\zeta))|_{\zeta_1=0} &= -\mathcal{L}\Big(\frac{\partial\omega}{\partial s_1}\Big)(0,\zeta') = -\int_{\mathbf{R}^{n-1}}\Big\{\int_{\mathbf{R}}\frac{\partial\omega}{\partial s_1}\,ds_1\Big\}e^{\zeta' s'}\,ds' \\ &= -\int_{\mathbf{R}^{n-1}}\omega'(s')e^{\zeta' s'}\,ds' = -\int_{\mathbf{R}^{n-1}} e^{\zeta' s'}k'(e^{-s'})\,ds' \\ &= -\mathcal{M}'(k')(\zeta')\end{aligned}$$

and hence by (30) there exists $\widetilde{K} \in \mathcal{O}(\mathbf{C}^n)$ such that

$$\mathcal{M}(\kappa \circ S)(\zeta) = -\frac{\mathcal{M}'k'(\zeta')}{\zeta_1} + \widetilde{K}(\zeta). \tag{31}$$

Since $\kappa \in M'_{(0)}$ (see Remark 5) by Theorem 8.4

$$(\mathcal{M}\kappa) \circ \mathbf{A}^{-1}(\zeta) = \mathcal{M}(\kappa \circ S)(\zeta) \qquad \text{for} \quad \operatorname{Re}\zeta < 0$$

and the unicity theorem for analytic functions yields assertion i).

As just observed $\mathcal{L}'(\omega')(\zeta') \in \mathcal{O}(\mathbf{C}^{n-1})$ hence from

$$\mathcal{L}'(\omega')(\zeta') = \int_{\mathbf{R}^{n-1}} e^{\zeta's'}\omega'(s')\,ds' = \int_{\mathbf{R}^{n-1}} e^{\zeta's'}k'(e^{-s'})\,ds' = \mathcal{M}'(k')(\zeta') \tag{32}$$

we get assertion (i').

For the proof of (ii) and (ii') note that by (30) and (32) we have

$$\begin{aligned}(a_1 + ib_1)\mathcal{M}(\kappa \circ S)(a+ib) &= -\int_{\mathbf{R}^n} e^{(a+ib)s}\frac{\partial\omega}{\partial s_1}\,ds = -\int_{\mathbf{R}^n}\left(e^{as}\frac{\partial\omega}{\partial s_1}\right)e^{ibs}\,ds\\ &= -(2\pi)^{\frac{n}{2}}\mathcal{F}^{-1}\left(e^{as}\frac{\partial\omega}{\partial s_1}\right)(b),\end{aligned}$$

$$\begin{aligned}\mathcal{M}'(k')(a'+ib') &= \int_{\mathbf{R}^{n-1}} (e^{a's'}\omega'(s'))e^{ib's'}\,ds'\\ &= (2\pi)^{\frac{n-1}{2}}\mathcal{F}^{-1}(e^{a's'}\omega'(s'))(b')\end{aligned}$$

and the rest of the proof runs as the proof of assertion (ii) of Proposition 7.4.

Remark 6. It follows at once from assertions (i) and (i') of Proposition 7 that

$$(\mathcal{M}\kappa) \circ \mathbf{A}^{-1} \in \mathcal{O}((\mathbf{C} \setminus \{0\}) \times \mathbf{C}^{n-1}).$$

Moreover we have the following explicit expressions for $K(\zeta) = \mathcal{M}(\kappa \circ S)(\zeta)$:

$$K(\zeta) = \begin{cases} \mathcal{L}\omega(\zeta) & \text{for } \operatorname{Re}\zeta_1 < 0\\ \mathcal{L}(\omega(s) - \omega'(s'))(\zeta) & \text{for } \operatorname{Re}\zeta_1 > 0.\end{cases} \tag{33}$$

In fact by (30) and (29)

$$\zeta_1\mathcal{M}(\kappa \circ S)(\zeta) = -\int_{\mathbf{R}^{n-1}} e^{\zeta's'}\left(\int_{-\infty}^{+\infty} e^{\zeta_1 s_1}\frac{\partial\omega}{\partial s_1}\,ds_1\right)ds' = \zeta_1(\mathcal{L}\omega)(\zeta)$$
$$\text{for} \quad \operatorname{Re}\zeta_1 < 0,$$

and

$$\begin{aligned}\zeta_1\mathcal{M}(\kappa\circ S)(\zeta) &= -\int_{\mathbf{R}^{n-1}} e^{\zeta' s'}\Big(\int_{-\infty}^{+\infty} e^{\zeta_1 s_1}\frac{\partial}{\partial s_1}(\omega(s_1,s')-\omega'(s'))\,ds_1\Big)ds' \\ &= \zeta_1(\mathcal{L}(\omega(s)-\omega'(s'))(\zeta) \qquad\qquad \text{for} \quad \operatorname{Re}\zeta_1>0.\end{aligned}$$

Exercises

1. Deduce from the definitions from Subsection 10.2 that:

(i) If $\mathring{\alpha},\omega\in\mathbf{R}^n$ and $\mathring{\alpha}<_K\omega$ then there exist $\overset{1}{\alpha},\overset{2}{\alpha}\in\mathbf{R}^n$ such that $\mathring{\alpha}<\overset{1}{\alpha}<_K\omega$, $\mathring{\alpha}<_K\overset{2}{\alpha}<\omega$.

(ii) For $\alpha\in\mathbf{R}^n$, $1\le j\le n$ there exists $\gamma\in\mathbf{R}^n$ such that $\gamma\le_K\alpha<\gamma+e^j$ where e^j is defined by (1.1).

(iii) If $K=\operatorname{conv}S$, the convex hull of the points in S, then $\alpha\le_K 0$ if and only if $\alpha\le_S 0$. In particular if K is a convex polyhedron the S may consist of a finite number of (corner) points.

(iv) For every fixed $t\in\mathbf{R}^n$ the set $\{\varepsilon\in\mathbf{R}^n\colon 0<_K\varepsilon<_K t\}$ is bounded.

2. Let $0<_K\eta$, $r<1$, $\tau\in\mathbf{N}_0^n$ and κ_r^K be given by (17). Show that $\sup\limits_{x\in\operatorname{supp}\kappa_r^K}|x^\eta(\ln x)^\tau|$ is bounded.

HINT. Show the boundedness of the function

$$(0,r]\times K\ni(t,\varrho)\mapsto t^{l(\eta,\varrho)}(\ln t)^{|\tau|}\varrho_1^{\tau_2}\cdot\ldots\cdot\varrho_{n-1}^{\tau_n}.$$

3. For the proof of the last assertion of Proposition 3 show the following inequalities:

$$\begin{aligned}&\int_{\operatorname{supp}\kappa_r^K}\big|(\ln x)^\tau x^{\eta-1}\big|\,dx\\ &\qquad\le\int_0^r t^{\min l(\eta,\varrho)-1}\,|\ln t|^{|\tau|+n-1}\,dt\cdot\int_K\varrho_1^{\tau_2}\cdot\ldots\cdot\varrho_{n-1}^{\tau_n}\,d\varrho<\infty.\end{aligned}$$

4. Let κ_r^K be given by (17), let $\psi=ax^\varepsilon\kappa_r^K$ with $0<_K\varepsilon$ and a smooth in the neighbourhood of zero. Show that $\psi\circ\mu\in S$ and hence $\psi\in\mathfrak{M}_{-1}(\mathbf{R}^n)$.

HINT. Apply Exercise 2 and Proposition 3 with $\eta=\tau=0$.

5. Let κ_r be a conical cut-off function from Subsection 1. Suppose $u\in\mathfrak{M}'_\alpha(\mathbf{R}^n_+)$ and a be a smooth function in a neighbourhood of zero. Show that $a\kappa_r u\in M'_{(\omega)}((0,\mathbf{r}])$ for every ω with $\sum\omega_j<\sum\alpha_j$ (r small enough) and that the multiplication by $a\kappa_r$ is continuous in the respective topologies.

HINT. See the proof of Proposition 5 and apply Proposition 1 instead of Proposition 3.

6. Let κ_r be a conical cut-off function and a be smooth in a neighbourhood of zero. Let $\psi = ax^{\varepsilon}\kappa_r$ with $\sum \varepsilon_j > 0$ and let $u \in \mathfrak{M}'_\alpha$. Show that then

$$(2\pi)^n \mathcal{M}(\psi u)(\alpha + i\beta) = (\mathcal{M}\psi(i\,\cdot\,) * \mathcal{M}_\alpha u)(\beta), \qquad \beta \in \mathbf{R}^n,$$

where the right hand side is the convolution of a function $\mathcal{M}\psi(i\,\cdot\,) \in S$ with a tempered distribution $\mathcal{M}_\alpha u$.

HINT. See the proof of Proposition 6.

7. Let $\omega, \mathring{\alpha} \in \mathbf{R}^n$, $\sum \mathring{\alpha}_j < \sum \omega_j$ and let $v \in M'_{(\omega)}$. Show that for every cut-off function κ_r defined in Subsection 1 $\kappa_r v \in \mathfrak{M}'_{\mathring{\alpha}}$ and $\mathcal{M}_{\mathring{\alpha}}(\kappa_r v)(\beta) = \mathcal{M}(\kappa_r v)(\alpha + i\beta)$ for $\beta \in \mathbf{R}^n$.

HINT. Apply Proposition 5.5 and Exercise 5.

8. Let $\omega, \mathring{\alpha} \in \mathbf{R}^n$, $\mathring{\alpha} <_K \omega$ and let $v \in M'_{(\omega)}$. Show that for every cut-off function κ^K (cf. (17)) $\kappa^K v \in \mathfrak{M}'_{\mathring{\alpha}}$ and $\mathcal{M}_{\mathring{\alpha}}(\kappa^K v)(\beta) = \mathcal{M}(\kappa^K v)(\mathring{\alpha} + i\beta)$ for $\beta \in \mathbf{R}^n$.

HINT. Apply Exercise 1, Propositions 5.5 and 5.

§11. IMPORTANT SUBSPACES OF MELLIN DISTRIBUTIONS

The basic analysis of Fuchsian differential operators of Sections 13 and 14 is performed in the spaces $M'^{s}_{(\omega)}$ of Mellin distributions of finite Mellin order, which can be regarded as a variant of weighted Sobolev spaces SP(s, s') (recalled below: see also [B]) and which allow one to measure the growth order of the Mellin transforms along the imaginary planes (see [Zie–K] for a connection with 2-microlocal regularity). The more refined regularity analysis of Section 15 takes place in the spaces $M(\Omega; \varrho)$, $Z_d(\Omega; s)$ which combine information on the extendibility of the Mellin transforms (see Appendix and [Zie6], [Zie5] for a connection with the theory of generalized Taylor expansions) with the control over their growth order along the imaginary planes.

1. Subspaces $M'^{s}_{(\omega)}$

Let $u \in M'_{(\omega)}(I)$, $\omega \in (\mathbf{R} \cup \{\infty\})^n$, $I = (0, t]$ be of Mellin order $\leq m$, $m \in \mathbf{N}_0$. Then by Theorem 9.4* (cf. Exercise 9.7) for every $a < \omega$ and $\tilde{t} > t$ there exists $C = C(a, \tilde{t}) < \infty$ such that

$$|\mathcal{M}u(\alpha + i\beta)| \leq C\langle\beta\rangle^m \tilde{t}^{-\alpha} \qquad \text{for } \alpha \leq a, \ \beta \in \mathbf{R}^n.$$

Definition 1. Let $\omega \in (\mathbf{R} \cup \{\infty\})^n$, $s \in \mathbf{R}$. We denote by $M'^{s}_{(\omega)}(I)$ the space of all distributions $u \in M'_{(\omega)}(I)$ such that for every $a < \omega$ and $\tilde{t} > t$ there exist a constant $C = C(a, \tilde{t}) < \infty$ such that

$$(1) \qquad |\mathcal{M}u(\alpha + i\beta)| \le C\langle\beta\rangle^s \tilde{t}^{-\alpha} \qquad \text{for } \alpha \le a, \ \beta \in \mathbf{R}^n.$$

It follows from the preceding remarks that every distribution $u \in M'_{(\omega)}(I)$ of Mellin order $\le m$ belongs to $M'^{m}_{(\omega)}$ and that $M'^{s}_{(\omega)} \subset M'^{\tilde{s}}_{(\omega)}$ for $s \le \tilde{s}$. Thus if we denote

$$(2) \qquad M'^{\infty}_{(\omega)} = \bigcup_{s \in \mathbf{R}} M'^{s}_{(\omega)}$$

we also have $M'^{\infty}_{(\omega)} = \bigcup_{m \in \mathbf{N}_0} M'^{m}_{(\omega)}$ and by Remark 9.1 it is the space of Mellin distributions of finite Mellin order. Hence by Remark 8.3 $M'^{\infty}_{(\omega)}$ is a proper subset of $M'_{(\omega)}(I)$ (cf. also Exercise 1).

Proposition 1. *Let $\varphi \in C_0^\infty(\mathbf{R}^n)$, $\operatorname{supp}\varphi \cap \mathbf{R}^n_+ \subset (0, t]$ and let $u \in M'_{(\omega)}$ satisfy the estimation*

$$(3) \qquad |\mathcal{M}u(\alpha + i\beta)| \le \tilde{C}(\alpha)\langle\beta\rangle^s \qquad \text{for } \beta \in \mathbf{R}^n$$

and α arbitrarily close to ω, $\alpha < \omega$. Then for any $a < \omega$ there exists a constant $C = C(a)$ such that

$$(4) \qquad |\mathcal{M}(\varphi u)(\alpha + i\beta)| \le C\langle\beta\rangle^s t^{-\alpha} \qquad \text{for } \alpha \le a, \ \beta \in \mathbf{R}^n.$$

Proof: We may assume that a is close enough to ω so that (3) holds. Let $\varepsilon = \frac{1}{2}(\omega - a)$ if $\omega < \infty$ and $\varepsilon = 1$ if $\omega = \infty$. By Remark 7.1 we have for $\alpha \le a$

$$\mathcal{M}(\varphi u)(\alpha + i\beta) = (2\pi)^{-n} \int_{\mathbf{R}^n} \mathcal{M}u(a + \varepsilon + i\gamma)\mathcal{M}\varphi(\alpha - a - \varepsilon + i(\beta - \gamma))\, d\gamma.$$

By (3)

$$|\mathcal{M}u(a + \varepsilon + i\gamma)| \le \tilde{C}(a + \varepsilon)\langle\gamma\rangle^s \qquad \text{for } \gamma \in \mathbf{R}^n.$$

Further by Theorem 9.5 for every $m \in \mathbf{N}_0$ there is a constant $C_m < \infty$ such that for $\alpha \le a$

$$|\mathcal{M}\varphi(\alpha - a - \varepsilon + i(\beta - \gamma))| \le C_m \frac{1}{\langle\beta - \gamma\rangle^m} t^{-\alpha + a + \varepsilon}.$$

Taking $m = |s| + n + 1$ we get from the above estimate

$$\begin{aligned} |\mathcal{M}(\varphi u)(\alpha + i\beta)| &\le (2\pi)^{-n}\tilde{C}(a + \varepsilon)C_m t^{-\alpha + a + \varepsilon} \int_{\mathbf{R}^n} \frac{\langle\gamma\rangle^s}{\langle\beta - \gamma\rangle^{|s| + n + 1}}\, d\gamma \\ &\le C(a)\langle\beta\rangle^s t^{-\alpha} \qquad \text{for } \alpha \le a \end{aligned}$$

in view of the Petree inequality

$$\frac{\langle\gamma\rangle^s}{\langle\beta-\gamma\rangle^{|s|}} \le \langle\beta\rangle^s \qquad \text{for any } s \in \mathbf{R}. \tag{5}$$

Remark 1. It follows from the proposition that the estimations (1) and (3) are equivalent.

Proof: Suppose (3), take $\varphi \in C_0^\infty(\mathbf{R}^n)$, $\operatorname{supp}\varphi \cap \mathbf{R}_+^n \subset (0,\tilde{t}]$, $\tilde{t} > t$, $\varphi \equiv 1$ on $\operatorname{supp} u \subset (0,t]$. Since we may consider u in $M'_{(\omega)}((0,\tilde{t}])$, (see Theorem 8.3), we get $\varphi u = u$ and (1) follows by Proposition 1 (with $\tilde{t}$ instead of t).

The implication (1) $\Rightarrow$ (3) is obvious.

Remark 2. Proposition 1 and Remark 1 imply that the spaces $M'^s_{(\omega)}$ are *local* i.e. for any $\varphi \in C_0^\infty(\mathbf{R}^n)$ and $u \in M'^s_{(\omega)}$, $\varphi u \in M'^s_{(\omega)}$. Therefore one may say that $u \in M'^s_{(\omega)}$-locally if for some $\varphi \in C_0^\infty(\mathbf{R}^n)$ $\varphi u \in M'^s_{(\omega)}$.

Proof: Let $u \in M'^s_{(\omega)}$. By Definition 1, Remark 1 and Proposition 1 we get (4) i.e.

$$|\mathcal{M}(\varphi u)(\alpha+i\beta)| \le \tilde{\tilde{C}}(\alpha)\langle\beta\rangle^s \qquad \text{for } \beta \in \mathbf{R}^n, \ \alpha < \omega$$

with $\tilde{\tilde{C}}(\alpha) = C(a)t^{-\alpha}$, $\alpha \le a < \omega$. Thus by Remark 1 $\varphi u \in M'^s_{(\omega)}((0,t])$.

If instead of the smooth cut-off functions φ as in Proposition 1 we use the conical cut-off functions κ (considered in Subsections 10.1 and 10.4) or the cut-off functions κ^K considered in Subsection 10.3 we are led to the following definition:

Definition 2. Let $u \in M'^\infty_{(\omega)}$, $\delta\mathring{x} \in \mathbf{R}_+^n$, $s \in \mathbf{R}$. We say that u belongs to $M'^s_{(\omega)}$ *2-locally* at $(0,\delta\mathring{x})$ if there exists a conical cut-off function $\kappa = \tilde{\kappa}\cdot\varphi$ as in Definition 10.1 with $\tilde{\kappa}(x) \equiv 1$ in a conical neighbourhood of $\delta\mathring{x}$ and $\varphi(x) \equiv 1$ in a neighbourhood of 0, such that $\kappa u \in M'^s_{(\omega)}$.

If instead of a conical cut-off function κ at $(0,\delta\mathring{x})$ we take a cut-off function κ_r^K related to a "tangent cone" Ξ_K (see Subsection 10.3) we get the following

Definition 3. Let $u \in M'^\infty_{(\omega)}$, $s \in \mathbf{R}$. We say that u belongs to $M'^s_{(\omega)}$ Ξ_K*-locally* (at zero) if there exists a cut-off function $\tilde{\kappa}_r^K$, $\tilde{\kappa}_r^K \equiv 1$ on Ξ_K close to zero, $\operatorname{supp}\tilde{\kappa}_r^K \subset \Xi_{K_1}$ for some K_1, $K \subset \operatorname{Int} K_1 \subset\subset \mathbf{R}_+^{n-1}$, such that for r small enough $\tilde{\kappa}_r^K u \in M'^s_{(\omega)}$.

The spaces defined by Definitions 2 and 3 are *2-local* and *Ξ_K-local*, respectively. In the case of Definition 3 this follows at once from the estimation (6) (with $\delta = 0$) of Lemma 1 below. The analogous proof for the case of Definition 2 is left to the reader (see Exercise 6). The more general setting of Lemma 1 will be of interest in Section 14.

Lemma 1. *Suppose $u \in M'^{s}_{(\omega)}$ Ξ_K-locally, and let $\tilde{\kappa}_r^K$ be such that $\tilde{\kappa}_r^K u \in M'^{s}_{(\omega)}$ (see Definition 3). Let $u_r = \kappa_r^K \tilde{\kappa}_r^K u$ with an arbitrary cut-off function (10.17) κ_r^K with support in Ξ_{K_1}. Fix $\mathring{\alpha} <_{K_1} \omega$. Then for every $\delta \in \mathbf{N}_0^n$ there exists a constant $C = C(\delta, \mathring{\alpha})$ such that*

$$\left|\left(\frac{\partial}{\partial\beta}\right)^{\delta}(\mathcal{M}u_r)(\mathring{\alpha}+i\beta)\right| < C\langle\beta\rangle^{s} \tag{6}$$

for r small enough. Hence $\mathcal{M}u_r(\mathring{\alpha}+i\,\cdot\,) \in O_M^s$ for r small (cf. Definition 4.2).

Proof: By Definition 3 there exists $\tilde{\kappa}_r^K$ such that $\tilde{\kappa}_r^K u \in M'^{s}_{(\omega)}$ and hence by Definition 1 and Remark 1 $\tilde{\kappa}_r^K u \in M'_{(\omega)}$ and

$$\left|\mathcal{M}(\tilde{\kappa}_r^K u)(\alpha+i\beta)\right| \le C\langle\beta\rangle^{s} \qquad \text{for } \beta \in \mathbf{R}^n \tag{7}$$

with a suitable constant $C = C(\alpha)$ and r small. We shall show that $\mathcal{M}(\kappa_r^K\tilde{\kappa}_r^K u)$ is holomorphic for $\operatorname{Re} z <_{K_1} \omega$. In fact, by Proposition 5.5 $\tilde{\kappa}_r^K u \in \mathfrak{M}'_{\alpha}$ for every $\alpha < \omega$ hence by Proposition 10.5 $\kappa_r^K\tilde{\kappa}_r^K u \in M'_{(\gamma)}$ for every $\gamma <_{K_1} \alpha$ and consequently $\kappa_r^K\tilde{\kappa}_r^K u \in M'_{(\gamma)}$ for $\gamma <_{K_1} \omega$. Thus $\mathcal{M}(\kappa_r^K\tilde{\kappa}_r^K u)(z)$ is holomorphic for $\operatorname{Re} z < \gamma$ if $\gamma <_{K_1} \omega$ and hence for $\operatorname{Re} z <_{K_1} \omega$. Choose $\tilde{\alpha}$ such that $\mathring{\alpha} <_{K_1} \tilde{\alpha} < \omega$ (see Exercise 10.1). By (7.9), Propositions 5.5, 10.6 and by Exercise 10.8 we have

$$\begin{aligned}
&\left|\left(\frac{\partial}{\partial\beta}\right)^{\delta}\mathcal{M}(\kappa_r^K\tilde{\kappa}_r^K u)(\mathring{\alpha}+i\beta)\right| \\
&\quad = \left|\left(\frac{\partial}{\partial\beta}\right)^{\delta}\mathcal{M}(x^{\tilde{\alpha}-\mathring{\alpha}}\kappa_r^K\tilde{\kappa}_r^K u)(\tilde{\alpha}+i\beta)\right| \\
&\quad = (2\pi)^{-n}\left|\left(\frac{\partial}{\partial\beta}\right)^{\delta}\int_{\mathbf{R}^n}\mathcal{M}(\tilde{\kappa}_r^K u)(\tilde{\alpha}+i\gamma)\mathcal{M}(x^{\tilde{\alpha}-\mathring{\alpha}}\kappa_r^K)(i(\beta-\gamma))\,d\gamma\right| \\
&\quad = (2\pi)^{-n}\left|\int_{\mathbf{R}^n}\mathcal{M}(\tilde{\kappa}_r^K u)(\tilde{\alpha}+i(\beta-\gamma))\mathcal{M}(x^{\tilde{\alpha}-\mathring{\alpha}}(\ln x)^{\delta}\kappa_r^K)(i\gamma)\,d\gamma\right|.
\end{aligned} \tag{8}$$

Now, from (10.22), for $h_1 = \ldots = h_n = |s|+2$, from Remark 10.4 and (7), (8) we get

$$\left|\left(\frac{\partial}{\partial\beta}\right)^{\delta}\mathcal{M}(\kappa_r^K\tilde{\kappa}_r^K u)(\mathring{\alpha}+i\beta)\right| \le C\int_{\mathbf{R}^n}\frac{\langle\beta-\gamma\rangle^{s}}{(\langle\gamma_1\rangle\cdot\ldots\cdot\langle\gamma_n\rangle)^{|s|+2}}\,d\gamma$$

which by Exercise 3 ends the proof of (6).

Remark 3. Definition 3 introduces a local regularity scale in the space $M'^{\infty}_{(\omega)}$. In fact, if $u \in M'^{\infty}_{(\omega)}$ then by (2) $u \in M'^{\tilde{s}}_{(\omega)}$ with some $\tilde{s} \in \mathbf{R}^n$. After a convenient localization of u by multiplication by a function $\tilde{\kappa}^K_r$ the distribution $\tilde{\kappa}^K_r u$ belongs to $M'^{s}_{(\omega)}$ which in the case $s < \tilde{s}$ means a higher regularity.

2. Subspaces $\mathrm{SP}_\Gamma(s, s')$ of Mellin Distributions

Definition 4 (cf. Def. 2.5 [B]) Let $u \in D'(\mathbf{R}^n \setminus \{0\})$ and let u vanish outside a unit ball in $\mathbf{R}^n$. Let s, $s' \in \mathbf{R}$ be such that $s + s'$ is a non-negative integer. We say that $u \in \mathrm{SP}(s, s')$ if

$$(9) \qquad \|x\|^{-s+|\lambda|} \left(\frac{\partial}{\partial x}\right)^{\lambda} u \in L^2(\mathbf{R}^n) \qquad \text{for } 0 \le |\lambda| \le s + s', \quad \lambda \in \mathbf{N}_0^n.$$

In our case it will be more convenient to use, instead of the space $L^2(\mathbf{R}^n)$, the space $L^2(\mathbf{R}^n_+)$ with the weight x^{-1} i.e. $u \in L^2(\mathbf{R}^n_+, x^{-1})$ if and only if $\int_{\mathbf{R}^n_+} |u(x)|^2 x^{-1}\, dx < \infty$ (actually the integral is over $\widetilde{B}_+ = \{x \in \mathbf{R}^n_+ \colon \|x\| \le 1\}$).

For $u \in D'(\mathbf{R}^n_+)$, (9) can be written as $u \in \mathrm{SP}(s, s')$ if and only if

$$(10) \qquad x^{1/2} \|x\|^{-s+|\lambda|} \left(\frac{\partial}{\partial x}\right)^{\lambda} u \in L^2(\mathbf{R}^n_+, x^{-1}), \qquad 0 \le |\lambda| \le s + s'.$$

If $\operatorname{supp} u \subset \Gamma$, where Γ is a *proper cone* in $\mathbf{R}^n_+$ (i.e. such that $\overline{\Gamma} \cap \overline{\mathbf{R}^n_+} = \{0\}$, overbar denotes closure in $\mathbf{R}^n$) then there exist constants $d_j > 0$ such that for $x \in \Gamma$ $x_j < \|x\| < d_k x_k$, $k = 1, 2, \dots, n$. Then (10) means that

$$(10') \qquad x^{1/2-\varrho} \left(x \frac{\partial}{\partial x}\right)^{\lambda} u \in L^2(\mathbf{R}^n_+, x^{-1})$$

for $\lambda \in \mathbf{N}_0^n$, $\varrho \in \mathbf{R}^n$, $0 \le |\lambda| \le s + s'$, with $\varrho_1 + \dots + \varrho_n = s$. Since $\operatorname{supp} u \subset \widetilde{B}_+$ this can be written as

$$(11) \qquad x^{\gamma} \left(x \frac{\partial}{\partial x}\right)^{\lambda} u \in L^2(\mathbf{R}^n_+, x^{-1}) \qquad \text{for } \lambda \in \mathbf{N}_0^n, \quad \gamma \in \mathbf{R}^n, \quad |\lambda| \le s + s'$$

where $\gamma_1 + \dots + \gamma_n \ge -s + \frac{n}{2}$, which is equivalent to

$$(11') \qquad x^{\mathring{\gamma}} \left(x \frac{\partial}{\partial x}\right)^{\lambda} u \in L^2(\mathbf{R}^n_+, x^{-1}) \qquad \text{for } \lambda \in \mathbf{N}_0^n, \quad |\lambda| \le s + s'$$

and some $\mathring{\gamma} \in \mathbf{R}^n$ such that $\sum \mathring{\gamma}_j = -s + \frac{n}{2}$.

We intend to describe the Mellin image of the space

$$\mathrm{SP}_\Gamma(s, s') = \{u \in \mathrm{SP}(s, s') \colon \operatorname{supp} u \subset \Gamma\},$$

where Γ is a proper cone as above. We start with

Lemma 2. *Let $u \in \mathrm{SP}(s,s')$, $s+s' \in \mathbf{N}_0$, $\operatorname{supp} u \subset \Gamma$ where Γ is a proper cone in $\mathbf{R}^n_+$. Then $u \in M'_{(\varrho)}$ for $\varrho \in \mathbf{R}^n$ with $\sum \varrho_j \le s - \frac{n}{2}$.*

Proof: From (11) for $\lambda = 0$ we have $x^\gamma u \in L^2(\mathbf{R}^n_+; x^{-1})$ for $\sum \gamma_j \ge -s + \frac{n}{2}$, $\gamma \in \mathbf{R}^n$. Since multiplication by x^γ, $\gamma \in \mathbf{R}^n$ maps $M'_{(\omega)} \to M'_{(\omega+\gamma)}$ it is enough to show that if $u \in L^2(\mathbf{R}^n_+, x^{-1})$ then $u \in M'_{(0)}$. Let $\varphi \in M_a$, $a < 0$; we have

$$\left|\int_{B_+} u\varphi\, dx\right| \le \left(\int_{B_+} |u|^2 x^{-1}\, dx\right)^{1/2} \cdot \left(\int_{B_+} x^1 |\varphi|^2\, dx\right)^{1/2}$$
$$\le C\left(\int_{B_+} x^{-2a-1}\, dx\right)^{1/2} \varrho_{a,0}(\varphi) < \infty$$

thus u is a continuous functional on the space $M_{(0)} = \bigcup_{a<0} M_a$.

Lemma 3. *Suppose $u \in D'_{\bar{B}_+}(\mathbf{R}^n_+)$. $u \in L^2(\mathbf{R}^n_+, x^{-1})$ if and only if $u \in M'_{(0)}$ and $(\mathcal{M}u)(z) \in L^2(\mathbf{R}^n)$ as a function of $\operatorname{Im} z$ for each fixed $\operatorname{Re} z \le 0$ (i.e. $\mathcal{M}u(z)$ defined for $\operatorname{Re} z < 0$ has a boundary value for $\operatorname{Re} z \to 0_-$, and the boundary value is in $L^2(\mathbf{R}^n)$).*

Proof: The condition $u \in L^2(\mathbf{R}^n_+, x^{-1})$ is equivalent to the condition that the function $h(y) = u(e^{-y}) \in L^2(\mathbf{R}^n)$. Since by (4.6) and (7.12)

$$\mathcal{M}u(\alpha + i\beta) = (2\pi)^{n/2} \mathcal{F}^{-1}(e^{\alpha y} h)(\beta) \qquad \text{for } \alpha < 0$$

we see that the latter condition is equivalent, in view of the continuity of the Fourier transformation in L^2, to the condition that $\mathcal{M}u(z) \in L^2(\mathbf{R}^n)$ for $\operatorname{Re} z \le 0$ (see Exercise 7).

From Lemma 2 and 3 and operational rules (7.9), (7.11) the following characterization of the space $\mathrm{SP}(s,s')$ can be derived:

Proposition 2. *Let $u \in D'_{B_+}(\mathbf{R}^n_+)$, $\operatorname{supp} u \subset \Gamma$— a proper cone, then the following conditions are equivalent:*

(i) $u \in \mathrm{SP}(s,s')$,

(ii) $u \in M'_{(\varrho)}$, $\varrho_1 + \cdots + \varrho_n \le s - \frac{n}{2}$, $\mathcal{M}u \in \mathcal{O}(\{z \in \mathbf{C}^n\colon \sum_{j=1}^n \operatorname{Re} z_j < s - \frac{n}{2}\})$

and for every z with $\sum_{j=1}^n \operatorname{Re} z_j \le s - \frac{n}{2}$

$$\mathcal{M}u(\operatorname{Re} z + i\beta) \in L^2(\mathbf{R}^n, \langle\beta\rangle^{2(s+s')}),$$

(iii) $u \in M'_{(\varrho)}$, $\varrho_1 + \cdots + \varrho_n \leq s - \frac{n}{2}$, $\mathcal{M}u \in \mathcal{O}(\{z \in \mathbf{C}^n\colon \sum_{j=1}^{n} \operatorname{Re} z_j < s - \frac{n}{2}\})$ *and for some* $\mathring{\alpha}$ *such that* $\sum_{j=1}^{n} \mathring{\alpha}_j = s - \frac{n}{2}$

$$\mathcal{M}u(\mathring{\alpha} + i\,\cdot) \in L^2(\mathbf{R}^n, \langle\beta\rangle^{2(s+s')}).$$

The spaces $\mathrm{SP}(s, s')$ turned out to be very useful in a whole range of non-linear and singular partial differential equations (see [B]). Inspired by the characterization of the space $\mathrm{SP}(s, s')$ in terms of the Mellin transformation given by Proposition 2 we introduce below the space $M(\Omega; \varrho)$ of *distributions with continuous radial asymptotics*. Instead of the weighted L^2 norms, as in Proposition 2, we use in the definition of $M(\Omega; \varrho)$ weighted supremum norms and extend the set of holomorphy of the Mellin transforms from the half-space $\{z \in \mathbf{C}\colon \sum_{i=1}^{n} \operatorname{Re} z_j < s - \frac{n}{2}\}$ to a set Ω.

3. Spaces $M(\Omega; \varrho)$ and $Z_d(\Omega; \varrho)$ of Distributions with Continuous Radial Asymptotics

Again as in Section 10.4, the singularities of the Mellin transforms considered in Proposition 2 are related to the variable $z_1 + \ldots + z_n$ and we use the change of variables $\zeta = \mathbf{A}z$, where

$$\begin{aligned} \zeta_1 &= z_1 + \ldots + z_n, \\ \zeta_j &= z_j \qquad \text{for } j = 2, \ldots, n. \end{aligned} \tag{12}$$

Recall that the change of variables $\zeta = \mathbf{A}z$ in the Mellin transform corresponds, according to Exercise 8.6, to the change $x = S(y)$ of the initial variable, where $S\colon \mathbf{R}^n_+ \to \mathbf{R}^n_+$ is given by

$$\begin{aligned} x_1 &= y_1, \\ x_j &= y_1 y_j \qquad \text{for } j = 2, \ldots, n. \end{aligned} \tag{13}$$

More precisely Theorem 8.4 gives in this case

Theorem 1. *Let* $u \in M'_{(\omega)}((0, t])$ *for some* $\omega \in (\mathbf{R} \cup \{\infty\})^n$ *and* $t \in \mathbf{R}^n_+$ *and suppose* u *is supported in a proper cone* $\Gamma \subset \mathbf{R}^n_+$. *Then* $u \circ S \in M'_{(\mathbf{A}\omega)}((0, \tilde{t}])$ *for some* $\tilde{t} \in \mathbf{R}^n_+$ *and*

$$(\mathcal{M}u) \circ \mathbf{A}^{-1}(\zeta) = \mathcal{M}(u \circ S)(\zeta) \qquad \text{for } \operatorname{Re}\zeta < \mathbf{A}\omega,$$

where $u \circ S$ *is defined by (8.19).*

Let

$$\Omega = \mathbf{A}^{-1}\left(\Omega^1 \times \mathbf{C}^{n-1}\right) \tag{14}$$

where Ω^1 is an $\mathbf{R}_-$ *connected* open subset of $\mathbf{C}$, i.e. a subset such that together with any point $\mathring{\zeta}_1 \in \Omega^1$ it contains the half line $\mathring{\zeta}_1 + \mathbf{R}_-$. We also assume that for any $\tau \in \operatorname{Re}\Omega^1 \stackrel{\text{def}}{=} \{a_1 \in \mathbf{R}\colon a_1 + ib_1 \in \Omega^1 \text{ for some } b_1\}$ the set

$$\Lambda_\tau = \{\zeta_1 \in \mathbf{C}\colon \zeta_1 \notin \Omega^1,\ \operatorname{Re}\zeta_1 \le \tau\}$$

is compact in $\mathbf{C}$.

Remark 4. Since Ω^1 is $\mathbf{R}_-$ connected it follows that for every $\nu \in \mathbf{R}^n$ with $\sum_{j=1}^n \nu_j \ge 0$ we have $\Omega \subset \Omega + \nu$.

Next we impose conditions on the function ϱ which measures the growth order of the Mellin transforms along the imaginary planes. By comparing (ii) and (iii) in Proposition 2 we note that ϱ should be a function of only $a_1 = \sum_{j=1}^n \operatorname{Re} z_j$ and should be non-decreasing i.e.

$$\varrho\colon \operatorname{Re}\Omega^1 \to \mathbf{R}$$

and $\varrho(a_1) \ge \varrho(\tilde{a}_1)$ if $a_1 \ge \tilde{a}_1$.

Definition 5. Let Ω and ϱ be as above. We say that a Mellin distribution u with support in a proper cone $\Gamma \subset \mathbf{R}^n_+$ belongs to $M(\Omega;\varrho)$ if the function $H(\zeta) = \mathcal{M}u \circ \mathbf{A}^{-1}(\zeta)$ satisfies the conditions

(i) $H \in \mathcal{O}(\Omega^1 \times \mathbf{C}^{n-1})$;

(ii) for any open neighbourhood W of $\Lambda = \mathbf{C} \setminus \Omega^1$

$$|H(a+ib)| \le C\langle b\rangle^{\varrho(a_1)} \qquad \text{for } a+ib \in (\mathbf{C}\setminus W) \times \mathbf{C}^{n-1},$$

where $C = C(W,a)$ is locally bounded in $a \in \operatorname{Re}(\mathbf{C}\setminus W) \times \mathbf{R}^{n-1}$.

We denote $M(\Omega,-\infty) = \bigcap_{s\in\mathbf{R}} M(\Omega;s)$ where s is the constant function $\varrho(\tau) = s$.

Next we impose more constraints on the set Ω^1 and on the behaviour of the Mellin transforms near the boundary of Ω^1.

Definition 6. Let $s \in \mathbf{R}$ and let $\Omega^1 = \mathbf{C} \setminus \bigcup_{j=1}^{k} L^j$ where $L^j = \theta_j + \overline{\mathbf{R}_+}$ where $\theta_j \in \mathbf{C}$, $j = 1, \ldots, k$.

Denote $\mathbf{B} = \{\operatorname{Im} \theta_j,\ j = 1, \ldots, k\}$ and set $\Omega = \mathbf{A}^{-1}(\Omega^1 \times \mathbf{C}^{n-1}\}$. We write $u \in Z_d(\Omega; s)$ if $u \in M'$ with support in some proper cone $\Gamma \subset \mathbf{R}^n_+$ and the function $H(\zeta) = \mathcal{M}u \circ \mathbf{A}^{-1}(\zeta)$ satisfies the conditions

(i) $H \in \mathcal{O}(\Omega^1 \times \mathbf{C}^{n-1})$;

(ii) for any open neighbourhood W of $\bigcup_{j=1}^{k} L^j$

$$|H(a + ib)| \le C\langle b\rangle^s \qquad \text{for } a + ib \in (\mathbf{C} \setminus W) \times \mathbf{C}^{n-1}$$

holds locally uniformly with respect to $a \in \mathbf{R}^n$;

(iii) $|H(a + ib)| \le \widetilde{C} \dfrac{\langle b'\rangle^s}{(\operatorname{dist}(b_1, \mathbf{B}))^m}$ for b_1 close to $\mathbf{B}$ and $\zeta' = a' + ib' \in \mathbf{C}^{n-1}$, with some constants $0 < \widetilde{C} = \widetilde{C}(a)$, $m = m(a) \in \mathbf{R}$ locally bounded for $a \in \mathbf{R}^n$.

We also denote $Z_d(\Omega; -\infty) = \bigcap_{s \in \mathbf{R}} Z_d(\Omega; s)$.

Let H be a function satisfying (i)–(iii) in Definition 6. For a fixed $\zeta' \in \mathbf{C}^{n-1}$ and every $j = 1, \ldots, k$ define

$$\Xi^j_{\zeta'} = \lim_{\substack{b_1 \to \operatorname{Im} \theta_j \\ b_1 > \operatorname{Im} \theta_j}} H(\,\cdot\, + ib_1, \zeta') - \lim_{\substack{b_1 \to \operatorname{Im} \theta_j \\ b_1 < \operatorname{Im} \theta_j}} H(\,\cdot\, + ib_1, \zeta'),$$

where the limits are taken in the sense of distributional convergence in $D'(\theta_j + \mathbf{R})$ as in Theorem 3.3. By that theorem and (iii) the limits clearly exist and define a holomorphic function

$$\mathbf{C}^{n-1} \ni \zeta' \mapsto \Xi^j_{\zeta'} \in D'(\theta_j + \mathbf{R}),$$

which satisfies the estimates

$$\left|\Xi^j_{a'+ib'}[\varphi]\right| \le C \sum_{|\alpha| \le m+1} \sup |D^\alpha \varphi|\, \langle b'\rangle^s \qquad \text{for } \varphi \in C_0^\infty(\theta_j + \mathbf{R})$$

locally uniformly in $a' \in \mathbf{R}^{n-1}$.

Remark 5. The spaces $M(\Omega;\varrho)$ and $Z_d(\Omega;s)$ are called the spaces of *distributions with "continuous radial asymptotics"* since the set Ω controls their asymptotic expansion at zero. Namely, the set $\mathbf{C}\setminus\Omega^1$ may be regarded as the set of those exponents which enter into an asymptotic expansion at zero with respect to the radial variable (see Appendix). In applications to Fuchsian PDE-s we choose Ω^1 as big as possible; usually $\mathbf{C}\setminus\Omega^1$ consists of a finite number of closed half lines parallel to the real axis as in Definition 6.

Definition 7. Let $\delta\mathring{x}\in\mathbf{R}^n_+$ and suppose $u\in\mathfrak{M}'_\omega$ for some $\omega\in\mathbf{R}^n$. We say that $u\in M(\Omega;\varrho)$ ($Z_d(\Omega;s)$ resp.) 2-locally at $(0;\delta\mathring{x})$ if there exists a conical cut-off function κ at $(0;\delta\mathring{x})$, such that $\kappa u\in M(\Omega;\varrho)$ ($Z_d(\Omega;s)$ resp.).

In Subsection 12.2 we shall prove that, similarly to the case of $M'^s_{(\omega)}$ (see Subsection 1), the spaces $M(\Omega;\varrho)$ and $Z_d(\Omega;s)$ are 2-local.

Exercises

1. Prove that $M'_{(\omega)}=\bigcap_{\varepsilon>0}M'^{\infty}_{(\omega-\varepsilon)}$. This will be used in Remark 13.4.

2. Let $\omega\in(\mathbf{R}\cup+\infty)^n$ and let F be a holomorphic function on the set $\operatorname{Re}z<\omega$ such that for every $\alpha<\omega$ there exists $\widetilde{C}(\alpha)<\infty$, $s\in\mathbf{N}_0$ such that $|F(\alpha+i\beta)|<\widetilde{C}(\alpha)\langle\beta\rangle^s$ for $\beta\in\mathbf{R}^n$. Show that then there exists a unique $u\in M'_{(\omega)}((0,t])$ such that $F(z)=\mathcal{M}u(z)$ for $\operatorname{Re}z<\omega$; u is a distribution of finite Mellin order $\le n(s+2)$. (cf. Theorem 9.3, Remark 9.1 and Remark 1).

3. Let $s\in\mathbf{R}$. Show that there exists $C<\infty$ such that

$$\int_{\mathbf{R}^n}\frac{\langle\beta-\gamma\rangle^s}{(\langle\gamma_1\rangle\cdot\ldots\cdot\langle\gamma_n\rangle)^{|s|+2}}\,d\gamma\le C\langle\beta\rangle^s.$$

4. Let $u\in M'^{\infty}_{(\omega)}$, $\delta\mathring{x}\in\mathbf{R}^n_+$. Show that the following statements are equivalent:

(i) $u\in M'^s_{(\omega)}$ 2-locally at $(0,\delta\mathring{x})$;

(ii) there exists a conical cut-off function $\kappa=\tilde\kappa\varphi$ as in Remark 10.1 with $\tilde\kappa(\delta\mathring{x})\ne0$ and $\varphi(0)\ne0$, such that $\kappa u\in M'^s_{(\omega)}$;

(iii) for every conical cut-off function κ, $\kappa u\in M'^s_{(\omega)}$.

5. Show that Definition 3 is equivalent to the following one: we say that $u\in M'^{\infty}_{(\omega)}$ belongs to $M'^s_{(\omega)}$ Ξ_K-locally if for every cut-off function κ^K_r, $\kappa^K_r\equiv1$ on Ξ_K close to zero, $\operatorname{supp}\kappa^K_r\subset\Xi_{K_1}$ for some K_1, $K\subset\subset\operatorname{Int}K_1\subset\subset\mathbf{R}^{n-1}_+$ for r small enough $\kappa^K_r u\in M'^s_{(\omega)}$.

6. Suppose $u \in M'^{s}_{(\omega)}$ 2-locally at δx for $\delta x \in \Gamma$, $\omega \in \mathbf{R}^n$ and let κ_0 be such that $\kappa_0 u \in M'^{s}_{(\omega)}$ (see Definition 2). Let $\tilde{u} = \kappa\kappa_0 u$ with an arbitrary conical cut-off function κ (see Subsection 10.1). Fix $\mathring{\alpha} \in \mathbf{R}^n$ with $\sum \mathring{\alpha}_j < \sum \omega_j$. Then for $\delta \in \mathbf{N}_0^n$ there exists a constant $C = C(\delta, \mathring{\alpha})$ such that (6) holds for r small.

7. Suppose $u \in D'_{\bar{B}_+}(\mathbf{R}^n_+)$. Show that

(a) if $u \in L^2(\mathbf{R}^n_+, x^{-1})$ then $u \in M'_{(0)}$ and

$$\lim_{\alpha \to 0_-} \mathcal{M}u(\alpha + i\,\cdot) = (2\pi)^{\frac{n}{2}} \mathcal{F}^{-1} h \qquad \text{in } L^2(\mathbf{R}^n_+)$$

where $h(y) = u(e^{-y})$;

(b) if $u \in M'_{(0)}$ and $\lim_{\alpha \to 0_-} \mathcal{M}u(\alpha + i\,\cdot)$ exists in $L^2(\mathbf{R}^n_+)$ then $h \in L^2(\mathbf{R}^n_+)$ (apply lemma Fatou).

§12. THE MODIFIED CAUCHY TRANSFORMATION

This section presents fundamental tools for the study of local properties of the Mellin transforms. Namely if κ is a cut-off function (of the type considered in Subsection 7.3 and Section 10) then for a Mellin distribution u computing the Mellin transform $\mathcal{M}\kappa u$ amounts (cf. Proposition 7.3, Remark 7.1) to finding a convolution of $\mathcal{M}\kappa$ and $\mathcal{M}u$. Since $\mathcal{M}\kappa$ bears a resemblance to the Cauchy kernel, $\mathcal{M}\kappa u$ is in fact a "modified" Cauchy transform of $\mathcal{M}u$. Below we present details in the case of one variable and of one variable with parameters. The reader interested in the case of several variables and its applications is invited to consult [Zie3]. The exposition of the classical Cauchy transformation is found in [St], [Rei]; see also [K].

1. Modified Cauchy and Hilbert Transformations in Dimension 1

We start with certain well-known facts about the classical Cauchy and Hilbert transformations in $L^2(\mathbf{R})$ in a slightly changed setting; namely the Cauchy transformation is not considered relative to the real axis $\mathbf{R}$ but relative to a fixed pure imaginary line $\mathring{\alpha} + i\mathbf{R}$ for some $\mathring{\alpha} \in \mathbf{R}$.

Let $T \in L^2(\mathbf{R})$ and fix $\mathring{\alpha} \in \mathbf{R}$. The *right* and the *left Cauchy transforms* of T (relative to the line $\mathring{\alpha} + i\mathbf{R}$) are defined as

$$C^+T(z) = -\frac{1}{2\pi} \int_{\mathbf{R}} \frac{T(\gamma)}{z - \mathring{\alpha} - i\gamma}\, d\gamma \qquad \text{for } \operatorname{Re} z > \mathring{\alpha},$$

$$C^-T(z) = -\frac{1}{2\pi} \int_{\mathbf{R}} \frac{T(\gamma)}{z - \mathring{\alpha} - i\gamma}\, d\gamma \qquad \text{for } \operatorname{Re} z < \mathring{\alpha}.$$

The following is a classical fact in the theory of the Hilbert transformation (see Exercise 2):

Theorem 1. *If* $T \in L^2(\mathbf{R})$ *then* $C^{\pm}T(\alpha + i\,\cdot) \in L^2(\mathbf{R})$ *and the following limits exist in* $L^2(\mathbf{R})$:

$$H^{\pm}T(\beta) = \lim_{\alpha \to \mathring{\alpha}\pm} C^{\pm}T(\alpha + i\beta)$$

and

$$H^{-}T - H^{+}T = T.$$

The functions $H^{+}T$, $H^{-}T \in L^2(\mathbf{R})$ are called the *right* and the *left Hilbert transforms* of T.

We want to establish analogues of these facts in S' instead of L^2. To this end we introduce modified Cauchy kernels which replace the *classical Cauchy kernel* $-\frac{1}{z}$. To justify the definition below recall that by Example 7.1

$$-\frac{1}{z} = (\mathcal{M}\chi_{(0,1]})(z) \qquad \text{for } z \neq 0,$$

where $\chi_{(0,1]}$ is the characteristic function of the interval $(0,1]$.

Further, recall that in Proposition 7.4 we considered the Mellin transforms of functions χ which are restrictions to $\mathbf{R}_{+}$ of functions $\tilde{\chi} \in C_0^{\infty}(\mathbf{R})$, $\tilde{\chi} \equiv 1$ in a neighbourhood of zero. We established there that $\mathcal{M}\chi$ differed from $-\frac{1}{z}$ by an entire function and $z\mathcal{M}\chi(z)$ was of rapid decrease along the imaginary axis. In Remark 7.3 we labelled $G = \mathcal{M}\chi$ — a *modified Cauchy kernel* (determined by χ).

Definition 1. Fix $\mathring{\alpha} \in \mathbf{R}$ and let $T \in S'(\mathbf{R})$. The (holomorphic) functions

$$C^{\pm}\,T(z) = \frac{1}{2\pi} T\big[G(z - \mathring{\alpha} - i\gamma)\big] \qquad \text{for} \quad \pm \operatorname{Re} z > \pm\mathring{\alpha} \tag{1}$$

are called the *right* and the *left modified Cauchy transforms* of T. In the sequel we often omit the word *modified* since the function G and the point $\mathring{\alpha}$ will remain fixed.

Theorem 2. *Let* $T \in S'(\mathbf{R})$ *and fix* $\mathring{\alpha} \in \mathbf{R}$. *Then in the sense of the convergence in* $S'(\mathbf{R})$ *the following limits exist*

$$\mathcal{H}^{\pm}T = \lim_{\alpha \to \mathring{\alpha}\pm} C^{\pm}T(\alpha + i\,\cdot)$$

and

$$\mathcal{H}^{-}T - \mathcal{H}^{+}T = T. \tag{2}$$

We call the distributions $\mathcal{H}^{\pm}T \in S'(\mathbf{R})$ the *right* and the *left (modified) Hilbert transforms* of $T \in S'(\mathbf{R})$.

Proof: By translation of the variable z we may assume that $\mathring{\alpha} = 0$. In view of Remark 7.2 (with $\kappa = \chi \circ \mu$) we have

$$G(\alpha + i\beta) = \begin{cases} \sqrt{2\pi}\mathcal{F}^{-1}\big(e^{\alpha y}\kappa(y)\big)(\beta) & \text{for } \alpha < 0 \\ \sqrt{2\pi}\mathcal{F}^{-1}\big(e^{\alpha y}(\kappa(y) - 1)\big)(\beta) & \text{for } \alpha > 0. \end{cases}$$

Now, by the formula for the Fourier transform of a convolution

$$\begin{aligned} \mathcal{C}^{-}T(\alpha + i\,\cdot\,) &= \mathcal{F}^{-1}\big(e^{\alpha y}\kappa(y)\,\mathcal{F}T\big) && \text{for } \alpha < 0 \\ \mathcal{C}^{+}T(\alpha + i\,\cdot\,) &= \mathcal{F}^{-1}\big(e^{\alpha y}(\kappa(y) - 1)\mathcal{F}T\big) && \text{for } \alpha > 0. \end{aligned}$$

Since $\mathcal{F}$ and $\mathcal{F}^{-1}$ are topological isomorphisms of S onto S it is enough, in order to prove the first part of the theorem, to show that for every $\sigma \in S(\mathbf{R})$

$$\begin{aligned} e^{\alpha y}\kappa(y)\sigma(y) &\to \kappa(y)\sigma(y) && \text{in } S(\mathbf{R}) \text{ as } \alpha \to 0_-, \\ e^{\alpha y}\big(\kappa(y) - 1\big)\sigma(y) &\to \big(\kappa(y) - 1\big)\sigma(y) && \text{in } S(\mathbf{R}) \text{ as } \alpha \to 0_+. \end{aligned}$$

But this is simple since $\operatorname{supp}\kappa$ is contained in a half line $y \geq -a$, and $\operatorname{supp}(\kappa - 1)$ in $y \leq a$ for some $a \in \mathbf{R}_+$, κ is bounded together with derivatives and $e^{\alpha y} \to 1$ as $\alpha \to 0$ uniformly on bounded sets.

From the above we get

$$\begin{aligned} \mathcal{H}^{-}T &= \mathcal{F}^{-1}(\kappa \cdot \mathcal{F}T) \\ \mathcal{H}^{+}T &= \mathcal{F}^{-1}\big((\kappa - 1) \cdot \mathcal{F}T\big) \end{aligned}$$

which gives the relation (2).

We end this subsection by an application of Theorem 2 to the study of holomorphic extensions of the left Cauchy transform, which is the leading motif of this section.

Let F be holomorphic in an open set $U \subset \mathbf{C}$. Let $\mathring{\alpha} \in \mathbf{R}$ and suppose that the function

$$\beta \mapsto F(\mathring{\alpha} + i\beta),$$

defined for $\beta \in \mathbf{R}$ such that $\mathring{\alpha} + i\beta \in U$, extends to a distribution in $S'(\mathbf{R})$ which we denote by $F_{\mathring{\alpha}}$.

Corollary 1. *Under the assumptions on F given above the function*

$$\psi(z) = \begin{cases} \mathcal{C}^- F_{\mathring{\alpha}}(z) & \text{for } \operatorname{Re} z < \mathring{\alpha} \\ \mathcal{C}^+ F_{\mathring{\alpha}}(z) + F(z) & \text{for } z \in \{\operatorname{Re} z > \mathring{\alpha}\} \cap U \end{cases}$$

extends to a holomorphic function on $\{\operatorname{Re} z < \mathring{\alpha}\} \cup U$.

Proof: It is easy to see that $\mathcal{C}^\pm F_{\mathring{\alpha}}(\alpha + i\beta)$ satisfies (3.2) in α. The rest follows immediately from (2) in view of the distributional Painlevé Theorem 3.3.

2. The Case with Parameters

We return now to the setting of Subsection 10.4. With the notation introduced before the statement of Proposition 10.7 we have

Theorem 3. *Let $T \in S'(\mathbf{R}^n)$ and fix $\mathring{a} \in \mathbf{R}^n$. Fix a cut-off function κ of the form (10.28) and denote*

$$K(\zeta) = (\mathcal{M}\kappa)(\mathbf{A}^{-1}\zeta) \qquad \text{for } \zeta \in (\mathbf{C} \setminus \{0\}) \times \mathbf{C}^{n-1}, \tag{3}$$

$$K'(\zeta') = \mathcal{M}'k'(\zeta') \qquad \text{for } \zeta' \in \mathbf{C}^{n-1}. \tag{4}$$

Denote

$$\widetilde{\mathcal{C}}^\pm(\zeta) = \widetilde{\mathcal{C}}^\pm T(\zeta) = \frac{1}{(2\pi)^n} T\,[K(\zeta - \mathring{a} - i\gamma)] \qquad \text{for } \pm \operatorname{Re} \zeta_1 > \pm \mathring{a}_1,\ \zeta' \in \mathbf{C}^{n-1} \tag{5}$$

and

$$\widetilde{\mathcal{C}}'_{\mathring{a}_1}(\zeta') = (\widetilde{\mathcal{C}}'T)_{\mathring{a}_1}(\zeta') = \frac{1}{(2\pi)^{n-1}} T\,[K'(\zeta' - \mathring{a}' - i\gamma')] \in S'(\mathbf{R}) \qquad \text{for } \zeta' \in \mathbf{C}^{n-1} \tag{6}$$

(in (6) T is regarded as an element of $S'(\mathbf{R}^{n-1}; S'(\mathbf{R}))$ under a canonical isomorphism $S'(\mathbf{R}^n) \simeq S'(\mathbf{R}^{n-1}; S'(\mathbf{R}))$ given by Theorem 4.3).

Then

$$\widetilde{\mathcal{C}}^\pm T \in \mathcal{O}\left(\{\pm \operatorname{Re} \zeta_1 > \pm \mathring{a}_1\} \times \mathbf{C}^{n-1}\right),$$

$$(\widetilde{\mathcal{C}}'T)_{\mathring{a}_1} \in \mathcal{O}\left(\mathbf{C}^{n-1};\ S'(\mathbf{R})\right)$$

and in the sense of convergence in $S'(\mathbf{R}^n)$

$$\lim_{\substack{a \to \mathring{a} \\ a_1 < \mathring{a}_1}} \widetilde{\mathcal{C}}^- T(a + i\cdot) - \lim_{\substack{a \to \mathring{a} \\ a_1 > \mathring{a}_1}} \widetilde{\mathcal{C}}^+ T(a + i\cdot) = (\widetilde{\mathcal{C}}'T)_{\mathring{a}_1}(\mathring{a}' + i\cdot) \tag{7}$$

(here $(\widetilde{\mathcal{C}}'T)_{\mathring{a}_1}(\mathring{a}' + i\cdot) \in S'(\mathbf{R}^{n-1};\ S'(\mathbf{R}))$ is regarded as an element of $S'(\mathbf{R}^n)$).

The proof given below can be regarded as a parameter version of the proof of Theorem 2.

Proof: By translation of the variable ζ we may assume $\mathring{a} = 0$. In view of formula (10.33) we have

$$K(a+ib) = \begin{cases} (2\pi)^{n/2}\mathcal{F}^{-1}\big(e^{as}\omega(s)\big)(b) & \text{for } a_1 < 0 \\ (2\pi)^{n/2}\mathcal{F}^{-1}\big(e^{as}(\omega(s)-\omega'(s'))\big)(b) & \text{for } a_1 > 0. \end{cases} \tag{8}$$

Now by the formula for the Fourier transform of a convolution

$$\widetilde{\mathcal{C}}^{-}T(a+i\cdot) = \mathcal{F}^{-1}\big(e^{as}\omega(s)\mathcal{F}T\big) \qquad \text{for } a_1 < 0,$$
$$\widetilde{\mathcal{C}}^{+}T(a+i\cdot) = \mathcal{F}^{-1}\big(e^{as}(\omega(s)-\omega'(s'))\mathcal{F}T\big) \qquad \text{for } a_1 > 0.$$

First we prove that the limits in (7) exist. To this end it is enough to show, since $\mathcal{F}$ and $\mathcal{F}^{-1}$ are topological isomorphism of S onto S, that for every $\psi \in S(\mathbf{R}^n)$

(9)

$$e^{as}\omega(s)\psi(s) \to \omega(s)\psi(s) \qquad \text{in } S(\mathbf{R}^n) \text{ as } a \to 0 \text{ with } a_1 < 0,$$

$$e^{as}(\omega(s)-\omega'(s'))\psi(s) \to (\omega(s)-\omega'(s'))\psi(s) \quad \text{in } S(\mathbf{R}^n) \text{ as } a \to 0 \text{ with } a_1 > 0.$$

But this is simple in view of the properties of the support of $\omega(s)$ and $\omega(s) - \omega'(s')$ and the fact that all derivatives of ω and ω' are bounded on $\mathbf{R}^n$ and $\mathbf{R}^{n-1}$, respectively.

From (9) we get

$$\lim_{\substack{a'\to 0\\ a_1\to 0_-}} \widetilde{\mathcal{C}}^{-}T(a+i\cdot) - \lim_{\substack{a'\to 0\\ a_1\to 0_+}} \widetilde{\mathcal{C}}^{+}T(a+i\cdot) = \mathcal{F}^{-1}\big(\omega'(s')\mathcal{F}T\big)$$

where δ_0 is the Dirac delta at zero in the variable s_1 and $\mathcal{F}'^{-1}$ denotes the inverse Fourier transformation in variables $s' = (s_2, \ldots, s_n)$. Now from (10.32) we see that

$$K'(ib') = (2\pi)^{\frac{n-1}{2}}\mathcal{F}'^{-1}\omega'(b') \tag{10}$$

and therefore

$$T * \big(\delta_0 \otimes \mathcal{F}'^{-1}\omega'\big) = \big(\widetilde{\mathcal{C}}'T\big)_0(i\cdot)$$

which proves (7). The proof of the holomorphicity of $\widetilde{\mathcal{C}}^{\pm}T$ and $(\widetilde{\mathcal{C}}'T)_{\mathring{a}_1}$ is left to the reader.

Corollary 2. *Let H be a function holomorphic on an open set $U \subset \mathbf{C}^n$. Fix $\mathring{a} \in \mathbf{R}^n$ and suppose that the function $b \mapsto H(\mathring{a} + ib)$, defined for $b \in \mathbf{R}^n$ such that $\mathring{a} + ib \in U$, extends to a distribution in $S'(\mathbf{R}^n)$ which we denote by $H_{\mathring{a}}$. Further suppose that there exists an open set $U^1 \subset \mathbf{C}$ such that for every $\zeta_1 \in U^1$ the function $b' \mapsto H_{\zeta_1}(\mathring{a}' + ib')$, defined for $b' \in \mathbf{R}^{n-1}$ such that $(\zeta_1, \mathring{a}' + ib') \in U$, extends to a distribution $H_{\zeta_1,\mathring{a}'}$ in $S'(\mathbf{R}^{n-1})$ and the distribution valued function*

$$U^1 \ni \zeta_1 \mapsto H_{\zeta_1,\mathring{a}'} \in S'(\mathbf{R}^{n-1}) \tag{11}$$

is holomorphic on U^1. Finally assume that there exists a regularization $\widetilde{H}_{\mathring{a}_1,\mathring{a}'} \in S'(\mathbf{R}; S'(\mathbf{R}^{n-1}))$ of the function $b_1 \mapsto H_{\mathring{a}_1+ib_1,\mathring{a}'} \in S'(\mathbf{R}^{n-1})$, defined for $b_1 \in \mathbf{R}$ with $\mathring{a}_1 + ib_1 \in U^1$, such that $\widetilde{H}_{\mathring{a}_1,\mathring{a}'} = H_{\mathring{a}}$ under the canonical isomorphism $S'(\mathbf{R}, S'(\mathbf{R}^{n-1})) \simeq S'(\mathbf{R}^n)$ of Theorem 4.3.

Then the function

$$\widetilde{\mathcal{C}}'_{\zeta_1}(\zeta') = \frac{1}{(2\pi)^{n-1}} H_{\zeta_1,\mathring{a}'}\left[K'(\zeta' - \mathring{a}' - i\gamma')\right], \quad (\zeta_1, \zeta') \in U^1 \times \mathbf{C}^{n-1} \tag{12}$$

is holomorphic on $U^1 \times \mathbf{C}^{n-1}$, and for every fixed $\zeta' \in \mathbf{C}^{n-1}$ the distribution $\widetilde{\mathcal{C}}'_{\mathring{a}_1}(\zeta') \in S'(\mathbf{R})$ is a regularization of the function

$$b_1 \mapsto \widetilde{\mathcal{C}}'_{\mathring{a}_1+ib_1}(\zeta')$$

defined for $b_1 \in \mathbf{R}$ such that $\mathring{a}_1 + ib_1 \in U^1$. Moreover the function

$$\tilde{\psi}(\zeta) = \begin{cases} \widetilde{\mathcal{C}}^-(\zeta) & \text{for } \operatorname{Re}\zeta_1 < \mathring{a}_1, \ \zeta' \in \mathbf{C}^{n-1} \\ \widetilde{\mathcal{C}}^+(\zeta) + \widetilde{\mathcal{C}}'_{\zeta_1}(\zeta') & \text{for } \operatorname{Re}\zeta_1 > \mathring{a}_1, \ \zeta_1 \in U^1, \ \zeta' \in \mathbf{C}^{n-1} \end{cases} \tag{13}$$

extends to a holomorphic function ψ on $\left(\{\operatorname{Re}\zeta_1 < \mathring{a}_1\} \cup U^1\right) \times \mathbf{C}^{n-1}$ (here $\widetilde{\mathcal{C}}^\pm(\zeta) = (\widetilde{\mathcal{C}}^\pm H_{\mathring{a}})(\zeta)$ as in Theorem 3).

Proof: Since the function (11) is holomorphic on U^1 it is immediate that $\widetilde{\mathcal{C}}'_{\zeta_1}(\zeta')$ is holomorphic on $U^1 \times \mathbf{C}^{n-1}$. The second assertion is also clear because $\widetilde{H}_{\mathring{a}_1,\mathring{a}'} = H_{\mathring{a}}$. Finally the extendibility of the function $\tilde{\psi}$ follows from (7) and a version of the Painlevé Theorem 3.3 with a holomorphic parameter.

Lemma 1. *Fix $\mathring{a} \in \mathbf{R}^n$. Let $T \in S'(\mathbf{R}^n)$ and for large $\|b\|$ let T be a function $T(b)$ such that for some $C > 0$ and $m \in \mathbf{R}$*

$$|T(b)| \le C\langle b\rangle^m \quad \text{for large } \|b\|. \tag{14}$$

Take $\mathring{a} \in \mathbf{R}^n$ and $r \in \mathbf{R}_+$ such that $0 < r < \mathring{a}_1 - \mathring{a}_1$. Then there exists $\widetilde{C} = \widetilde{C}(r) < \infty$ such that

$$\left|\widetilde{\mathcal{C}}^+ T(a + ib)\right| \le \widetilde{C}\langle b\rangle^m \qquad \text{for } |a_1 - \mathring{a}_1| \le r, \ \|a' - \mathring{a}'\| \le r.$$

Proof: Let φ be a $C_0^\infty(\mathbf{R}^n)$-function such that $\widetilde{T} = (1-\varphi)T$ is a function and satisfies (14) for $b \in \mathbf{R}^n$. Thus

$$\widetilde{\mathcal{C}}^+T(\zeta) = I_1(\zeta) + I_2(\zeta),$$

where

$$I_1(\zeta) = \frac{1}{(2\pi)^n}\int_{\mathbf{R}^n} \widetilde{T}(\gamma)K(\zeta - \mathring{a} - i\gamma)\,d\gamma,$$
$$I_2(\zeta) = \frac{1}{(2\pi)^n}\varphi T[K(\zeta - \mathring{a} - i\gamma)].$$

Since the function $b \mapsto K(a+ib)$ is in $S(\mathbf{R}^n)$ locally uniformly with respect to a for $a_1 > 0$, it follows from Proposition 10.7(ii) that for every $h > 0$ there exists a constant $C_1 = C_1(r,h)$ such that

$$|K(a - \mathring{a} + ib)| \le \frac{C_1}{\langle b\rangle^h} \qquad \text{for } |a_1 - \mathring{a}_1| \le r,\ \ \|a' - \mathring{a}'\| \le r.$$

From this for $h = |m| + n + 1$ and from (14) we get

$$|I_1(a+ib)| \le C_2 \int_{\mathbf{R}^n} \frac{\langle\gamma\rangle^m}{\langle b-\gamma\rangle^{|m|+n+1}}\,d\gamma \le C_2\int_{\mathbf{R}^n}\frac{d\gamma}{\langle\gamma\rangle^{n+1}}\langle b\rangle^m = C_3\langle b\rangle^m$$

in view of the Petree inequality (11.5).

Concerning $I_2(a+ib)$ observe that it is uniformly in S for $|a_1 - \mathring{a}_1| \le r$, $\|a' - \mathring{a}'\| \le r$. Indeed, under the Fourier transformation (with respect to b) this amounts in view of (8) to the fact that the product of the function $\mathcal{F}(\varphi T) \in O_M$ by the function

$$s \mapsto e^{(a-\mathring{a})s}(\omega(s) - \omega'(s'))$$

is in S uniformly for a varying as above.

Lemma 2. *Fix $\mathring{a} \in \mathbf{R}^n$. Let $\mathring{a} \in \mathbf{R}^n$ and suppose that*

$$\zeta_1 \mapsto T_{\zeta_1} \in S'(\mathbf{R}^{n-1})$$

is a family of tempered distributions holomorphic in an open neighbourhood of the set

(15) $$\mathcal{T}_r^1(\mathring{a}_1) = \{\zeta_1 \in \mathbf{C}\colon |\mathrm{Re}\,\zeta_1 - \mathring{a}_1| \le r,\ \mathrm{Im}\,\zeta_1 \le \tilde{b}_1 \text{ or } \mathrm{Im}\,\zeta_1 \ge \tilde{b}_2\}$$

for some $\tilde{b}_1, \tilde{b}_2 \in \mathbf{R}$, $\tilde{b}_1 \le \tilde{b}_2$ *and* $r > 0$. *Further suppose that*

i) *There exist* $C < \infty$ *and* $k \in \mathbf{N}_0$ *independent of* $\zeta_1 \in \mathcal{T}_r^1(\mathring{a}_1)$ *such that for every* $\varphi' \in C_0^\infty(\mathbf{R}^{n-1})$

$$(16) \qquad |T_{a_1+ib_1}[\varphi']| \le C\langle b_1\rangle^{\varrho(a_1)} |||\varphi'|||_k$$

where ϱ *is an non-decreasing bounded function of* a_1 *and*

$$|||\varphi'|||_k = \sum_{|\nu|\le k} \sup \left|\left(\frac{\partial}{\partial b'}\right)^\nu \varphi'\right|.$$

ii) *For large* $\|b'\|$, T_{ζ_1} *is a function* $T_{\zeta_1}(b')$ *such that for some constant* C *independent of* $\zeta_1 \in \mathcal{T}_r^1(\mathring{a}_1)$ *and* b'

$$|T_{a_1+ib_1}(b')| \le C\langle b\rangle^{\varrho(a_1)} \qquad \text{for } a_1 + ib_1 \in \mathcal{T}_r^1(\mathring{a}_1),\ \|b'\| \text{ large enough.}$$

Then for every $\tilde{r} > 0$ *there exists* $\widetilde{C} = \widetilde{C}(r, \tilde{r}) < \infty$ *such that*

$$\left|\widetilde{\mathcal{C}}'_{a_1+ib_1}(a' + ib')\right| \le \widetilde{C}\langle b\rangle^{\varrho(a_1)}$$

for $a_1 + ib_1 \in \mathcal{T}_r^1(\mathring{a}_1)$, $\|a' - \mathring{a}'\| < \tilde{r}$, $b' \in \mathbf{R}^{n-1}$, *where (see (12))*

$$\widetilde{\mathcal{C}}'_{\zeta_1}(\zeta') = \frac{1}{(2\pi)^{n-1}} T_{\zeta_1}\left[K'(\zeta' - \mathring{a}' - i\gamma')\right].$$

Proof: As in the proof of Lemma 1 we choose $\varphi' \in C_0^\infty(\mathbf{R}^{n-1})$ such that

$$\widetilde{T}_{\zeta_1} = (1 - \varphi')T_{\zeta_1}$$

is a function and satisfies

$$\left|\widetilde{T}_{a_1+ib_1}(b')\right| \le C\langle b\rangle^{\varrho(a_1)} \qquad \text{for } a_1 + ib_1 \in \mathcal{T}_r^1(\mathring{a}_1).$$

Write

$$\widetilde{\mathcal{C}}'_{\zeta_1}(\zeta') = \widetilde{I}_1(\zeta) + \widetilde{I}_2(\zeta) \qquad \text{where} \qquad \begin{aligned} \widetilde{I}_1(\zeta) &= \frac{1}{(2\pi)^{n-1}} \widetilde{T}_{\zeta_1}\left[K'(\zeta' - \mathring{a}' - i\gamma')\right], \\ \widetilde{I}_2(\zeta) &= \frac{1}{(2\pi)^{n-1}} \varphi' T_{\zeta_1}\left[K'(\zeta' - \mathring{a}' - i\gamma')\right]. \end{aligned}$$

Then in the same way as in the proof of Lemma 1 using the estimation

$$\frac{(|b_1| + \langle\gamma'\rangle)^m}{\langle\gamma' - b'\rangle^{|m|}(|b_1| + \langle b'\rangle)^m} \le 1 \qquad \text{for } m \in \mathbf{R},$$

we get for any $\tilde{r} > 0$

$$(17)\qquad \left|\widetilde{I}_1(a+ib)\right| \le C_1\langle b\rangle^{\varrho(a_1)} \qquad \text{for } a_1+ib_1 \in \mathcal{T}^1_r(\mathring{a}_1),\ \|a'-\mathring{a}'\| < \tilde{r}.$$

To deal with $\widetilde{I}_2$, denote $\tilde{\tilde{T}}_{\zeta_1} = \varphi' T_{\zeta_1}$ and observe that in view of (16) we have

$$(18)\qquad \left|\mathcal{F}'\tilde{\tilde{T}}_{a_1+ib_1}(s')\right| = (2\pi)^{-\frac{n-1}{2}}\left|T_{a_1+ib_1}\left[\varphi' e^{-ib's'}\right]\right| \le C_1\langle b_1\rangle^{\varrho(a_1)}\langle s'\rangle^k.$$

In view of (10)

$$(19)\qquad \mathcal{F}'\left(\widetilde{I}_2(\zeta_1, a'+i\cdot)\right)(s') = \mathcal{F}'\tilde{\tilde{T}}_{\zeta_1}(s')\cdot e^{(a'-\mathring{a}')s'}\omega'(s'),$$

where $\omega'(s') = k'(e^{-s'})$, with k' defined before the statement of Proposition 10.7. From Proposition 10.7 we know that $k' \in C_0^\infty(\mathbf{R}^{n-1}_+)$ hence $\omega' \in C_0^\infty(\mathbf{R}^{n-1})$. Thus from (18) and (19) we easily find that for every $p_1, p_2 \in \mathbf{N}_0$

$$\left|\Delta^{p_1}_{n-1}\mathcal{F}'\left(\widetilde{I}_2(a_1+ib_1,\ a'+i\cdot)\right)(s')\right| \le C_2\frac{\langle b_1\rangle^{\varrho(a_1)}}{\langle s'\rangle^{p_2}}$$

for $a_1+ib_1 \in \mathcal{T}^1_r(\mathring{a}_1)$, $|a'-\mathring{a}'| < \tilde{r}, s' \in \mathbf{R}^{n-1}$ (Δ_{n-1} the Laplace operator in $\mathbf{R}^{n-1}$). Hence $\mathcal{F}'\widetilde{I}_2 \in \mathcal{S}(\mathbf{R}^{n-1})$. Since $\mathcal{F}'(\Delta_{n-1}\sigma)(b') = -\|b'\|^2\mathcal{F}\sigma(b')$ for $\sigma \in \mathcal{S}(\mathbf{R}^{n-1})$, we get for every $p_1 \in \mathbf{N}_0$

$$(20)\qquad \left|\widetilde{I}_2(a+ib)\right| \le C_3\frac{\langle b_1\rangle^{\varrho(a_1)}}{\langle b'\rangle^{2p_1}}$$

for $a_1+ib_1 \in \mathcal{T}^1_r(\mathring{a}_1)$, $|a'-\mathring{a}'| < \tilde{r}$, $b' \in \mathbf{R}^{n-1}$. If $\varrho(a_1) \ge 0$ then we take $p_1 = 0$ and in view of (17) we get the desired estimation for $\widetilde{C}'_{\zeta_1}(\zeta')$. If $\varrho(a_1) < 0$, we take p_1 such that $2p_1 \ge |\varrho(a_1)|$.

Then since

$$\langle b_1\rangle\langle b'\rangle \ge \langle b\rangle$$

we get from (20)

$$\left|\widetilde{I}_2(a+ib)\right| \le C_3\langle b\rangle^{\varrho(a_1)}$$

which combined with (17) ends the proof.

Theorem 4. *If $u \in M(\Omega;\varrho)$ ($Z_d(\Omega;s)$ resp.) then for any conical cut-off function κ, $\kappa u \in M(\Omega;\varrho)$ ($Z_d(\Omega;s)$ resp.). In other words the spaces $M(\Omega;\varrho)$ and $Z_d(\Omega;s)$ are 2-local.*

Proof: We consider the case of $M(\Omega;\varrho)$; that of $Z_d(\Omega;s)$ is left to the reader as an exercise (see Exercise 1). Since $\mathcal{M}u \in \mathcal{O}(\Omega)$, it follows from the properties of the set Ω that for some $\tilde{\alpha} \in \mathbf{R}^n$ the set $V = \{z \in \mathbf{C}^n\colon \sum_{j=1}^n \operatorname{Re} z_j < \sum_{j=1}^n \tilde{\alpha}_j\}$ is contained in Ω. Thus $\mathcal{M}u \in \mathcal{O}(V)$ and from (ii) in Definition 11.5

$$|\mathcal{M}u(\alpha+i\beta)| \le C\langle\beta\rangle^{\varrho(\sum \alpha_j)} \qquad \text{if } \textstyle\sum\alpha_j < \sum\tilde{\alpha}_j. \tag{21}$$

Let $\mathring{\alpha} < \tilde{\alpha}$. From Remark 7.1 we have

$$\mathcal{M}(\kappa u)(z) = \frac{1}{(2\pi)^n}\int_{\mathbf{R}^n} \mathcal{M}u(\mathring{\alpha}+i\xi)\mathcal{M}\kappa(z-\mathring{\alpha}-i\xi)\,d\xi \tag{22}$$

for $\sum_{j=1}^n \operatorname{Re} z_j < \sum_{j=1}^n \mathring{\alpha}_j$, and the integral exists in view of (21) and Proposition 10.7. Thus denoting $\mathring{a} = \mathbf{A}\mathring{\alpha}$ we get under notation (3), (4), (5)

$$\begin{aligned}\mathcal{M}(\kappa u)(\mathbf{A}^{-1}\zeta) &= \frac{1}{(2\pi)^n}\int_{\mathbf{R}^n} \mathcal{M}u(\mathbf{A}^{-1}(\mathring{a}+i\gamma))K(\zeta-\mathring{a}-i\gamma)\,d\gamma \\ &= \widetilde{C}^{-}T(\zeta) \qquad\qquad \text{for } \operatorname{Re}\zeta_1 < \mathring{a}_1,\end{aligned} \tag{23}$$

where

$$T(b) = \mathcal{M}u\big(\mathbf{A}^{-1}(\mathring{a}+ib)\big) \qquad \text{for } b \in \mathbf{R}^n.$$

Now, we intend to apply Corollary 2 to the function

$$H(\zeta) = \mathcal{M}u(\mathbf{A}^{-1}\zeta).$$

Since $\mathcal{M}u \in \mathcal{O}(\Omega)$ it follows that $H \in \mathcal{O}(\Omega^1 \times \mathbf{C}^{n-1})$. Thus the function

$$\Omega^1 \ni \zeta_1 \mapsto H_{\zeta_1,\mathring{a}'} = H(\zeta_1, \mathring{a}' + i\cdot)$$

is a holomorphic function on Ω^1 with values in $O_M(\mathbf{R}^{n-1}) \subset S'(\mathbf{R}^{n-1})$. We easily check that all remaining assumptions of Corollary 2 are also satisfied with $U^1 = \Omega^1$. Since $T = H_{\mathring{a}}$, the assertion of Corollary 2 states that $\widetilde{C}^{-}T$ extends to a holomorphic function on $\big(\{\operatorname{Re}\zeta_1 < \mathring{a}_1\}\cup\Omega^1\big)\times\mathbf{C}^{n-1} = \Omega^1\times\mathbf{C}^{n-1}$ and consequently in view of (23), $\mathcal{M}(\kappa u) \in \mathcal{O}(\Omega)$.

We shall now estimate the growth order of $\mathcal{M}(\kappa u)$ along the imaginary planes. Take $\mathring{a} \in \operatorname{Re}\Omega_1$ and let $T_r^1(\mathring{a}_1)$ be a set of the form (15) such that (ii) in Definition 11.5 holds for $a_1 + ib_1 \in T_r^1(\mathring{a}_1)$. Take $\mathring{a}$ as in the proof above and such that $\mathring{a}_1 < \mathring{a}_1 - r$. It is easy to see that the assumptions of Lemma 1 are satisfied for

$T = H_{\mathring{a}}$ and $m = \varrho(\mathring{a}_1)$. Next in Lemma 2 we put $T_{\zeta_1} = H_{\zeta_1,\mathring{a}'}$, and the only non obvious fact to check is the estimate (16) if $\varrho(a_1) < 0$. Since $H_{\zeta_1,\mathring{a}'} \in O_M(\mathbf{R}^{n-1})$ we have for $a_1 + ib_1 \in \Omega^1$ and $\psi' \in C_0^\infty(\mathbf{R}^{n-1})$

$$\begin{aligned} |T_{a_1+ib_1}[\psi']| &= \left| \int_{\mathbf{R}^{n-1}} H(a_1 + ib_1, \mathring{a}' + i\gamma')\psi'(\gamma')d\gamma' \right| \\ &\leq C_1 \int_{\mathbf{R}^{n-1}} \frac{|\psi'(\gamma')|}{(|b_1| + \langle\gamma'\rangle)^{|\varrho(a_1)|}} d\gamma' \\ &\leq \langle b_1\rangle^{\varrho(a_1)} C_1 C_2^{\varrho(a_1)} \int_{\mathbf{R}^{n-1}} \frac{|\psi'(\gamma')|}{\langle\gamma'\rangle^{|\varrho(a_1)|}} d\gamma' \\ &\leq C_3 \langle b_1\rangle^{\varrho(a_1)} \sup|\psi'| \end{aligned}$$

in view of the estimate

$$|b_1| + \langle\gamma'\rangle \geq C_2\langle b_1\rangle\langle\gamma'\rangle \qquad \text{for } \gamma' \in \operatorname{supp}\psi',$$

where the constants C_2 and C_3 depend on the support of ψ'.

Now in view of the representation (13) of the function $\mathcal{M}(\kappa u)(\mathbf{A}^{-1}\zeta)$ it follows from Lemmas 1 and 2 that

$$\left|\mathcal{M}(\kappa u)(\mathbf{A}^{-1}(a + ib))\right| \leq C\langle b\rangle^{\varrho(a_1)} \qquad \text{for } a + ib \in \mathcal{T}_r^1(\mathring{a}) \times \{\|a' - \mathring{a}'\| < r\}$$

since $\max(\varrho(\mathring{a}_1), \varrho(a_1)) = \varrho(a_1)$ for $a_1 > \mathring{a}_1$ because ϱ is non-decreasing.

We end this section with a property of the classical Cauchy transformation which will find application in Section 15.

Proposition 1. *Let T be a distribution in $E'(\mathbf{R})$ of order $\tilde{p}$ with bounded support. Suppose T restricted to an interval $(0, \mathring{b})$, $\mathring{b} > 0$ is a differentiable function such that for $j = 0, 1$*

$$\left|\left(\frac{\partial}{\partial\gamma}\right)^j T(\gamma)\right| \leq \frac{C}{\gamma^p} \qquad \text{for } \gamma \in (0, \mathring{b}), \tag{24}$$

where $C > 0$, $p \geq 0$ are some constants. Then the (classical) left Cauchy transform

$$C^-T(z) = -\frac{1}{2\pi}T\left[\frac{1}{z - i\gamma}\right]$$

defined and holomorphic for $\operatorname{Re} z < 0$ extends to a continuous function on the set $\{z \in \mathbf{C}\colon \operatorname{Re} z \leq 0, 0 < \operatorname{Im} z < \mathring{b}\}$ and for every $\mathring{\alpha} < 0$ there exists $\widetilde{C} > 0$ such that

$$\left|C^-T(\alpha + i\beta)\right| \leq \frac{\widetilde{C}}{\beta^{\hat{p}}} \qquad \text{for } \mathring{\alpha} \leq \alpha < 0 \text{ and small } \beta > 0,$$

where $\hat{p} = \max(p, \tilde{p} + 1)$.

Proof: Let $\psi \in C_0^\infty(\mathbf{R})$ be such that $\psi \equiv 0$ on $\mathbf{R} \setminus (-\frac{2}{3}, \frac{2}{3})$, $|\psi| \le 1$, $\psi \equiv 1$ on $(-\frac{1}{2}, \frac{1}{2})$ and $\psi(-\gamma) = \psi(\gamma)$ for $\gamma \in \mathbf{R}$. Let $\beta > 0$ be so small that T is a differentiable function $T(\gamma)$ on $\operatorname{supp} \psi_\beta \subset [\frac{1}{3}\beta, \frac{5}{3}\beta]$ where $\psi_\beta(\gamma) = \psi(\frac{\gamma-\beta}{\beta})$. Write

$$T\left[\frac{1}{z - i\gamma}\right] = I_1(z) + I_2(z),$$

where

$$I_1(\alpha + i\beta) = \int_{\mathbf{R}} \frac{T(\gamma)\psi_\beta(\gamma)}{\alpha + i(\beta - \gamma)}\, d\gamma,$$

$$I_2(\alpha + i\beta) = (1 - \psi_\beta) T\left[\frac{1}{\alpha + i(\beta - \gamma)}\right].$$

We first estimate I_1. To this end we write for $\gamma \in \mathbf{R}$

$$\psi_\beta(\gamma)T(\gamma) = \psi_\beta(\gamma)T(\beta) + \psi_\beta(\gamma)(\gamma - \beta) \int_0^1 T'(\beta + \theta(\gamma - \beta))\, d\theta.$$

Observe that since ψ is symmetric we have for $\alpha < 0$

$$\begin{aligned}\int_{\mathbf{R}} \frac{\psi_\beta(\gamma)}{\alpha + i(\beta - \gamma)}\, d\gamma &= \int_{\mathbf{R}} \frac{\psi(\frac{\delta}{\beta})}{\alpha + i\delta}\, d\delta = \int_{\mathbf{R}} \frac{(\alpha - i\delta)\psi(\frac{\delta}{\beta})}{\alpha^2 + \delta^2}\, d\delta \\ &= \alpha \int_{\mathbf{R}} \frac{\psi(\frac{\delta}{\beta})}{\alpha^2 + \delta^2}\, d\delta = -\int_{\mathbf{R}} \frac{\psi(\frac{\alpha\varrho}{\beta})}{1 + \varrho^2}\, d\varrho.\end{aligned}$$

For fixed β and $\alpha \to 0_-$, the last integral tends to $\int \frac{1}{1+\varrho^2}\, d\varrho = \pi$. Since $\left|\psi(\frac{\alpha\varrho}{\beta})\right| \le 1$, this together with (24) for $j = 0$ gives

$$\left| T(\beta) \int_{\mathbf{R}} \frac{\psi_\beta(\gamma)}{\alpha + i(\beta - \gamma)}\, d\gamma \right| \le C_1 \frac{1}{|\beta|^p} \qquad \text{for } \mathring{a} \le \alpha < 0 \text{ and small } \beta > 0.$$

Next for $\mathring{\alpha} \le \alpha < 0$ and small $\beta > 0$ we have in view of (24) for $j = 1$

$$\begin{aligned}&\left| \int_{\mathbf{R}} \frac{(\gamma - \beta)\psi(\frac{\gamma-\beta}{\beta}) \int_0^1 T'(\beta + \theta(\gamma - \beta))\, d\theta}{\alpha + i(\beta - \gamma)}\, d\gamma \right| \\ &\quad \le C_1 \int_{\mathbf{R}} \left(\int_0^1 |T'(\beta + \theta(\gamma - \beta))|\, d\theta \right) \left| \psi\left(\frac{\gamma - \beta}{\beta}\right) \right| d\gamma \\ &\quad \le C_2 \int_{\mathbf{R}} \int_0^1 \frac{\left|\psi(\frac{\gamma-\beta}{\beta})\right|}{|\beta + \theta(\gamma - \beta)|^p}\, d\theta\, d\gamma \le C_2 \int_0^1 \int_{-\frac{2}{3}\beta}^{\frac{2}{3}\beta} \frac{d\delta\, d\theta}{|\beta + \theta\delta|^p} \le C_3 \frac{1}{|\beta|^p}.\end{aligned}$$

Now, consider $I_2(z)$ for $\mathring{\alpha} \le \alpha < 0$ and β small. Let $\operatorname{supp} T \subset K$ —a bounded interval in $\mathbf{R}$. Since $\operatorname{supp}(1-\psi_\beta)T \subset K \setminus (\frac{1}{2}\beta, \frac{3}{2}\beta)$ and T is of order $\tilde{p}$, we have by Theorem 3.1

$$|I_2(\alpha + i\beta)| \le \sum_{j=0}^{\tilde{p}} \sup_{\gamma \in K \setminus (\frac{1}{2}\beta, \frac{3}{2}\beta)} \left| \frac{d^j}{d\gamma^j} \frac{1}{\alpha + i(\beta - \gamma)} \right|$$

$$\le C_1 \sup_{\gamma \in K \setminus (\frac{1}{2}\beta, \frac{3}{2}\beta)} \frac{1}{|\beta - \gamma|^{\tilde{p}+1}} \le C_2 \frac{1}{|\beta|^{\tilde{p}+1}}.$$

We shall also need the following parameter version of Proposition 1

Corollary 3. *Let $\mathbf{C}^{n-1} \ni \zeta' \mapsto T_{\zeta'} \in E'(\mathbf{R})$ be a distribution valued holomorphic function which is rapidly decreasing as a function of $\operatorname{Im} \zeta'$, locally uniformly in $\operatorname{Re} \zeta'$ and such that the orders of $T_{\zeta'}$ are commonly bounded. Suppose that $T_{\zeta'}$ restricted to an interval $(0, \mathring{b})$, $\mathring{b} > 0$ is a function $T_{\zeta'}(\gamma_1)$ for $\zeta' \in \mathbf{C}^{n-1}$ and for $j = 0, 1$ and some $l \in \mathbf{N}_0$*

$$\left|\left|\left| \frac{\partial^j}{\partial \gamma_1^j} T_{a'+i\cdot}(\gamma_1) \right|\right|\right|_{S;l} \le \frac{C_l}{\gamma_1^p} \qquad \gamma_1 \in (0, \mathring{b})$$

locally uniformly with respect to $a' \in \mathbf{R}^{n-1}$, where

$$|||\sigma|||_{S;l} = \sup_{x \in \mathbf{R}^{n-1}} \langle x \rangle^l \Big(\sum_{|\alpha| \le l} \Big| \Big(\frac{\partial}{\partial x}\Big)^\alpha \sigma(x) \Big| \Big)$$

for $\sigma \in S(\mathbf{R}^{n-1})$. Then for $\mathring{a}_1 \le a_1 < 0$ and small $b_1 > 0$

$$\left|\left|\left| C^- T_{a'+i\cdot}(a_1 + ib_1) \right|\right|\right|_{S;l} \le \frac{\widetilde{C}_l}{b_1^{\hat{p}}}$$

locally uniformly in $a' \in \mathbf{R}^{n-1}$, where $\hat{p} = \max(p, 1 + \tilde{p})$, $\tilde{p} = \sup_{\zeta' \in \mathbf{C}^{n-1}} \operatorname{order} T_{\zeta'}$ and

$$C^- T_{\zeta'}(\zeta_1) = -\frac{1}{2\pi} T_{\zeta'} \left[\frac{1}{\zeta_1 - i\gamma_1} \right] \qquad \text{for } \zeta' \in \mathbf{C}^{n-1},\ \operatorname{Re} \zeta_1 < 0.$$

The corollary follows easily from Proposition 1 since C^- commutes with differentiations in variable b'.

Exercises

1. In addition to the assumptions of Lemma 2 suppose that for $\zeta_1 \in \mathbf{C} \setminus \overline{\mathbf{R}_+}$, $\zeta_1 \mapsto T_{\zeta_1} \in S'(\mathbf{R}^{n-1})$ is a holomorphic function such that for $|b_1|$ close to zero and $\sigma \in S(\mathbf{R}^{n-1})$

$$|T_{a_1+ib_1}[\sigma]| \le \frac{C}{|b_1|^m}$$

with $C = C(a_1, \sigma)$, $m = m(a_1)$ locally bounded in $a_1 \in \mathbf{R}$. Show that $\widetilde{C}'_{\zeta_1}(\zeta') \in \mathcal{O}((\mathbf{C} \setminus \overline{\mathbf{R}_+}) \times \mathbf{C}^{n-1})$ and for $|b_1|$ close to zero

$$|\widetilde{C}'_{a_1+ib_1}(a' + ib')| \leq \frac{\widetilde{C}}{|b_1|^m}$$

with $\widetilde{C} = \widetilde{C}(a_1, a')$ locally bounded in $a \in \mathbf{R}^n$.

2. Prove Theorem 1.

HINT. Modify the proof of Theorem 2. Use the fact that $\mathcal{F}$ and $\mathcal{F}^{-1}$ are continuous mappings of $L^2(\mathbf{R}) \to L^2(\mathbf{R})$.

Chapter III

FUCHSIAN TYPE SINGULAR OPERATORS

§13. FUCHSIAN TYPE ORDINARY DIFFERENTIAL OPERATORS

Before passing to the study of Fuchsian type differential operators with smooth coefficients we must acquaint ourselves with the notion of an asymptotic expansion of a function into polyhomogeneous terms (cf. [J]). It can be regarded as a generalization of the notion of smoothness.

1. Asymptotic Expansions

Definition 1. Let $\{\gamma_k\}_{k\in\mathbf{N}_0}$ be a sequence of complex numbers such that $\lim\limits_{k\to\infty} \operatorname{Re}\gamma_k = +\infty$. Let $f \in C^\infty((0,t))$ for some $t \in \mathbf{R}_+$. We say that f admits as $x \to 0_+$ an *asymptotic expansion*

$$f(x) \sim \sum_{k=0}^{\infty}\sum_{j=0}^{a} c_{kj} x^{\gamma_k}(\ln x)^j, \qquad c_{kj} \in \mathbf{C}, \quad a \in \mathbf{N}_0 \tag{1}$$

infinitely differentiable term by term, if for any $\nu \in \mathbf{N}_0$ and any $r \in \mathbf{R}$ the remainder term

$$R_r^\nu(x) = \frac{d^\nu}{dx^\nu} f(x) - \sum_{j=0}^{a} \sum_{\operatorname{Re}\gamma_k<r} c_{kj} \frac{d^\nu}{dx^\nu}(x^{\gamma_k}(\ln x)^j)$$

has the property that $x^{-b}R_r^\nu(x)$ is bounded at zero for any $b \le r-\nu$.

Remark 1. The Taylor series at zero of a smooth function in a neighbourhood of zero gives its asymptotic expansion infinitely differentiable term by term. The example of a function equal to its asymptotic expansion is given in Example 7.4.

Our purpose is to establish a characterization of the Mellin transforms of functions having the asymptotic expansion (1) (cf. Example 7.5).

Proposition 1. *Let $\{\gamma_k\}_{k\in\mathbb{N}_0}$ be a sequence of complex numbers such that $\lim_{k\to\infty} \operatorname{Re}\gamma_k = +\infty$ and let $f \in M'_{(\omega)}$ for some $\omega \in \mathbb{R}$. Then f is smooth on $(0,t)$, $t>0$ and admits an asymptotic expansion (1) infinitely differentiable term by term if and only if for any cut-off function $\chi \in C^\infty(\mathbb{R}_+)$, $\operatorname{supp}\chi \subset (0,t)$, $\chi \equiv 1$ close to zero:*

(i) *$\mathcal{M}(\chi f)$ is meromorphic on $\mathbb{C}$ with poles at most at the points γ_k, $k \in \mathbb{N}_0$ of order at most $a+1$,*

(ii) *for any $m \in \mathbb{N}$, $\varepsilon > 0$ and any bounded set $W \subset \mathbb{R}$ there exists a constant $C_{\varepsilon,m,W}$ such that*

$$|\mathcal{M}(\chi f)(\alpha + i\beta)| \le \frac{C_{\varepsilon,m,W}}{\langle\beta\rangle^m} \qquad \text{for } \alpha \in W,\ |\beta| \ge \varepsilon + \sup|\operatorname{Im}\gamma_k|,$$

where the sup is taken over the indices k such that $\operatorname{Re}\gamma_k \in W$ ($\beta \in \mathbb{R}$ if $\{\operatorname{Re}\gamma_k\} \cap W = \emptyset$ for $k \in \mathbb{N}_0$).

Proof: Suppose $f \in C^\infty((0,t))$ and admits asymptotic expansion (1). Choose arbitrarily $b \in \mathbb{R}$ and $r > b$. We shall show that the function

$$x^{-b}\left(x\frac{d}{dx}\right)^{\nu}(\chi R_r^0) \qquad \text{with } \nu \in \mathbb{N}_0$$

is bounded on $(0,t]$. This is clear for $\nu = 0$. To prove it for $\nu = 1$ write

$$(\chi R_r^0)(x) = (\chi f)(x) - \chi(x) \sum_{\operatorname{Re}\gamma_k < r} \sum_{j=0}^{a} c_{kj} x^{\gamma_k} (\ln x)^j \tag{2}$$

and observe that near zero

$$x^{-b}\left(x\frac{d}{dx}\right)(\chi R_r^0)(x) = x^{-b+1}\frac{d}{dx}R_r^0 = x^{-b+1}R_r^1(x),$$

which by assumption is bounded at zero since $b < r$. We end the proof by induction. Thus $\chi R_r^0 \in M_{-b-1} \subset M'_{(b)}$ and hence $\mathcal{M}(\chi R_r^0)(z)$ is holomorphic for $\operatorname{Re} z < b$. Moreover the function $x^{-b}\chi R_r^0 \in \mathring{M}_{[-1]}$ and hence by Theorem 9.5* for every $m \in \mathbb{N}_0$ and $h < 0$ there exists a constant $C_{m,h}$ such that

$$\left|\mathcal{M}(x^{-b}\chi R_r^0)(z)\right| \le C_{m,h}\frac{t^{-\operatorname{Re} z}}{\langle\beta\rangle^m} \qquad \text{for } \operatorname{Re} z \le h.$$

Thus, by the operation rule (7.9) for the Mellin transformation we get

$$\left|\mathcal{M}(\chi R_r^0)(\alpha + i\beta)\right| \le \frac{C_{m,h}}{\langle\beta\rangle^m} t^{-\alpha+b} \qquad \text{for } \alpha \le b + h. \tag{3}$$

From Corollary 7.3 follows that the function

$$\mathcal{M}(x^{\gamma_k}\chi(\ln x)^j)(z) = \mathcal{M}(\chi(\ln x)^j)(z-\gamma_k)$$

is meromorphic on $\mathbf{C}$ with a pole at the point γ_k of order $j+1$. Thus, (i) follows from (2) since $b \in \mathbf{R}$ could be chosen arbitrarily and $\mathcal{M}(\chi R_r^0)(z)$ is holomorphic for $\operatorname{Re} z < b$.

For the proof of the assertion (ii) take $m \in \mathbf{N}$, $\varepsilon > 0$ and a bounded set $W \subset \mathbf{R}$. Let $b > \sup_{x\in W} x + 1$, $r > b$ and let $V_k = W - \operatorname{Re}\gamma_k$ for $\operatorname{Re}\gamma_k < r$. By Corollary 7.3 there exist constants C_{ε,m,V_k} such that

$$\left|\mathcal{M}(\chi x^{\gamma_k}(\ln x)^j)(z)\right| = \left|\mathcal{M}(\chi(\ln x)^j)(z-\gamma_k)\right| < \frac{C_{\varepsilon,m,V_k}}{\langle \beta - \operatorname{Im}\gamma_k\rangle^m}$$

for $\alpha \in W$, $|\beta| \geq \varepsilon + |\operatorname{Im}\gamma_k|$. Thus, there exists a constant $C_{\varepsilon,m,W,r}$ such that

$$\left|\sum_{\operatorname{Re}\gamma_k<r}\sum_{j=0}^{a} c_{kj}\mathcal{M}(\chi x^{\gamma_k}(\ln x)^j)(z)\right| \leq \sum_{\operatorname{Re}\gamma_k<r}\sum_{j=0}^{a}|c_{kj}|\frac{C_{\varepsilon,m,V_k}}{\langle\beta-\operatorname{Im}\gamma_k\rangle^m} \leq \frac{C_{\varepsilon,m,W,r}}{\langle\beta\rangle^m}$$
$$\text{for}\quad \alpha\in W,\ |\beta| \geq \varepsilon + \sup_{\operatorname{Re}\gamma_k<r}|\operatorname{Im}\gamma_k|,$$

which together with (3) yields assertion (ii) since W is bounded.

Conversely, suppose that $f \in M'_{(\omega)}$ and that (i), (ii) hold for any cut-off function χ. Then $\chi f \in M'_{(\omega)}$ and $\mathcal{M}(\chi f)(z)$ is holomorphic for $\operatorname{Re} z < \omega$. Take an arbitrary $r \neq \operatorname{Re}\gamma_k$, $k \in \mathbf{N}_0$ and $s < \omega$. Then it follows from the Cauchy formula and the estimation (ii) for $m = 2$ and $W = [s,r]$ that

$$\frac{1}{2\pi i}\int_{s+i\mathbf{R}}\mathcal{M}(\chi f)(z)x^z\,dz$$
$$= \frac{1}{2\pi i}\int_{r+i\mathbf{R}}\mathcal{M}(\chi f)(z)x^z\,dz - \sum_{\operatorname{Re}\gamma_k<r}\operatorname{Res}_{z=\gamma_k}(\mathcal{M}(\chi f)(z)x^z).$$

Since the orders of the poles of γ_k are not greater than $a+1$ it follows that

$$-\operatorname{Res}_{z=\gamma_k}(\mathcal{M}(\chi f)(z)x^z) = \sum_{j=0}^{a} c_{kj}x^{\gamma_k}(\ln x)^j$$

for some complex constants c_{kj}.

Let

$$R_r(x) = \frac{1}{2\pi i}\int_{r+i\mathbf{R}}\mathcal{M}(\chi f)(z)x^z\,dz.$$

Then we get from the last formulas

$$(4)\qquad \frac{1}{2\pi i}\int_{s+i\mathbf{R}} \mathcal{M}(\chi f)(z)x^z\,dz = R_r(x) + \sum_{\operatorname{Re}\gamma_k<r}\sum_{j=0}^{a} c_{kj}x^{\gamma_k}(\ln x)^j.$$

By assumption (ii) for $W=\{s\}$ we have for any $m\in\mathbf{N}$

$$|\mathcal{M}(\chi f)(s+i\beta)| \le \frac{C_m}{\langle\beta\rangle^m} \qquad \text{for } \beta\in\mathbf{R}.$$

Since $\operatorname{supp}(\chi f)\subset(0,t)$ by Remark 11.1 we get as in the proof of Theorem 9.5* that for every $b<\omega$ there exists a constant $\widetilde{C}=\widetilde{C}_m(b)$ such that

$$|\mathcal{M}(\chi f)(\alpha+i\beta)| \le \frac{\widetilde{C}}{\langle z\rangle^m}t^{-\alpha} \qquad \text{for } \alpha\le b,\ \beta\in\mathbf{R}.$$

From this estimation with $m=2$ we get by Lemma 9.1 and Theorem 9.2 that

$$(5)\qquad (\chi f)(x) = \frac{1}{2\pi i}\int_{s+i\mathbf{R}} \mathcal{M}(\chi f)(z)x^z\,dz$$

is continuous for $x>0$.

Moreover, we see that for every $p\in\mathbf{N}$ (by applying the same estimation with $m=p+2$)

$$\left(\frac{d}{dx}\right)^p(\chi f)(x) = \frac{1}{2\pi i}\int_{s+i\mathbf{R}} \mathcal{M}(\chi f)(z)z(z-1)\cdot\ldots\cdot(z-p+1)x^{z-p}\,dz$$

is continuous and hence f is smooth near zero for $x>0$. Since χ is an arbitrary function $\chi\equiv 1$ close to zero, $\chi\in C^\infty(\mathbf{R}_+)$, $\operatorname{supp}\chi\subset(0,t)$, f is smooth on $(0,t)$. Now from (4) and (5) results $(\chi f)(x) = R_r(x) + \sum_{\operatorname{Re}\gamma_k<r}\sum_{j=0}^{a} c_{kj}x^{\gamma_k}(\ln x)^j$ and

$$R_r(x) = f(x) - \sum_{\operatorname{Re}\gamma_k<r}\sum_{j=0}^{a} c_{kj}x^{\gamma_k}(\ln x)^j = R_r^0 \qquad \text{near zero.}$$

Observe that $R_r^\nu = \frac{d^\nu}{dx^\nu}R_r^0$ and by assumption (ii)

$$\begin{aligned}\left|x^{-b}R_r^\nu(x)\right| &\le \frac{1}{2\pi}x^{-b}\int_{r+i\mathbf{R}} |z|\,|z-1|\cdot\ldots\cdot|z-\nu+1|\,|\mathcal{M}(\chi f)(z)|\,x^{r-\nu}\,|dz|\\ &\le Cx^{-b+r-\nu} \qquad\qquad \text{with some } C<\infty.\end{aligned}$$

Hence $x^{-b}R_r^\nu(x)$ is bounded at zero if $r-b-\nu\ge 0$, $(\nu=0,1,\ldots)$.

The assumption $r\ne\operatorname{Re}\gamma_k$ for $k\in\mathbf{N}_0$ was unessential. In fact if $r=\operatorname{Re}\gamma_{\mathring{k}}$ for some $\mathring{k}\in\mathbf{N}_0$ we choose $\tilde{r}<r$ such that $\operatorname{Re}\gamma_k\notin[\tilde{r},r)$ for $k\in\mathbf{N}_0$ and we have $R_r^\nu(x)=R_{\tilde{r}}^\nu(x)$.

The following lemma will be applied in Subsection 3.

Lemma 1. *Let $u \in M'_{(b)}((0,t])$ for some $b \in \mathbf{R}$, $t > 0$. Let a be a smooth function (in a neighbourhood) on $[0,t]$. Suppose $\mathcal{M}u$ defined for $\operatorname{Re} z < b$ extends holomorphically to a connected set $\Omega \subset \mathbf{C}$. Then $au \in M'_{(b)}((0,t])$ and $\mathcal{M}(au)$ extends holomorphically to the set $\bigcap_{j=0}^{\infty} \Omega + j$.*

Moreover, if for any open neighbourhood W of $\mathbf{C} \setminus \Omega$

$$|\mathcal{M}u(\alpha + i\beta)| \leq C(\alpha, W)\langle\beta\rangle^p \qquad \text{for } \alpha + i\beta \in \mathbf{C} \setminus W \tag{6}$$

for some $p \in \mathbf{R}$ and $C(\alpha, W)$ locally bounded in α, then for any open neighbourhood $\widetilde{W}$ of $\mathbf{C} \setminus (\bigcap_{j=0}^{\infty} \Omega + j)$

$$|\mathcal{M}(au)(\alpha + i\beta)| \leq \widetilde{C}(\alpha, \widetilde{W})\langle\beta\rangle^p \qquad \text{for } \alpha + i\beta \in \mathbf{C} \setminus \widetilde{W} \tag{7}$$

with $\widetilde{C}(\alpha, \widetilde{W})$ locally bounded in α.

Proof: Take $r \in \mathbf{N}$ and write

$$a(x) = \sum_{j=0}^{r} \frac{a^{(j)}(0)}{j!} x^j + x^{r+1} f(x),$$

where f is smooth on $[0,t]$. Then

$$\mathcal{M}(au)(z) = \sum_{j=0}^{r} \frac{a^{(j)}(0)}{j!} \mathcal{M}u(z-j) + \mathcal{M}(fu)(z-r-1). \tag{8}$$

Since $fu \in M'_{(b)}$ it follows that $\mathcal{M}(fu)(z-r-1)$ is holomorphic for $\operatorname{Re} z < b+r+1$. Clearly $\mathcal{M}u(z-j)$ is holomorphic on $\Omega + j$ and the first assertion follows since r can be arbitrarily large.

To prove the estimate (7) we again make use of (8) and observe that it is enough to show that

$$|\mathcal{M}(fu)(\alpha + i\beta)| \leq C(\alpha)\langle\beta\rangle^p \qquad \text{for } \alpha < b,\ \beta \in \mathbf{R},$$

where $C(\alpha)$ is locally bounded for $\alpha < b$. The last assertion follows from Propositions 7.3 and 7.6. To apply those propositions we extend f to a smooth function $\tilde{f}$ on $\mathbf{R}_+$ with bounded support and note that $fu = \tilde{f}u$ (see Theorem 8.3).

Remark 2. Observe that if $\mathcal{M}u$ is meromorphic on $\mathbf{C}$ with poles of order less than k then so is $\mathcal{M}(au)$ with poles of order also less than k.

2. The Equation $P(x\frac{d}{dx})u = f$ and Definition of Ordinary Fuchsian Type Differential Operators

Let P be a polynomial with constant coefficients of degree m. The operator $P(x\frac{d}{dx})$ is the *constant coefficient ordinary Fuchsian differential operator.*

Choose q such that $P(z) \neq 0$ for $\operatorname{Re} z \leq q$. Let $s \in \mathbf{R}$ and $f \in M'^{s}_{(q)}((0,t_0])$ (see Definition 11.1). Since $f \in M'_{(q)}$, $\mathcal{M}f(z)$ is holomorphic for $\operatorname{Re} z < q$ and to every $b < q$, $\tilde{t}_0 > t_0$ there exists a constant $C(b,\tilde{t}_0) < \infty$ such that

$$|\mathcal{M}f(z)| < C(b,\tilde{t}_0)\langle\beta\rangle^s \tilde{t}_0^{-\alpha} \qquad \text{for} \quad \alpha \leq b, \quad z = \alpha + i\beta, \quad \beta \in \mathbf{R}. \tag{9}$$

We seek a solution to the equation

$$P\Big(x\frac{d}{dx}\Big)u = f \tag{10}$$

in the class $M'_{(q)}((0,t_0])$. Suppose first that there exists such a solution u. Applying the Mellin transformation to both sides of (10) we get

$$P(z)(\mathcal{M}u)(z) = (\mathcal{M}f)(z) \qquad \text{for} \quad \operatorname{Re} z < q.$$

Let

$$F(z) = \frac{\mathcal{M}f(z)}{P(z)} \qquad \text{for} \quad \operatorname{Re} z < q.$$

We see that F is holomorphic for $\operatorname{Re} z < q$. By (9) (and the assumption on the polynomial P) it follows that for every $b < q$ and $\tilde{t}_0 > t_0$ there exists a constant $\widetilde{C}(b,\tilde{t}_0)$ such that

$$|F(z)| < \widetilde{C}(b,\tilde{t}_0)\langle\beta\rangle^{s-m}\tilde{t}_0^{-\alpha} \qquad \text{for} \quad \operatorname{Re} z = \alpha \leq b.$$

By Theorem 9.4* there exists a unique distribution $u \in M'_{(q)}((0,t_0])$, such that $(\mathcal{M}u)(z) = F(z)$ for $\operatorname{Re} z < q$ and hence $u \in M'^{s-m}_{(q)}((0,t_0])$, $\mathcal{M}(P(x\frac{d}{dx})u - f)(z) = P(z)F(z) - (\mathcal{M}f)(z) = 0$ for $\operatorname{Re} z < q$. By Theorem 9.2 $P(x\frac{d}{dx})u - f = 0$. Thus, we have shown that there exists a unique distribution u which verifies the equation (10) on $\mathbf{R}_+$ and belongs to $M'^{s-m}_{(q)}((0,t_0])$. By Theorem 8.1 such u can be considered as a restriction to $\mathbf{R}_+$ of a distribution in $D'(\mathbf{R})$ with support in $[0,t_0]$.

In the sequel we consider operators of the form $R(x, x\frac{d}{dx})$ where $R(x,z)$ is a polynomial in z of order m, whose coefficients are smooth (analytic) functions of x in a neighbourhood of zero i.e.

$$R(x,z) = \sum_{\nu=0}^{m} a_\nu(x) z^\nu, \qquad (a_m(0) \neq 0).$$

We consider the equation

$$R\Big(x, x\frac{d}{dx}\Big)u = f, \tag{11}$$

where $f \in M'^{s}_{(q)}(I_0)$, $I_0 = (0, t_0]$, $t_0 > 0$, $q \in \mathbf{R}$, $s \in \mathbf{R}$ (or $f \in M'^{s}_{(q)}$-locally(cf. Remark 11.2)).

We may assume that $a_m(x) \equiv 1$ (dividing both sides by $a_m(x)$, if necessary). Writing $a_\nu(x) = a_\nu(0) - x\tilde{a}_\nu(x)$ for smooth (analytic) $\tilde{a}_\nu(x)$, $\nu = 0, 1, \dots, m-1$ we may write $R(x, z)$ in the form

$$R(x, z) = P(z) - xQ(x, z),$$

where $P(z) = R(0, z)$ and Q has the same properties as R and is of order $m-1$. In other words

$$R\Big(x, x\frac{d}{dx}\Big) = P\Big(x\frac{d}{dx}\Big) - xQ\Big(x, x\frac{d}{dx}\Big) \tag{12}$$

so that $R\big(x, x\frac{d}{dx}\big)$ can be regarded as a perturbation of the operator $P\big(x\frac{d}{dx}\big)$ with $-xQ\big(x, x\frac{d}{dx}\big)$ being the perturbation term. Such operators R are called *ordinary Fuchsian type differential operators* or *ordinary singular differential operators with regular singularity* at zero. Putting $c_\nu = a_\nu(0)$ $(\nu = 0, 1, \dots, m-1)$ we write (11) in the following form

$$P\Big(x\frac{d}{dx}\Big)u = xQ\Big(x, x\frac{d}{dx}\Big)u + f, \tag{11$'$}$$

where

$$P\Big(x\frac{d}{dx}\Big) = \Big(x\frac{d}{dx}\Big)^m + \sum_{\nu=0}^{m-1} c_\nu \Big(x\frac{d}{dx}\Big)^\nu,$$

$$xQ\Big(x, x\frac{d}{dx}\Big) = x\sum_{\nu=0}^{m-1} \tilde{a}_\nu(x)\Big(x\frac{d}{dx}\Big)^\nu.$$

The polynomial $P(z) = R(0, z)$ is called the *characteristic polynomial* of the operator R and zeros of $P(z)$ are called the *characteristic roots* of R at zero. By $\operatorname{Char} P$ we denote the set of all different complex roots of P. We shall solve the equation (11′) and establish the regularity of the solution. The case of smooth (analytic) coefficients a_ν $(\nu = 0, 1, \dots, m)$ is examined in Subsection 3 (Subsection 4) (see also [Tou] for related questions).

3. Case of Smooth Coefficients

Suppose that the coefficients a_ν ($\nu = 0, 1, \ldots, m$) in (12) are smooth in a closed right-sided neighbourhood of zero: $0 \le x \le \mathring{r}$, $\mathring{r} > 0$. It is easy to show that there exist functions b_ν ($\nu = 0, 1, \ldots, m-1$) smooth for $0 \le x \le \mathring{r}$ and such that (cf. Exercise 5.1)

$$x \sum_{\nu=0}^{m-1} \tilde{a}_\nu(x)\Big(x\frac{d}{dx}\Big)^\nu u = \sum_{\nu=0}^{m-1} \Big(x\frac{d}{dx}\Big)^\nu (xb_\nu u).$$

Thus the equation (11′) can be equivalently written in the form

$$P\Big(x\frac{d}{dx}\Big)u = \sum_{\nu=0}^{m-1} \Big(x\frac{d}{dx}\Big)^\nu (xb_\nu u) + f. \tag{13}$$

Fix numbers q, K such that

$$P(z) \ne 0, \qquad \frac{|z|^\nu}{|P(z)|} \le K \qquad \text{for } \operatorname{Re} z \le q, \quad \nu = 0, 1, \ldots, m-1. \tag{14}$$

Let $s \in \mathbf{R}$, $0 < t_0 < \frac{\mathring{r}}{2}$ and fix a function $\varphi \in C_0^\infty(\mathbf{R})$, $\varphi \equiv 1$ for $0 \le x \le \frac{1}{2}$ and $\varphi(x) = 0$ for $x \ge 1$. Let $\chi_{t_0}(x) = \varphi(\frac{x}{2t_0})$ for $x \in \mathbf{R}_+$. Let $f \in M'^s_{(q)}((0, t_0])$. We see that $\chi_{t_0} f = f$.

First we consider the equation

$$P\Big(x\frac{d}{dx}\Big)u = f + \sum_{\nu=0}^{m-1} \Big(x\frac{d}{dx}\Big)^\nu (xb_\nu \chi_{t_0} u) \tag{15}$$

and solve it for sufficiently small t by the method of successive approximations according to the scheme:

$$P\Big(x\frac{d}{dx}\Big)u_0 = f, \tag{16}$$

$$P\Big(x\frac{d}{dx}\Big)u_j = f + \sum_{\nu=0}^{m-1} \Big(x\frac{d}{dx}\Big)^\nu (xb_\nu \chi_{t_0} u_{j-1}) \qquad (j = 1, 2, \ldots). \tag{17}$$

Since $f \in M'^s_{(q)}((0, t_0])$ by Definition 11.1 for every $\tilde{t}_0 > t_0$ and every $b < q$ there exists a constant $\widetilde{C}(b, \tilde{t}_0) < \infty$ such that

$$|\mathcal{M}f(\alpha + i\beta)| \le \widetilde{C}(b, \tilde{t}_0)\langle\beta\rangle^s \tilde{t}_0^{-\alpha} \qquad \text{for } \alpha \le b, \quad \beta \in \mathbf{R} \tag{18}$$

and, as it was shown in Subsection 2, there exists a unique $u_0 \in M'^{s-m}_{(q)}((0,t_0])$ satisfying (16). Put

$$p = s - m, \qquad h_0(z) = (\mathcal{M}f)(z)/P(z)$$

and observe that $(\mathcal{M}u_0)(z) = h_0(z)$ and

$$\text{(19)} \qquad |h_0(z)| \le B(b,\tilde{t}_0)\langle\beta\rangle^p \tilde{t}_0^{-\alpha} \qquad \text{for } \alpha \le b, \ \beta \in \mathbf{R},$$

where $B(b,\tilde{t}_0) = K\widetilde{C}(b,\tilde{t}_0) < \infty$.

To get u_j $(j = 1,2,\ldots)$ we apply the Mellin transformation to both sides of (17), define

$$h_j(z) = \frac{1}{P(z)}\Big\{(\mathcal{M}f)(z) + \sum_{\nu=0}^{m-1} z^\nu \mathcal{M}(b_\nu \chi_{t_0} u_{j-1})(z-1)\Big\} \qquad (j = 1,2,\ldots)$$

and consider the equations

$$\mathcal{M}u_j = h_j \qquad (j = 1,2,\ldots).$$

We begin with $j = 1$ and we shall prove for h_1 an estimation analogous to (19). By Proposition 7.3 we get for $0 < \varepsilon < 1$

$$\begin{aligned}\text{(20)} \quad & \mathcal{M}(b_\nu \chi_{t_0} u_0)(\alpha - 1 + i\beta) \\ & = \frac{1}{2\pi}\int_{\mathbf{R}} \mathcal{M}u_0(\alpha - 1 + \varepsilon + i\gamma)\mathcal{M}(b_\nu\chi_{t_0})(-\varepsilon + i(\beta-\gamma))\,d\gamma \\ & \qquad\qquad \text{for } \alpha \le b - \varepsilon + 1, \ \beta \in \mathbf{R}.\end{aligned}$$

Put $\varepsilon = \frac{1}{2}$. By (19) and Proposition 7.6 we get from (20) the following estimation (in which $C_{|p|+2}$ does not depend on t small enough)

$$\begin{aligned}\text{(21)} \quad & |\mathcal{M}(b_\nu\chi_{t_0}u_0)(\alpha - 1 + i\beta)| \le \frac{1}{\pi}B(b,\tilde{t}_0)\tilde{t}_0^{-\alpha+\frac{1}{2}} C_{|p|+2}(2t_0)^{\frac{1}{2}} \int_{\mathbf{R}} \frac{\langle\gamma\rangle^p}{\langle\beta-\gamma\rangle^{|p|+2}}\,d\gamma \\ & \qquad\qquad \text{for } \alpha \le b + \frac{1}{2}, \ \beta \in \mathbf{R}.\end{aligned}$$

By applying the inequality (11.5) we get from (21) that

$$|\mathcal{M}(b_\nu\chi_{t_0}u_0)(\alpha - 1 + i\beta)| \le C^*\tilde{t}_0 B(b,\tilde{t}_0)\tilde{t}_0^{-\alpha}\langle\beta\rangle^p \qquad \text{for } \alpha \le b + \frac{1}{2}, \ \beta \in \mathbf{R},$$

where

$$C^* = \frac{\sqrt{2}}{\pi} C_{|p|+2} \int_{\mathbf{R}} \frac{d\gamma}{\langle\gamma\rangle^2} < \infty. \tag{22}$$

Thus by (19) and (14) we get the estimation

$$|h_1(z)| \leq B(b, \tilde{t}_0)\tilde{t}_0^{-\alpha}\langle\beta\rangle^p(1 + mKC^*\tilde{t}_0) \qquad \text{for } \alpha = \operatorname{Re} z \leq b, \quad \beta \in \mathbf{R}. \tag{23}$$

By Theorem 9.4* there exists a unique $u_1 \in M'_{(q)}((0, t_0])$ such that $\mathcal{M}u_1(z) = h_1(z)$ and hence by (23) $u_1 \in M'^p_{(q)}((0, t_0])$.

Let

$$a = \min\left(\frac{1}{mKC^*}, \frac{\mathring{r}}{2}\right)$$

and note that it depends only on R, s and on the choice of φ, and that $0 < a < \infty$. By induction we show that

$$|h_j(z)| \leq B(b, \tilde{t}_0)\tilde{t}_0^{-\alpha}\langle\beta\rangle^p \sum_{\nu=0}^{j} \left(\frac{\tilde{t}_0}{a}\right)^\nu \qquad \text{for } \operatorname{Re} z \leq b, \quad \beta \in \mathbf{R}, \quad (j = 0, 1, \ldots).$$

We conclude (similarly as for $j = 1$) that there exist

$$u_j \in M'^p_{(q)}((0, t_0]), \qquad \mathcal{M}u_j(z) = h_j(z) \qquad \text{for } \operatorname{Re} z < q, \quad (j = 1, 2, \ldots).$$

Fix arbitrarily $t_0 < a$, choose $\tilde{t}_0 \in (t_0, a)$ and put $C = \sum_{\nu=0}^{\infty} \left(\frac{\tilde{t}_0}{a}\right)^\nu$. Then $C < \infty$ and the functions $\mathcal{M}u_j$ holomorphic for $\operatorname{Re} z \leq b$ fulfil the estimations

$$\begin{aligned} &|\mathcal{M}u_j(z)| \leq CB(b, \tilde{t}_0)\tilde{t}_0^{-\alpha}\langle\beta\rangle^p && \text{for } \operatorname{Re} z \leq b, \quad \beta \in \mathbf{R}, \quad (j = 0, 1, \ldots), \\ &|\mathcal{M}u_{j+1}(z) - \mathcal{M}u_j(z)| \leq \left(\frac{\tilde{t}_0}{a}\right)^{j+1} \langle\beta\rangle^p B(b, \tilde{t}_0)\tilde{t}_0^{-\alpha} && (j = 0, 1, \ldots). \end{aligned} \tag{24}$$

Hence

$$g = \lim_{j \to \infty} \mathcal{M}u_j$$

is holomorphic for $\operatorname{Re} z < q$ (as a limit of an almost uniformly convergent sequence of holomorphic functions). By (24) we get

$$|g(z)| \leq CB(b, \tilde{t}_0)\tilde{t}_0^{-\alpha}\langle\beta\rangle^p \qquad \text{for } \operatorname{Re} z = \alpha \leq b$$

and consequently there exists a unique $u \in M'^p_{(q)}((0, t_0])$ such that

$$g(z) = (\mathcal{M}u)(z) \qquad \text{for } \operatorname{Re} z < q.$$

Applying Proposition 7.3 (as in (20)) we write $\mathcal{M}u_j = h_j$ in the form

$$\mathcal{M}u_j(z) = \frac{1}{P(z)}\Big\{\mathcal{M}f(z)+ + \frac{1}{2\pi}\sum_{\nu=0}^{m-1} z^\nu \int_{\mathbf{R}} \mathcal{M}u_{j-1}(\alpha - \tfrac{1}{2} + i\gamma)\mathcal{M}(b_\nu\chi_{t_0})(-\tfrac{1}{2} + i(\beta-\gamma))\,d\gamma\Big\} \qquad (j = 1, 2, \ldots),$$

and in virtue of Proposition 7.6 and estimation (24) we get by the Lebesgue dominated convergence theorem

$$\begin{aligned}\mathcal{M}u(z) &= \frac{1}{P(z)}\Big\{\mathcal{M}f(z)+ + \frac{1}{2\pi}\sum_{\nu=0}^{m-1} z^\nu \int_{\mathbf{R}} \mathcal{M}u(\alpha - \tfrac{1}{2} + i\gamma)\mathcal{M}(b_\nu\chi_{t_0})(-\tfrac{1}{2} + i(\beta-\gamma))\,d\gamma\Big\}\\ &= \frac{1}{P(z)}\Big\{\mathcal{M}f(z) + \sum_{\nu=0}^{m-1} z^\nu \mathcal{M}(b_\nu\chi_{t_0}u)(z-1)\Big\}.\end{aligned}$$

Thus

$$P(z)\mathcal{M}u(z) = \mathcal{M}\Big(f + \sum_{\nu=0}^{m-1}\Big(x\frac{d}{dx}\Big)^\nu (xb_\nu\chi_{t_0}u)\Big)(z)$$

and hence u fulfills equation (15) on $\mathbf{R}_+$ and, as just observed $u \in M'^{p}_{(q)}((0,t_0])$. To prove the uniqueness of the solution to the equation (15) in the class of Mellin distributions, suppose that there exists another distribution $v \in M'_{(c)}((0,t_0])$ with $c \le q$ satisfying (15). Hence $\tilde{g} = \mathcal{M}v$ is holomorphic in the half-space $\operatorname{Re} z < c$ and for every $b < c$ and every $\tilde{t}_0 > t_0$ there exist $A = A(b,\tilde{t}_0) > B(b,\tilde{t}_0)$ and $\tilde{p} > p$ such that $|\tilde{g}(z)| < A\tilde{t}_0^{-\alpha}\langle\beta\rangle^{\tilde{p}}$ for $\alpha \le b$. Since

$$\tilde{g}(z) = \frac{1}{P(z)}\Big\{\mathcal{M}f(z) + \sum_{\nu=0}^{m-1} z^\nu \mathcal{M}(b_\nu\chi_{t_0}v)(z-1)\Big\}$$

we get

$$|\tilde{g}(z) - h_j(z)| \le \Big(\frac{\tilde{t}_0}{a}\Big)^{j+1} A\tilde{t}_0^{-\alpha}\langle\beta\rangle^{\tilde{p}} \qquad \text{for } \alpha \le b,\ \beta \in \mathbf{R}, \qquad (j = 0, 1, \ldots)$$

and passing to the limit with $j \to \infty$ we see that $\tilde{g}(z) = g(z)$ for $\operatorname{Re} z < c$. Then by Theorem 9.2 $u = v$ in $M'_{(c)}$. Finally note that $\chi_{t_0} \equiv 1$ on $\operatorname{supp} u$ so that $\chi_{t_0}u = u$ and we have shown the following

Theorem 1. *Let R be a singular differential operator with regular singularity at zero and smooth coefficients in a neighbourhood of zero (see Subsection 2). Let $s \in \mathbf{R}$ and $P(z) \neq 0$ for $\operatorname{Re} z \leq q$. Then there exist a number $0 < a$ such that for every $t_0 < a$ the equation (13) (i.e. (11)) with $f \in M'^{s}_{(q)}((0, t_0])$ has a unique solution $u \in M'^{s-m}_{(q)}((0, t_0])$ on $(0, a)$.*

Remark 3. If we do not require that u belong to $M'_{(q)}((0, t_0])$ the uniqueness assertion in Theorem 1 fails to be true since the homogeneous equation has non-trivial solutions (but with unbounded support).

Remark 4. Note that (see Exercise 11.1) if one takes $f \in M'_{(q)}((0, t_0])$ in Theorem 1 then $u \in M'_{(q)}((0, t_0])$.

Corollary 1. *Let $R(x, x\frac{d}{dx}) = \sum_{j=0}^{m} a_j(x)(x\frac{d}{dx})^j$ be an operator with smooth coefficients in a neighbourhood of an interval $[0, t)$ for some $t > 0$ and let $a_m(x) \neq 0$ on $[0, t)$. Then for every distribution $f \in D'((0, t))$ extendible over zero every solution $u \in D'((0, t))$ of the equation $Ru = f$ on $(0, t)$ is extendible over zero.*

Proof: Let $u \in D'((0, t))$ be a solution of $Ru = f$ on $(0, t)$. Let $\chi \in C^\infty(\mathbf{R}_+)$ be such that $\chi \equiv 1$ on $(0, \tilde{t})$, $\operatorname{supp} \chi \subset I_0$, where $I_0 = (0, \mathring{t}]$, $\tilde{t} < \mathring{t} < t$. To see that u is extendible over zero it is enough by Theorem 8.1 to prove that $\chi u \in M'_{(b)}(I_0)$ for some $b \in \mathbf{R}$. Put

$$g = R(\chi u)$$

and observe that $\operatorname{supp} g \subset I_0$ and g is extendible over zero since coincides with f on $(0, \tilde{t})$. Thus from Theorem 8.1 $g \in M'_{(\omega)}(I_0)$ for some $\omega \in \mathbf{R}$. By Remark 4 there exists $\tilde{u} \in M'_{(\omega)}(I_0)$ such that $R\tilde{u} = g$ on $\mathbf{R}_+$. Thus by (5.16) and Proposition 3.1 $v = \chi u - \tilde{u} \in D'_{I_0}(\mathbf{R}_+)$ and is a distributional solution of the homogeneous equation on $(0, t)$. Since $a_m(x) \neq 0$ on $(0, t)$ it is a classical solution of this equation (see Exercise 1). Because $v \equiv 0$ on $(\mathring{t}, t)$ it follows from the classical theory of the uniqueness of the initial value problem that $v \equiv 0$ on $(0, t)$.

The following regularity theorem gives asymptotic expansions at zero for solutions to (11) with smooth right-hand side:

Theorem 2. *Let $R(x, x\frac{d}{dx}) = \sum_{j=0}^{m} a_j(x)(x\frac{d}{dx})^j$ be an operator with smooth coefficients in a neighbourhood of the interval $[0, t)$ for some $t > 0$ and $a_m(x) \neq 0$ on $[0, t)$. Suppose that $f \in C^\infty([0, t))$ and that $u \in D'((0, t))$ satisfies $Ru = f$ on*

$(0,t)$. Then u is smooth on $(0,t)$ and admits an asymptotic expansion at zero (in the sense of Definition 1) of the form

$$u(x) \sim \sum_{l=0}^{\infty}\sum_{j=0}^{m}\sum_{k=0}^{m} c_{ljk} x^{\gamma_k + l} (\ln x)^j, \tag{25}$$

where $\gamma_0 = 0$ and $\gamma_1, \ldots, \gamma_m$ are the characteristic roots of R at zero (see Subsection 2).

Proof: By Corollary 2 the distribution u is extendible over zero. Let $\chi \in C^\infty(\mathbf{R})$, $\chi \equiv 1$ on $[0, \tilde{t}]$, $\chi \equiv 0$ on $(\tilde{\tilde{t}}, +\infty)$ for some $0 < \tilde{t} < \tilde{\tilde{t}} < t$. Denote $v = \chi u$. Then $\operatorname{supp} v \subset (0, \tilde{\tilde{t}}]$, $v \in D'(\mathbf{R}_+)$ is extendible over zero and by Theorem 8.1 $v \in M'_{(b)}((0,\tilde{\tilde{t}}])$ for some $b \in \mathbf{R}$. Thus by Theorem 9.4* for every $r < b$ there exist $C = C(r, \alpha)$, locally bounded in α and $p = p(r)$ such that

$$|\mathcal{M}v(\alpha + i\beta)| \le C\langle \beta \rangle^p \qquad \text{for } \alpha \le r < b. \tag{26}$$

Write $R(x, x\frac{d}{dx}) = \sum\limits_{j=0}^{m} e_j(x) \frac{d^j}{dx^j}$. Then $e_m(x) \neq 0$ on $(0,t)$ and the equation $Ru = f$ is equivalent on $(0,t)$ to

$$\frac{d^m}{dx^m} u + \sum_{j=0}^{m-1} \tilde{e}_j(x) \frac{d^j}{dx^j} u = \tilde{f}, \tag{27}$$

where $\tilde{f}(x) = \frac{f(x)}{e_m(x)}$ and $\tilde{e}_j(x) = \frac{e_j(x)}{e_m(x)}$ $(j = 0, 1, \ldots, m-1)$ are smooth on $(0,t)$. Hence it follows that $u \in C^\infty((0,t))$ since $u \in D'((0,t))$ is a solution to equation (27) (see Exercise 1). To prove that u admits at zero an asymptotic expansion (25) we shall make use of Proposition 1.

Denote $g = Rv$ and observe that $\operatorname{supp} g \subset (0, \tilde{\tilde{t}}]$, $g = f$ on $(0, \tilde{t})$ and $g \in C^\infty([0,t))$ because $f \in C^\infty([0,t))$.

As it was mentioned in Subsection 2 we may assume $a_m(x) \equiv 1$ on $[0,t)$ and write operator R in the form (12). Computing the Mellin transform of $Rv = g$ we find

$$P(z)(\mathcal{M}v)(z) = \mathcal{M}\Big(xQ\Big(x, x\frac{d}{dx}\Big)v\Big)(z) + \mathcal{M}g(z) \qquad \text{for } \operatorname{Re} z < b \tag{28}$$

and

$$\mathcal{M}\Big(xQ\Big(x, x\frac{d}{dx}v\Big)\Big)(z) = \sum_{\nu=0}^{m-1} z^\nu \mathcal{M}(b_\nu v)(z-1).$$

By Lemma 1 we get the estimations

(29) $$\left|\mathcal{M}\Big(xQ\Big(x, x\frac{d}{dx}\Big)v\Big)(z)\right| \le C_1\langle\beta\rangle^{p+m-1} \quad \text{for} \quad \operatorname{Re} z \le r+1$$

with a new constant $C_1 = C_1(r, \alpha)$ locally bounded in α. Observe that $g\tilde{\chi} = g$. Since $g \in C^\infty([0,t))$, by Remark 1 and Proposition 1 $\mathcal{M}g \in \mathcal{O}(\mathbb{C} \setminus \mathbb{N}_0)$ and for any $j \in \mathbb{N}$, $\varepsilon > 0$ and a bounded set $W \subset \mathbf{R}$ there exists a constant $C_3 = C_3(\varepsilon, j, W)$ such that

(30) $$|\mathcal{M}g(\alpha + i\beta)| \le \frac{C_3}{\langle\beta\rangle^j} \quad \text{for} \quad \alpha \in W, \quad |\beta| \ge \varepsilon.$$

Taking $j \ge -m-p+1$, $j \in \mathbb{N}$ we obtain by (28)–(30) that for every ε and a bounded set W there exists a constant $C_4 = C_4(\varepsilon, W, p)$ such that

(31) $$|Mv(\alpha + i\beta)| \le C_4\langle\beta\rangle^{p-1} \quad \text{for} \quad \alpha \in W, \quad \alpha \le r+1$$

as a function of $\beta = \operatorname{Im} z$ for z outside a neighbourhood of the complex roots of P, and $|\beta| \ge \varepsilon$.

In this way starting from the estimation (26) we get the estimation (31) valid in a bigger domain and with power $p-1$.

Next we determine the set of holomorphy of $\mathcal{M}v$. To this end we consider again equation (28). Observe that $\mathcal{M}v$ is holomorphic for $\operatorname{Re} z < b$ and $\mathcal{M}g \in \mathcal{O}(\mathbb{C} \setminus \mathbb{N}_0)$. Since $v \in M'_{(b)}((0, \tilde{\tilde{t}}])$ it follows by Example 6.3 that $xQ(x, x\frac{d}{dx})v \in M'_{(b+1)}((0, \tilde{\tilde{t}}])$. Thus $\mathcal{M}(xQ(x, x\frac{d}{dx})v)$ is holomorphic for $\operatorname{Re} z < b+1$. Dividing both sides of (28) by $P(z)$ we hence find that $\mathcal{M}v$ is holomorphic for $\operatorname{Re} z < b+1$ outside the complex roots of $P(z)$ and the nonnegative integers. By successive applications of Lemma 1 we find by induction that $\mathcal{M}v$ is holomorphic outside the points $\mathbb{N}_0 \cup (\{z\colon P(z) = 0\} + \mathbb{N}_0)$ and (ii) in Proposition 1 is satisfied for v. Finally by (28) and Remark 2 we infer that at those points $\mathcal{M}v$ has poles of order not greater than $m+1$.

A similar reasoning leads to the following extension of Theorem 1:

Theorem 1*. If in Theorem 1 we take $f \in M(\Omega; \varrho)$ (see Definition 11.5) then $u \in (\widetilde{\Omega}; \varrho - m)$ where $\widetilde{\Omega} = \Omega \setminus (\operatorname{Char} P + \mathbb{N}_0)$.

4. Case of Analytic Coefficients

Before passing to the study of Fuchsian type differential operators with analytic coefficients we introduce the definition of a generalized analytic function. We begin with a proposition characterizing analytic functions in terms of their Mellin transforms. Below, by $\chi_{\tilde{r},r}$ we denote the cut-off function considered in Proposition 7.5.

Proposition 2. *Let f be a function analytic in a (right-sided) neighbourhood of zero of radius of convergence at zero not less than ϱ i.e. $f(x) = \sum_{j=0}^{\infty} a_j x^j$ for $0 \le x < \varrho$ and assume that $\varrho > r > 0$. Then for any cut-off function $\chi_{\tilde{r},r}$*

(i) *$\mathcal{M}(\chi_{\tilde{r},r} f)(z)$ is meromorphic on $\mathbf{C}$ with simple poles at the points of $\mathbf{N}_0$.*

(ii) *For any $0 < \varepsilon < \frac{1}{2}$ and any $p \in \mathbf{N}$ there exists a constant $0 < C^*_{\varepsilon p} < \infty$ (independent of $\tilde{r}, r$ for $\tilde{r} = r\delta < r < \mathring{r}$, with a fixed $\mathring{r} < \varrho$) such that*

$$(32) \qquad |\mathcal{M}(\chi_{\tilde{r},r} f)(\alpha + i\beta)| \le \begin{cases} C^*_{\varepsilon p} \dfrac{r^{-\alpha}}{\langle\beta\rangle^p} & \text{for } \alpha \le -1 \\ C^*_{\varepsilon p} \dfrac{\tilde{r}^{-\alpha}}{\langle\beta\rangle^p} & \text{for } \alpha > -1,\ \alpha + i\beta \notin (\mathbf{N}_0)_\varepsilon, \end{cases}$$

where $(\mathbf{N}_0)_\varepsilon = \{z \in \mathbf{C}\colon \operatorname{dist}(z, \mathbf{N}_0) \le \varepsilon\}$.

Conversely, if F is a meromorphic function on $\mathbf{C}$ with simple poles at $\mathbf{N}_0$ satisfying the condition:

(ii′) *there exist constants $C > 0$ and $s \in \mathbf{R}$ such that*

$$(33) \qquad |F(\alpha + i\beta)| \le \begin{cases} C\langle\beta\rangle^s r^{-\alpha} & \text{for } \alpha \le -1 \\ C\langle\beta\rangle^s \tilde{r}^{-\alpha} & \text{for } \alpha > -1,\ \alpha + i\beta \notin (\mathbf{N}_0)_\varepsilon \end{cases}$$

for a fixed ε, $0 < \varepsilon < \frac{1}{2}$ and some $0 < \tilde{r} < r$,

then F is the Mellin transform of a Mellin distribution with support in $(0, r]$ which on the set $(0, \tilde{r})$ coincides with a function analytic on $(-\tilde{r}, \tilde{r})$ with a radius of convergence not less than $\tilde{r}$.

Proof: Write

$$\chi_{\tilde{r},r}(x) f(x) = \sum_{j=0}^{\infty} a_j x^j \chi_{\tilde{r},r}(x) \qquad \text{for } 0 \le x \le r$$

with the series being absolutely convergent. Then

$$\mathcal{M}(\chi_{\tilde{r},r} f)(z) = \sum_{j=0}^{\infty} a_j (\mathcal{M}\chi_{\tilde{r},r})(z - j).$$

Fix $r < \mathring{r} < \varrho$. By Proposition 7.5 we get

$$|\mathcal{M}(\chi_{\tilde{r},r}f)(\alpha+i\beta)|$$

$$\leq \begin{cases} \dfrac{C_{\varepsilon p}}{\langle\beta\rangle^p}\sum\limits_{j=0}^{\infty}|a_j|\, r^{-\alpha+j} & \text{for } \alpha\leq -1 \\ \dfrac{C_{\varepsilon p}}{\langle\beta\rangle^p}\Big(\sum\limits_{j=0}^{k}|a_j|\,\tilde{r}^{-\alpha+j}+\sum\limits_{j=k+1}^{\infty}|a_j|\,r^{-\alpha+j}\Big) & \text{for } \alpha+i\beta\notin(\mathbf{N}_0)_\varepsilon \end{cases}$$

$$k-1<\alpha\leq k \quad (k=0,1,\ldots).$$

Put $c=\sum\limits_{j=0}^{\infty}|a_j|\,\mathring{r}^j$ and note that $\sum\limits_{j=0}^{\infty}|a_j|\,\tilde{r}^j<c$, $\sum\limits_{j=0}^{\infty}|a_j|\,r^j<c$. Then estimations (32) follow immediately from the above formulas with $C^*_{\varepsilon p}=cC_{\varepsilon p}(1+\delta^\alpha)$, since $\tilde{r}^{-\alpha}+r^{-\alpha}=\tilde{r}^{-\alpha}(1+\delta^\alpha)$.

Now let F be a meromorphic function on $\mathbf{C}$ with simple poles at $\mathbf{N}_0$ satisfying (33). One may assume that $s\in\mathbf{N}$. Fix $\mathring{\alpha}\leq -1$ and set

$$G(z)=\frac{F(z)}{(z-\mathring{\alpha})^{s+2}},$$

$$g(x)=\frac{1}{2\pi i}\int_{\mathring{\alpha}-1-i\infty}^{\mathring{\alpha}-1+i\infty}G(z)x^z\,dz \qquad \text{for } x>0.$$

G is a meromorphic function on $\mathbf{C}$ and is holomorphic for $\operatorname{Re} z<\mathring{\alpha}$. By (33) and Lemma 9.1 the function g is continuous, $\operatorname{supp} g\subset(0,r]$, $g\in M'_{(\mathring{\alpha}-1)}$ and $\mathcal{M}g(z)=G(z)$ for $\operatorname{Re} z\leq\mathring{\alpha}-1$. Let

$$f=\Big(x\frac{d}{dx}-\mathring{\alpha}\Big)^{s+2}g \qquad \text{on } \mathbf{R}_+.$$

Clearly $\operatorname{supp} f\subset(0,r]$ and

$$\mathcal{M}f(z)=(z-\mathring{\alpha})^{s+2}\mathcal{M}g(z)=F(z).$$

It remains to prove that f is an analytic function which extends analytically to a neighbourhood of zero with a radius of convergence not less than $\tilde{r}$. To this aim take c arbitrarily big and such that the line $c+i\mathbf{R}$ is contained in $\mathbf{C}\setminus(\mathbf{N}_0)_\varepsilon$. By (33) and the residue theorem we get

$$\frac{1}{2\pi i}\int_{c-i\infty}^{c+i\infty}G(z)x^z\,dz-g(x)=\sum_{j=0}^{[c]}-a_jx^j+\operatorname{Res}_{z=\mathring{\alpha}}(G(z)x^z),$$

where $[c]$ is the entire part of c and $-a_j$ is the residuum of G at the point j. Using once more the estimations (33) we find

$$\left| \int_{c-i\infty}^{c+i\infty} G(z)x^z \, dz \right| \le \widetilde{C} \left(\frac{x}{\tilde{r}}\right)^c \qquad \text{with } \widetilde{C} < \infty.$$

For $0 < x < \tilde{r}$ the integral can be made arbitrarily small if c is large enough. This proves that

$$g(x) + \mathrm{Res}_{z=\mathring{\alpha}}(G(z)x^z) = \sum_{j=0}^{\infty} a_j x^j \qquad \text{for } 0 < x < \tilde{r}$$

and hence $h(x) \stackrel{\text{def}}{=} \sum_{j=0}^{\infty} a_j x^j$ converges for $|x| < \tilde{r}$. Writing

$$\mathrm{Res}_{z=\mathring{\alpha}}(G(z)x^z) = \frac{1}{2\pi i} \int_\gamma G(z)x^z \, dz$$

(γ is a suitable curve encircling $\mathring{\alpha}$) we easily get

$$\left(x\frac{d}{dx} - \mathring{\alpha}\right)^{s+2} (\mathrm{Res}_{z=\mathring{\alpha}} \, G(z)x^z)) = 0$$

and hence

$$f(x) = \left(x\frac{d}{dx} - \mathring{\alpha}\right)^{s+2} h(x) \qquad \text{for } 0 < x < \tilde{r}$$

is analytic with the radius of convergence not less than $\tilde{r}$.

In the same way, with Proposition 7.5 replaced by Corollary 7.3(i) and by Exercise 7.13 one can prove the following generalization of Proposition 2.

Proposition 2′. *Let $a, b \in \mathbb{N}_0$, $\gamma_k \in \mathbb{C}$ for $k = 0, \ldots, b, \varepsilon > 0$. Let $\widetilde{Z} = \{\gamma_0 + \mathbb{N}_0\} \cup \cdots \cup \{\gamma_b + \mathbb{N}_0\}$, $\widetilde{Z}_\varepsilon = \{z \in \mathbb{C}\colon \mathrm{dist}(z, \widetilde{Z}) \le \varepsilon\}$. Suppose that the series $\sum_{l=0}^{\infty} c_{ljk} x^l$ $(j = 0, \ldots, a,\ k = 0, \ldots, b)$ have radius of convergence $\varrho > 0$ and let*

$$f(x) = \sum_{l=0}^{\infty} \sum_{j=0}^{a} \sum_{k=0}^{b} c_{ljk} x^{\gamma_k + l} (\ln x)^j \tag{34}$$

for $0 < x < \varrho$. Then for any cut-off function $\chi_{\tilde{r},r}$ with $r < \varrho$

(i) *$\mathcal{M}(\chi_{\tilde{r},r}f)(z)$ is meromorphic on $\mathbf{C}$ with poles at $\widetilde{Z}$ of order not greater than $a+1$,*

(ii) *For any $0 < \varepsilon < \frac{1}{2}$ and any $p \in \mathbf{N}$ there exists a constant $C_{\varepsilon p} = C_{\varepsilon p}(r) > 0$ such that*

$$(35) \qquad |\mathcal{M}(\chi_{\tilde{r},r}f)(\alpha+i\beta)| \le \begin{cases} C_{\varepsilon p}\dfrac{r^{-\alpha}}{\langle\beta\rangle^p} & \text{for } \alpha \le -1 \\ C_{\varepsilon p}\dfrac{\tilde{r}^{-\alpha}}{\langle\beta\rangle^p} & \text{for } \alpha > -1, \alpha+i\beta \notin \widetilde{Z}_\varepsilon. \end{cases}$$

Conversely, if F is a meromorphic function on $\mathbf{C}$ with poles at $\widetilde{Z}$ of order not greater than $a+1$ satisfying the condition:

(ii′) *there exist constants $C > 0$ and $s \in \mathbf{R}$ such that*

$$|F(\alpha+i\beta)| \le \begin{cases} C\langle\beta\rangle^s r^{-\alpha} & \text{for } \alpha \le -1 \\ C\langle\beta\rangle^s \tilde{r}^{-\alpha} & \text{for } \alpha > -1,\ \alpha+i\beta \ne \widetilde{Z}_\varepsilon \end{cases}$$

for a fixed $0 < \varepsilon$ sufficiently small and some $0 < \tilde{r} < r$,

then F is the Mellin transform of a Mellin distribution with support in $(0,r]$ which on the set $(0,\tilde{r})$ coincides with the function (34) where the series $\sum_{l=0}^{\infty} c_{ljk}x^l$ ($j = 0,\dots,a$, $k = 0,\dots,b$) have the radius of convergence not less than $\tilde{r}$.

Remark 5. If we assume that condition (ii′) in Proposition 2 ((ii′) in Proposition 2′, resp.) hold for any $\tilde{r} < r < \varrho$ with a constant $C = C(\tilde{r},r)$ then Proposition 2 (2′, resp.) so modified characterizes functions analytic on $(-\varrho,\varrho)$ of convergence radius not less than ϱ (functions (34), resp.).

Inspired by Propositions 2 and 2′ and Corollary 9.1 with Exercise 9.8 we introduce the definition of a generalized analytic function.

Let Z be a closed subset of $\mathbf{C}$ such that there exist points $\zeta_k \in \mathbf{C}$, $k = 0,\dots,b$ so that

$$Z \subset \mathbf{L} = \bigcup_{k=0}^{b} \zeta_k + \overline{\mathbf{R}_+},$$

and denote

$$\mathbf{L}_\varepsilon = \{z \in \mathbf{C}\colon \operatorname{dist}(z,\mathbf{L}) \le \varepsilon\}.$$

Definition 2. A distribution $u \in D'((0,t))$ extendible over zero is called a *generalized analytic function* (*of type* Z and order p) in a right-sided neighbourhood of zero, of convergence radius not less than ϱ, $0 < \varrho \leq t$, if for any cut-off function $\chi_{\tilde{r},r}$, $\tilde{r} < r < \varrho$

(i) $\mathcal{M}(\chi_{\tilde{r},r}u)$ extends[(1)] to a holomorphic function on $\mathbf{C} \setminus Z$,

(ii) there exists $s \in \mathbf{R}$ such that for any $\varepsilon > 0$ and $\kappa > 0$ if $\alpha + i\beta \notin \mathbf{L}_\varepsilon$ and $r_\kappa = e^\kappa r$, $\tilde{r}_{-\kappa} = e^{-\kappa}\tilde{r}$

$$|\mathcal{M}(\chi_{\tilde{r},r}u)(\alpha+i\beta)| \leq \begin{cases} C_{\varepsilon,\kappa}\langle\beta\rangle^s r_\kappa^{-\alpha} & \text{for } \alpha \leq -1 \\ C_{\varepsilon,\kappa}\langle\beta\rangle^s \tilde{r}_{-\kappa}^{-\alpha} & \text{for } \alpha \geq -1 \end{cases}$$

and some constants $C_{\varepsilon,\kappa}$ (which may depend on r),

(iii) $\mathcal{M}(\chi_{\tilde{r},r}u)$ is of *uniformly*[(2)]*polynomial growth* close to $\mathbf{L}$, i.e. there exist $\varepsilon > 0$ and some $p = p(\kappa) \in \mathbf{N}$ such that for any $\kappa > 0$

$$|\mathcal{M}(\chi_{\tilde{r},r}u)(\alpha+i\beta)| \leq C_\kappa \frac{\tilde{r}_{-\kappa}^{-\alpha}}{(\operatorname{dist}(\beta, \{\operatorname{Im}\zeta_0, \ldots, \operatorname{Im}\zeta_b\}))^p}$$

for $\alpha + i\beta \in \mathbf{L}_\varepsilon$, with some constant C_κ.

Remark 6. Every function u analytic at zero is a generalized analytic function of type $\mathbf{N}_0$ with $p = 1$. Indeed, by Proposition 2 it is enough to show that there exist $\varepsilon > 0$ and a constant $C < \infty$ such that

$$|\mathcal{M}(\chi_{\tilde{r},r}u)(\alpha+i\beta)| \leq C\frac{\tilde{r}^{-\alpha}}{|\beta|} \qquad \text{for } \alpha + i\beta \in (\mathbf{R}_+)_\varepsilon,$$

and this follows from Exercise 2. Moreover it can be proved that every generalized analytic function $u \in M'_{(0)}$ of type $\mathbf{N}_0$ with $p = 1$ is in fact an analytic function at zero. Analogously, generalized analytic functions of type $Z = \bigcup_{k=0}^{b} \{\gamma_k + \mathbf{N}_0\}$ with $p = a + 1$ coincide with the functions of Proposition 2′ (see Exercise 3).

Below we give a less evident example

Example 1. The function u

$$(0,1) \ni x \mapsto \frac{1}{(-\ln x)^j}$$

---

(1) i.e. it extends to a (single valued) holomorphic function in a simply connected domain $\mathbf{C} \setminus \mathbf{L}$, extendible by continuity to a holomorphic function on $\mathbf{C} \setminus Z$.

(2) The condition that the polynomial growth is uniform with respect to α is strictly connected with the "analyticity". In the "smooth" case the estimations are locally uniform with respect to α (see [Zie 4, 5, 6]).

is a generalized analytic function at zero of convergence radius $\mathring{r} = 1$ (of type $\overline{\mathbf{R}_+}$ with $p = 1$).

Proof: Observe that $u \in M'_{(0)}$ locally at zero and that for $0 < x < 1$

$$\frac{1}{(-\ln x)^j} = \int_0^\infty x^\varrho \frac{\varrho^{j-1}}{(j-1)!}\, d\varrho. \tag{36}$$

Take a cut off function $\chi_{\tilde{r},r}$, $r < 1$. Then it follows from (36) that

$$\mathcal{M}\Big(\frac{\chi_{\tilde{r},r}}{(-\ln x)^j}\Big)(z) = \int_0^\infty \frac{\varrho^{j-1}}{(j-1)!}(\mathcal{M}\chi_{\tilde{r},r})(z-\varrho)\, d\varrho$$

and hence by (7.34) we verify the properties (i), (ii) in Definition 2 with $\mathbf{Z} = \overline{\mathbf{R}_+}$, $s = 0$. For the proof of (iii) with $p = 1$, $\mathbf{L}_\varepsilon = (\overline{\mathbf{R}_+})_\varepsilon$ we proceed similarly as in Remark 7 (cf. Exercise 2 and Example 7.1).

We are just ready to examine equation (11) in which R is a singular differential operator with regular singularity at zero and analytic coefficients in a neighbourhood of zero. Obviously Theorem 1 holds in this case but our aim now is to establish that if f is a generalized analytic function then so is every solution u of (11) (see Theorem 3).

Fix q so that $P(z) \neq 0$ for $\operatorname{Re} z \leq q$. Denote by $z_1, \ldots, z_k$ $(1 \leq k \leq m)$ all different roots of the polynomial $P(z) = z^m + \sum_{\nu=0}^{m-1} c_\nu z^\nu$ i.e. $\operatorname{Char} P = \{z_1, \ldots, z_k\}$. Denote by $\mathring{r}$ the radius of convergence of the coefficients of operator R and let

$$\operatorname{Reg}^\varepsilon P = \{z \in \mathbf{C}\colon \operatorname{dist}(z, \operatorname{Char} P) \geq \varepsilon\}.$$

Let $\zeta_0, \ldots, \zeta_b \in \mathbf{C}$, $\min(\operatorname{Re}\zeta_0, \ldots, \operatorname{Re}\zeta_b) > q$ and let $\mathbf{L} = \bigcup_{j=0}^{b} (\zeta_j + \overline{\mathbf{R}_+})$.

Let $v \in M'_{(\omega)}((0, r])$, $q < \omega < \min(\operatorname{Re}\zeta_0, \ldots, \operatorname{Re}\zeta_b)$ and suppose $\mathcal{M}v$ satisfies the conditions (i)–(iii) in Definition 2 for some fixed $0 < \tilde{r} < r < \mathring{r}$. First we shall solve the equation $Ru = v$, which can be written in the form

$$P\Big(x\frac{d}{dx}\Big)u + \sum_{\nu=0}^{m-1}\sum_{j=0}^{\infty} a_j^\nu x^{j+1}\Big(x\frac{d}{dx}\Big)^\nu u = v, \tag{37}$$

where $\sum_{j=0}^{\infty} |a_j^\nu|\, \mathring{r}^j < \infty$ for $\nu = 0, 1, \ldots, m-1$.

Suppose that there exists a Mellin distribution $u \in M'_{(q)}((0,r])$ satisfying (37). Put

$$g(z) = \mathcal{M}u(z), \qquad h(z) = \frac{\mathcal{M}v(z)}{P(z)}.$$

Then by the Mellin transformation we get from (37)

$$g(z) = -\sum_{l=1}^{\infty}\sum_{\nu=0}^{m-1} a_{l-1}^{\nu}\frac{(z-l)^{\nu}}{P(z)}g(z-l) + h(z). \tag{38}$$

We solve (38) by the method of successive approximations according to the scheme

$$g_0(z) = h(z), \qquad g_{j+1}(z) = \sum_{l=1}^{\infty} p_l(z)g_j(z-l) + h(z) \qquad (j = 0, 1, \ldots), \tag{39}$$

where

$$p_l(z) = -\sum_{\nu=0}^{m-1} a_{l-1}^{\nu}\frac{(z-l)^{\nu}}{P(z)} \qquad (l = 1, 2, \ldots).$$

By the assumption on v we see that $g_0 = h$ satisfies the following conditions:

(i_0) g_0 is holomorphic on the set $\mathbf{C} \setminus (\mathbf{L} \cup \operatorname{Char} P)$,

(ii_0) there exists $s \in \mathbf{R}$ such that for any $\varepsilon > 0$ and $\kappa > 0$ there exist $C_{\varepsilon\kappa}$ such that

$$|g_0(\alpha + i\beta)| \le \begin{cases} C_{\varepsilon\kappa}\langle\beta\rangle^s r_{\kappa}^{-\alpha} & \text{for } \alpha \le -1 \\ C_{\varepsilon\kappa}\langle\beta\rangle^s \tilde{r}_{-\kappa}^{-\alpha} & \text{for } \alpha > -1 \end{cases}$$

for $\alpha + i\beta \in (\mathbf{C} \setminus \mathbf{L}_{\varepsilon}) \cap \operatorname{Reg}^{\varepsilon} P$.

We shall state similar conditions (i_j)–(ii_j) for g_j $(j = 1, 2, \ldots)$. We begin with the function

$$g_1(z) = \sum_{l=1}^{\infty} p_l(z)g_0(z-l) + h(z).$$

It is clear that

(i_1) g_1 is holomorphic on the set $\mathbf{C} \setminus (\mathbf{L} \cup (\operatorname{Char} P + \mathbf{N}_0))$.

To get the estimations for g_1 observe first that

$$|g_1(z)| \le \sum_{l=1}^{\infty}\sum_{\nu=0}^{m-1} |a_{l-1}^{\nu}|\frac{|z-l|^{\nu}}{|P(z)|}|g_0(z-l)| + |h(z)|$$

and that to every $\varepsilon > 0$ there exists $C_\varepsilon < \infty$ such that

$$\frac{|z-l|^\nu}{|P(z)|\, l^\nu} < C_\varepsilon \qquad \text{if } z \in \operatorname{Reg}^\varepsilon P \qquad (l = 1, 2, \ldots, \quad \nu = 0, 1, \ldots, m-1).$$

Hence from estimations (ii_0) for $g_0 = h$ we get (in a similar way as (32) was derived from (7.34)) for $z \in (\mathbf{C} \setminus \mathbf{L}_\varepsilon) \cap (\operatorname{Reg}^\varepsilon P + \mathbf{N}_0)$ the estimations

$$|g_1(\alpha + i\beta)| \le C_\varepsilon C_{\varepsilon\kappa} \sum_{l=1}^{\infty} \sum_{\nu=0}^{m-1} l^\nu \left|a_{l-1}^\nu\right| \langle\beta\rangle^s r_\kappa^{-\alpha+l} + C_{\varepsilon\kappa} \langle\beta\rangle^s r_\kappa^{-\alpha} \qquad \text{for } \alpha \le -1,$$

(40)

$$|g_1(\alpha + i\beta)| \le C_\varepsilon C_{\varepsilon\kappa} \sum_{l=1}^{\infty} \sum_{\nu=0}^{m-1} l^\nu \left|a_{l-1}^\nu\right| \langle\beta\rangle^s \tilde{r}_{-\kappa}^{-\alpha+l} + C_{\varepsilon\kappa} \langle\beta\rangle^s \tilde{r}_{-\kappa}^{-\alpha} \qquad \text{for } \alpha > -1.$$

Let

$$A(r) = \sum_{\nu=0}^{m-1} \left(\frac{d}{dr} r\right)^\nu \sum_{j=0}^{\infty} |a_j^\nu|\, r^j.$$

By assumption on the coefficients a_j^ν in (37), $A(r) < \infty$ since $r < \mathring{r}$. Moreover $A(r)$ is a non decreasing function and

$$A(r) = \sum_{l=1}^{\infty} \sum_{\nu=0}^{m-1} \left|a_{l-1}^\nu\right| r^{l-1} l^\nu.$$

Hence we can put the estimations (40) in the following form

$$|g_1(\alpha + i\beta)| \le \begin{cases} C_{\varepsilon\kappa}(1 + C_\varepsilon r_\kappa A(r_\kappa)) \langle\beta\rangle^s r_\kappa^{-\alpha} & \text{for } \alpha \le -1 \\ C_{\varepsilon\kappa}(1 + C_\varepsilon \tilde{r}_{-\kappa} A(\tilde{r}_{-\kappa})) \langle\beta\rangle^s \tilde{r}_{-\kappa}^{-\alpha} & \text{for } \alpha > -1, \\ & z \in (\mathbf{C} \setminus \mathbf{L}_\varepsilon) \cap (\operatorname{Reg}^\varepsilon P + \mathbf{N}_0). \end{cases}$$

By induction we show that for $j = 1, 2, \ldots$

(i_j) $\quad g_j$ is holomorphic in the set $C \setminus (\mathbf{L} \cup (\operatorname{Char} P + \mathbf{N}_0))$

and moreover that the following estimations hold

$$|g_j(\alpha + i\beta)| \le \begin{cases} C_{\varepsilon\kappa} \langle\beta\rangle^s r_\kappa^{-\alpha} \sum_{l=0}^{j} (C_\varepsilon r_\kappa A(r_\kappa))^l & \text{for } \alpha \le -1, \\ C_{\varepsilon\kappa} \langle\beta\rangle^s \tilde{r}_{-\kappa}^{-\alpha} \sum_{l=0}^{j} (C_\varepsilon \tilde{r}_{-\kappa} A(\tilde{r}_{-\kappa}))^l & \text{for } \alpha > -1, \\ & z \in (\mathbf{C} \setminus \mathbf{L}_\varepsilon) \cap (\operatorname{Reg}^\varepsilon P + \mathbf{N}_0). \end{cases}$$

Let r_1 be so small that $C_\varepsilon r_1 A(r_1) < 1$. Let $r_\kappa < r_1$. Since A in non-decreasing ($C_\varepsilon r A(r) < 1$ for $r < r_1$) then for some constant $\widetilde{C}_{\varepsilon\kappa}$ (independent of j) we have for $j = 1, 2, \ldots$

$$(\mathrm{ii}_j) \qquad |g_j(\alpha + i\beta)| \le \begin{cases} \widetilde{C}_{\varepsilon\kappa}\langle\beta\rangle^s r_\kappa^{-\alpha} & \text{for } \alpha \le -1, \\ \widetilde{C}_{\varepsilon\kappa}\langle\beta\rangle^s \tilde{r}_{-\kappa}^{-\alpha} & \text{for } \alpha > -1, \quad z \in (\mathbf{C} \setminus \mathbf{L}_\varepsilon) \cap (\operatorname{Reg}^\varepsilon P + \mathbf{N}_0). \end{cases}$$

In a similar way we derive, for $j = 1, 2, \ldots$, the estimations

$$|g_j(\alpha + i\beta) - g_{j-1}(\alpha + i\beta)| \le \begin{cases} C_{\varepsilon\kappa}\langle\beta\rangle^s r_\kappa^{-\alpha}(C_\varepsilon r_\kappa A(r_\kappa))^j & \text{for } \alpha \le -1, \\ C_{\varepsilon\kappa}\langle\beta\rangle^s \tilde{r}_{-\kappa}^{-\alpha}(C_\varepsilon \tilde{r}_{-\kappa} A(\tilde{r}_{-\kappa}))^j & \text{for } \alpha > -1, \\ & z \in (\mathbf{C} \setminus \mathbf{L}_\varepsilon) \cap (\operatorname{Reg}^\varepsilon P + \mathbf{N}), \end{cases}$$

which, due to our choice of r_1 prove that the series $g_0 + \sum_{j=1}^{\infty} g_j - g_{j-1}$ is almost uniformly convergent on the set $\mathbf{C} \setminus (\mathbf{L} \cup (\operatorname{Char} P + \mathbf{N}_0))$, and $g(z) = \lim_{j\to\infty} g_j(z)$ is a holomorphic function on that set. By (ii_j) we get for $\alpha + i\beta \in (\mathbf{C} \setminus \mathbf{L}) \cap (\operatorname{Reg}^\varepsilon P + \mathbf{N}_0)$

$$|g(\alpha + i\beta)| \le \begin{cases} \widetilde{C}_{\varepsilon\kappa}\langle\beta\rangle^s r_\kappa^{-\alpha} & \text{for } \alpha \le -1, \\ \widetilde{C}_{\varepsilon\kappa}\langle\beta\rangle^s \tilde{r}_{-\kappa}^{-\alpha} & \text{for } \alpha > -1. \end{cases}$$

Hence by Theorem 9.3 g is the Mellin transform of a distribution $u \in M'_{(q)}((0, r])$ which solves (37). We shall check that the function g satisfies the condition

$$(41) \qquad |g(\alpha + i\beta)| \le C\tilde{r}_{-\kappa}^{-\alpha} \cdot (\operatorname{dist}(\beta, \{\operatorname{Im}\zeta_0, \ldots, \operatorname{Im}\zeta_b, \operatorname{Im} z_1, \ldots, \operatorname{Im} z_k\}))^{-p-m}.$$

To this end observe that in establishing estimations (ii_j) we made no use of the term $\langle\beta\rangle^s$. Thus it can be replaced by $(\operatorname{dist}(\beta, \{\operatorname{Im}\zeta_0, \ldots, \operatorname{Im}\zeta_b, \operatorname{Im} z_1, \ldots, \operatorname{Im} z_k\}))^{-p-m}$ which gives (41).

Theorem 3. *Let $\mathring{r} > 0$ and let $R = R(x, x\frac{d}{dx})$ be an operator with coefficients analytic at zero whose radius of convergence is not less than $\mathring{r}$. Then there exists $0 < r_1 < \mathring{r}$ such that for any generalized analytic function f (of type Z and order p) at zero of convergence radius $\mathring{r}$, every Mellin distribution u satisfying $Ru = f$ on $(0, \mathring{r})$ is a generalized analytic function (of type $Z \cup \{\operatorname{Char} P + \mathbf{N}_0\}$ and order $p + m$) of convergence radius not less than r_1.*

Proof: Let $\chi_{\tilde{r},r}$ be the cut-off function with $r < r_1$ where r_1 was defined before the statement of Theorem 3. We have

$$R(\chi_{\tilde{r},r} u) = \chi_{\tilde{r},r} Ru + \tilde{f}.$$

$\tilde{f}$ is clearly a distribution on $(0, \mathring{r})$ supported by $[\tilde{r}, r]$. Then by Corollary 9.1 $\mathcal{M}\tilde{f}$ is entire and for every $\kappa > 0$ there exists $\tilde{s} \in \mathbf{R}$ such that

$$|\mathcal{M}\tilde{f}(\alpha + i\beta)| \le \begin{cases} \widetilde{C}_\kappa \langle\beta\rangle^{\tilde{s}} r_\kappa^{-\alpha} & \text{for } \alpha \le -1, \\ \widetilde{C}_\kappa \langle\beta\rangle^{\tilde{s}} \tilde{r}_{-\kappa}^{-\alpha} & \text{for } \alpha > -1. \end{cases} \tag{42}$$

We have $Ru = f$, thus $\mathcal{M}(\chi_{\tilde{r},r} Ru) = \mathcal{M}(\chi_{\tilde{r},r} f)$ fulfills the conditions (i)–(iii) in Definition 2. These conditions together with (42) imply that the function $R(\chi_{\tilde{r},r} u) \in M'_{(\omega)}((0, r])$ is such that its Mellin transform satisfies (i)–(iii) in Definition 2. It follows from the considerations preceding Theorem 3 that $\chi_{\tilde{r},r} u$ also fulfills (i)–(iii) in Definition 2 which proves that it is a generalized analytic function at zero.

From the proof of Theorem 3 and Proposition 2′ we easily get (see Exercise 4)

Corollary 2. *If $R = R\big(x, x\frac{d}{dx}\big)$ is an operator with coefficients analytic at zero then any solution $u \in D'((0,t))$ of $Ru = 0$ can be written in the form*

$$u(x) = \sum_{l=0}^{\infty} \sum_{j=0}^{m-1} \sum_{i=1}^{k} c_{lji} x^{z_i + l} (\ln x)^j,$$

where the series has a positive radius of convergence.

5. Special functions as generalized analytic functions

Generalized analytic functions of Subsection 13.4 are represented in the form (46) which can be regarded as a continuous version of convergent power series. It is shown that all important special functions (except the elliptic functions) are generalized analytic functions.

We start by recalling the definition of some of the special functions and transforming them to the form suitable for our purposes:
the error function

$$\mathrm{Erf}(y) = \int_0^y e^{-t^2}\, dt = \frac{y}{2} \int_0^1 e^{-y^2\alpha} \frac{d\alpha}{\sqrt{\alpha}} \qquad \text{for } y > 0;$$

the complementary error function

$$\mathrm{Erfc}(y) = \int_y^\infty e^{-t^2}\, dt = \frac{y}{2} \int_1^\infty e^{-y^2\alpha} \frac{d\alpha}{\sqrt{\alpha}} \qquad \text{for } y > 0;$$

the incomplete gamma functions

$$\gamma(a,y) = \int_0^y e^{-t} t^{a-1}\, dt = y^a \int_0^1 e^{-y\alpha} \alpha^{a-1}\, d\alpha \qquad \text{for } y > 0,$$

$$\Gamma(a,y) = \int_y^\infty e^{-t} t^{a-1}\, dt = y^a \int_1^\infty e^{-y\alpha} \alpha^{a-1}\, d\alpha \qquad \text{for } y > 0$$

(clearly $\gamma(a,y) + \Gamma(a,y) = \Gamma(a)$—the "usual" Euler Γ-function);
the exponential integral

$$-\operatorname{Ei}(-y) = \int_y^\infty e^{-t} \frac{dt}{t} = \int_1^\infty e^{-y\alpha} \frac{d\alpha}{\alpha} \qquad \text{for } y > 0;$$

the logarithmic integral

$$\operatorname{Li}(x) = \int_0^x \frac{dt}{\ln t} = -\int_1^\infty x^\alpha \frac{d\alpha}{\alpha} \qquad \text{for } 0 < x < 1.$$

Further examples of special functions are listed at the end of the subsection.

Observe that the function Li is the example of a function of the form

$$u(x) = \int_0^\infty x^\alpha T(\alpha)\, d\alpha \tag{43}$$

for a suitable function T. The representation (43) will be generalized below to cover the remaining special functions listed above (in variables $x = e^{-y^2}$ or $x = e^{-y}$, respectively). First, $T(\alpha)\, d\alpha$ may be replaced by a complex measure $T = d\mu(\alpha)$. Thus (43) becomes

$$u(x) = \int_0^\infty x^\alpha\, d\mu(\alpha)$$

which can equivalently be written as

$$u(x) = T[x^\alpha], \tag{44}$$

where $T = d\mu(\alpha)$ is regarded as a linear functional in α (distribution of order 0) with support in $\overline{\mathbf{R}_+}$ and x^α denotes the test function $\mathbf{R} \ni \alpha \mapsto x^\alpha$ for a fixed $x > 0$. For instance if

$$T = \sum_{j=0}^\infty a_j \delta_{(j)}$$

with $\overline{\lim_{j\to\infty}} \sqrt[j]{|a_j|} < \infty$, then (44) becomes

$$u(x) = T[x^\alpha] = \Big(\sum_{j=0}^{\infty} a_j\delta_{(j)}\Big)[x^\alpha] = \sum_{j=0}^{\infty} a_j x^j$$

i.e. u is an analytic function in a neighbourhood of zero. We can generalize further the expression (44) by taking as T the so called Laplace distributions (which are the Mellin distributions in the variable $\alpha = -\ln x$), which can be described as follows (see Theorem 8.2): Given $\omega \in \mathbf{R}$, for every $\kappa > 0$ T can be expressed as a distributional derivative

$$T = \sum_{k=0}^{m} \Big(\frac{d}{d\alpha}\Big)^k T_{k,\kappa}$$

where $T_{k,\kappa}$ is a measurable function on $\mathbf{R}$ with support in $\overline{\mathbf{R}_+}$ satisfying

$$(45) \qquad |T_{k,\kappa}(\alpha)| \le Ce^{(-\omega+\kappa)\alpha} \qquad \text{for } 0 \le \alpha < \infty$$

with $m = m(\kappa) \in \mathbf{N}_0$ and $C = C(\kappa) > 0$.

For T being a Laplace distribution we have

$$(46) \qquad u(x) = T[x^\alpha] = \sum_{k=0}^{n}(-\ln x)^k \int_0^\infty x^\alpha T_{k,\kappa}(\alpha)\, d\alpha \qquad \text{for } 0 < x < e^{\omega-\kappa}$$

with the integral converging absolutely. We then say that the (absolute) convergence radius of u is greater than or equal to $\varrho = e^\omega$.

In Theorem 4 below we establish that the generalized analytic functions introduced in Subsection 13.4 coincide with the above class of functions. Thus we call functions (46) the generalized analytic functions (abbreviated to GAF).

Theorem 4. *A distribution $u \in D'((0,t))$ is a generalized analytic function of type $\overline{\mathbf{R}_+}$ and convergence radius $\ge \varrho = e^\omega$ if and only if for every $\kappa > 0$ u can be represented in the form (46) for $0 < x < e^{\omega-\kappa}$.*

Proof: Suppose u is a function of the form (46) with T being a Laplace distribution. Denote

$$G_{\tilde{r},r} = \mathcal{M}\chi_{\tilde{r},r}$$

where $\chi_{\tilde{r},r}$ is the cut-off function as in Proposition 7.5. Let $r < \varrho = e^\omega$ and choose κ so that $r < \varrho e^{-2\kappa} = \mathring{r}$. Since

$$\Big(\frac{d}{dz}\Big)^k G_{\tilde{r},r}(z-\alpha) = \mathcal{M}((-\ln x)^k \chi_{\tilde{r},r})(z-\alpha),$$

it follows from (45) and (46), after entering with the Mellin transformation under the integral sign, that

$$(47)\qquad \mathcal{M}(\chi_{\tilde{r},r}u)(z) = \sum_{k=0}^{m} \int_0^\infty \left(\frac{d}{dz}\right)^k G_{\tilde{r},r}(z-\alpha)T_{k,\kappa}(\alpha)\,d\alpha \qquad \text{for} \quad \operatorname{Re} z < 0.$$

By (45) and Exercise 7.13 for every $\varepsilon > 0$ and $p \in \mathbf{N}$ there exists a constant $C_{\varepsilon,p,r}$ such that for $0 < \alpha < \infty$

$$\left|\left(\frac{d}{dz}\right)^k G_{\tilde{r},r}(z-\alpha)T_{k,\kappa}(\alpha)\right| \le \begin{cases} C_{\varepsilon,p,r}\dfrac{r^{-\operatorname{Re} z}}{\langle \operatorname{Im} z\rangle^p} e^{-\kappa\alpha} & \text{for } \operatorname{Re} z \le -1, \\ C_{\varepsilon,p,r}\dfrac{\tilde{r}^{-\operatorname{Re} z}}{\langle \operatorname{Im} z\rangle^p} e^{-\kappa\alpha} & \text{for } \operatorname{Re} z > -1,\ |z-\alpha| \ge \varepsilon \end{cases}$$

(since $\alpha(\ln r - \omega + \kappa) < -\kappa\alpha$). Hence for some new constant $C = C_{\varepsilon,p,r}$

$$|\mathcal{M}(\chi_{\tilde{r},r}u)(z)| \le \begin{cases} C\dfrac{r^{-\operatorname{Re} z}}{\langle \operatorname{Im} z\rangle^p} & \text{for } \operatorname{Re} z \le -1, \\ C\dfrac{\tilde{r}^{-\operatorname{Re} z}}{\langle \operatorname{Im} z\rangle^p} & \text{for } \operatorname{Re} z > -1,\ z \notin (\overline{\mathbf{R}_+})_\varepsilon. \end{cases}$$

Thus $\mathcal{M}(\chi_{\tilde{r},r}u)$ verifies the conditions (i) and (ii) in Definition 13.2 with $Z = L = \overline{\mathbf{R}_+}$, $s = -p$. To prove (iii) observe that by Proposition 7.5

$$(48)\qquad G_{\tilde{r},r}(z) = -\frac{\tilde{r}^{-z}}{z} + \widetilde{G}_{\tilde{r},r}(z)$$

with $\widetilde{G}_{\tilde{r},r}$ entire on $\mathbf{C}$. Hence in a similar way as above we get for $p = 1$ the estimation

$$\left|\left(\frac{d}{dz}\right)^k \widetilde{G}_{\tilde{r},r}(z-\alpha)T_{k,\kappa}(\alpha)\right| \le \begin{cases} C\dfrac{r^{-\operatorname{Re} z}}{\langle \operatorname{Im} z\rangle} e^{-\kappa\alpha} & \text{for } \operatorname{Re} z \le -1, \\ C\dfrac{\tilde{r}^{-\operatorname{Re} z}}{\langle \operatorname{Im} z\rangle} e^{-\kappa\alpha} & \text{for } \operatorname{Re} z > -1. \end{cases}$$

Observe that the function

$$H^{k,\kappa}_{\tilde{r},r}(z) \stackrel{\text{def}}{=} \int_0^\infty \left(\frac{d}{dz}\right)^k \widetilde{G}_{\tilde{r},r}(z-\alpha)T_{k,\kappa}(\alpha)\,d\alpha$$

is entire and for $\operatorname{Re} z > -1$

$$\left|H^{k,\kappa}_{\tilde{r},r}(z)\right| \le C\frac{\tilde{r}^{\operatorname{Re} z}}{\langle \operatorname{Im} z\rangle}$$

with a new constant $C > 0$.

Thus in the study of the behaviour of $\mathcal{M}(\chi_{\tilde{r},r}u)$ near $\overline{\mathbf{R}_+}$ we are reduced, by (47) and (48), to the integrals of the form

$$H_{\tilde{r}}^{k,\kappa}(z) = \int_0^\infty \left(\frac{d}{dz}\right)^k \left(\frac{\tilde{r}^{\alpha-z}}{\alpha - z}\right) T_{k,\kappa}(\alpha)\, d\alpha$$

for which condition (iii) in Definition 13.2 (with $p = k+1$) is easily verified since by (45)

$$\left|H_{\tilde{r}}^{k,\kappa}(z)\right| \le C \frac{\tilde{r}^{-\operatorname{Re} z}}{|\operatorname{Im} z|^{k+1}} \qquad \text{for } z \in (\overline{\mathbf{R}_+})_\varepsilon.$$

To prove the second part of Theorem 4 suppose that $u \in D'((0,t))$ and satisfies (i)–(iii) in Definition 13.2 with $Z = L = \overline{\mathbf{R}_+}$. Assume also that $\varrho \le 1$. Let

$$F(z) = \mathcal{M}(\chi_{\tilde{r},r}u)(z) \tag{49}$$

with $r < \varrho$. By (i) F is holomorphic on $\mathbf{C} \setminus \overline{\mathbf{R}_+}$ and since (iii) holds we deduce from Theorem 3.3 (with $\Omega = \mathbf{R}$, $V = \mathbf{C}$) that the limits $\lim\limits_{\beta \to 0_+} F(\cdot \pm i\beta)$ exist in the sense of $D'(\mathbf{R})$. Define

$$T = \frac{1}{2\pi i}\left(\lim_{\beta\to 0_+} F(\cdot + i\beta) - \lim_{\beta \to 0_+} F(\cdot - i\beta)\right) \tag{50}$$

and observe that T is a distribution on $\mathbf{R}$ with support in $\overline{\mathbf{R}_+}$ (by (i)) and is independent of the choice of the cut-off function $\chi_{\tilde{r},r}$. We want to prove that T is a Laplace distribution. To this end we want to represent T as a (iterated) primitive of a function satisfying (45). The reasoning given below slightly modifies that from the proof of Theorem 3.3. By assumption (iii) there exists $\varepsilon > 0$ such that for any $\kappa > 0$ and some $p = p(\kappa) \in \mathbf{N}$

$$|F(z)| \le C_\kappa \frac{1}{|\operatorname{Im} z|^p} e^{\operatorname{Re} z(\kappa - \ln \tilde{r})} \qquad \text{for } z \in (\overline{\mathbf{R}_+})_\varepsilon. \tag{51}$$

Let $\mathring{z} = -\eta$ where $\eta > 0$ is sufficiently small. For $z \in (\overline{\mathbf{R}_+})_\varepsilon$, $|\operatorname{Im} z| > 0$, let γ_z be a curve joining $\mathring{z}$ to z in $\mathbf{C} \setminus \overline{\mathbf{R}_+}$ (it can consist of some line segments as in the proof of Theorem 3.3). We define $JF(z) = \int_{\gamma_z} F(\zeta)\, d\zeta$. Observe that in contrast to the proof of Theorem 3.3 we choose one base point $\mathring{z} \in \mathbf{R}_-$ and define only one operation J (instead of $J_\pm$). This is possible because $F \in \mathcal{O}(\mathbf{C} \setminus \overline{\mathbf{R}_+})$ and gives us $JF \in \mathcal{O}(\mathbf{C} \setminus \overline{\mathbf{R}_+})$. It is easily checked that JF satisfies the estimation (51) with $|\operatorname{Im} z|^{-p}$ replaced by $|\operatorname{Im} z|^{-p+1}$ if $p > 1$ and by $C + |\ln|\operatorname{Im} z||$ if $p = 1$ (we use here the assumption that $\tilde{r} < 1$). Thus after $p+1$ iterations of the operation J we arrive at the function $J^{p+1}F \in \mathcal{O}(\mathbf{C} \setminus \overline{\mathbf{R}_+})$ satisfying the estimation

$$\left|J^{p+1}F(z)\right| \le Ce^{\operatorname{Re} z(\kappa - \ln \tilde{r})} \qquad \text{for } z \in (\overline{\mathbf{R}_+})_\varepsilon.$$

Moreover the limits of $J^{p+1}F(\alpha \pm \beta i)$ as $\beta \to 0_+$ exist locally uniformly and give continuous functions on $(-\varepsilon, \infty)$

$$F_\pm(\alpha) \stackrel{\text{def}}{=} \lim_{\beta \to 0_+} J^{p+1}F(\alpha \pm i\beta)$$

satisfying the estimation

$$|F_\pm(\alpha)| \le Ce^{\alpha(\kappa - \ln \tilde{r})} \qquad \text{for } \alpha > -\varepsilon. \tag{52}$$

Hence the function

$$T_\kappa(\alpha) \stackrel{\text{def}}{=} \frac{1}{2\pi i}(F_+(\alpha) - F_-(\alpha))$$

is supported by $\overline{\mathbf{R}_+}$, and satisfies estimation (52). Moreover we have $\left(\frac{d}{d\alpha}\right)^{p+1} T_\kappa = T$ which shows that T is a Laplace distribution.

Define

$$\tilde{u}(x) = T[x^\alpha] \qquad \text{for } 0 < x < \varrho.$$

Thus $\tilde{u}$ is a GAF of convergence radius $\ge \varrho$. It follows from the first part of Theorem 4, which we have already proved, that $\mathcal{M}(\chi_{\tilde{r},r}\tilde{u})$ (with $r < \varrho$) satisfies the conditions (i)–(iii) of Definition 13.2. Let

$$H = \mathcal{M}(\chi_{\tilde{r},r}u) - \mathcal{M}(\chi_{\tilde{r},r}\tilde{u}). \tag{53}$$

Since by assumption $\mathcal{M}(\chi_{\tilde{r},r}u)$ satisfies (ii) it follows that there exists $s \in \mathbf{R}$ such that for any $\varepsilon > 0$ and $\kappa > 0$

$$|H(z)| \le \begin{cases} C\langle \operatorname{Im} z\rangle^s r_\kappa^{-\operatorname{Re} z} & \text{for } \operatorname{Re} z \le -1 \\ C\langle \operatorname{Im} z\rangle^s \tilde{r}_{-\kappa}^{-\operatorname{Re} z} & \text{for } \operatorname{Re} z > -1,\ z \notin (\overline{\mathbf{R}_+})_\varepsilon. \end{cases} \tag{54}$$

Next we shall prove that H is an entire function. To this end it is enough to show, in view of Theorem 3.3 that

$$b(\mathcal{M}\chi_{\tilde{r},r}\tilde{u}) = 2\pi i T. \tag{55}$$

From (47) we have the formula

$$\mathcal{M}\chi_{\tilde{r},r}\tilde{u}(z) = T[G_{\tilde{r},r}(z - \alpha)] \tag{56}$$

which shows that if $\operatorname{supp} T \subset [a, \infty)$ ($a > 0$) then $\mathcal{M}\chi_{\tilde{r},r}\tilde{u} \in \mathcal{O}(\mathbf{C} \setminus [a, \infty))$ and consequently $b(\mathcal{M}\chi_{\tilde{r},r}\tilde{u}) \equiv 0$ on $(-\infty, a)$. Thus to prove (55) one can assume that $\operatorname{supp} T \subset [0, a]$. Then $\mathcal{M}\chi_{\tilde{r},r}\tilde{u} \in \mathcal{O}(\mathbf{C} \setminus [0, a])$.(1) Since both sides of (55) are distributions it suffices to check that (55) holds on a dense subset of $D([0, a])$, of functions φ analytic in a neighbourhood of $[0, a]$. For such φ we have

$$b(\mathcal{M}\chi_{\tilde{r},r}\tilde{u})[\varphi] = \int_C (\mathcal{M}\chi_{\tilde{r},r}\tilde{u})(\zeta)\varphi(\zeta)\, d\zeta$$

where C is a curve encircling $[0, a]$ (anti clockwise) and contained in a complex neighbourhood of $[0, a]$ to which φ extends as a holomorphic function.

By (48) $G_{\tilde{r},r}(z) = -1/z + \widetilde{G}(z)$, with $\widetilde{G}$ entire and we get from (56)

$$\int_C (\mathcal{M}\chi_{\tilde{r},r}\tilde{u})(\zeta)\varphi(\zeta)\, d\zeta = \frac{1}{2\pi i} \int_C T\left[\frac{1}{\zeta - z}\right] \varphi(\zeta)\, d\zeta.$$

Since T has compact support and integration is over a compact set it is possible to integrate under the sign of T and the Cauchy integral formula yields

$$\frac{1}{2\pi i} \int_C T\left[\frac{1}{\zeta - z}\right] \varphi(\zeta)\, d\zeta = T\Big[\frac{1}{2\pi i} \int_C \frac{\varphi(\zeta)}{\zeta - z}\, d\zeta\Big] = T[\varphi].$$

This proves (55).

Both $\mathcal{M}(\chi_{\tilde{r},r}u)$ and $\mathcal{M}(\chi_{\tilde{r},r}\tilde{u})$ satisfy (iii) in Definition 13.2 hence for every $\kappa > 0$

$$|H(\alpha + i\beta)| \le C \frac{\tilde{r}_{-\kappa}^{-\alpha}}{|\beta|^q} \qquad \text{for } \alpha + i\beta \in (\overline{\mathbf{R}_+})_\varepsilon$$

(with some $q \in \mathbf{N}$ and $\varepsilon > 0$). Proceeding as in the first part of the proof we may find a function $J^{q+1}H$ such that

$$\text{(57)} \qquad \left|J^{q+1}H(\alpha + i\beta)\right| \le \widetilde{C}\tilde{r}_{-\kappa}^{-\alpha} \qquad \text{for } a + i\beta \in (\overline{\mathbf{R}_+})_\varepsilon$$

and $(\frac{d}{dz})^{q+1}J^{q+1}H = H$. Since H is entire (57) gives (cf. Exercise 7.13)

$$|H(\alpha + i\beta)| \le C\tilde{r}_{-\kappa}^{-a} \qquad \text{for } \alpha + i\beta \in (\overline{\mathbf{R}_+})_\varepsilon$$

with a new constant C. This together with (54) yields the estimation

$$|H(\alpha + i\beta)| \le \begin{cases} C\langle\beta\rangle^s r_\kappa^{-\alpha} & \text{for } \alpha \le -1 \\ C\langle\beta\rangle^s \tilde{r}_{-\kappa}^{-\alpha} & \text{for } \alpha > -1. \end{cases}$$

(1) The technique used below is typical in hyperfunction theory (see e.g. [K] for details).

Now, it follows from Corollary 9.1 and Exercise 9.8 that H is the Mellin transform of a Mellin distribution with support in $[\tilde{r}, r]$. Thus $u = \tilde{u}$ on $(0, \tilde{r})$ for any $\tilde{r} < \varrho$ hence $u = \tilde{u}$ on $(0, \varrho)$.

Finally it remains only to get rid of the assumption $\varrho < 1$. To this end note that if u satisfies (i)–(iii) of Definition 13.2 with some $\varrho \in \mathbf{R}_+$ then $u^* = u \circ I_\varrho$, where $I_\varrho(x) = x\varrho$ satisfies (i)–(iii) with $\varrho = 1$. This is clear in view of the identity

$$\mathcal{M}u^*(z) = \varrho^z \mathcal{M}u(z).$$

The important feature of generalized analytic function is the following quasi-analyticity property:

Property A: If $T \equiv 0$ on $\mathbf{R}$ then u given by (46) is zero on $(0, \varrho)$. In other words if u given by (46) is flat of order r for an arbitrary r then $u \equiv 0$ on $(0, \varrho)$[(1)].

As remarked before generalized analytic functions can be regarded as a "continuous" version of convergent power series. The "continuous" counterpart of the operation of multiplication of power series is the operation of convolution: if

$$u_j(x) = T_j[x^\alpha], \qquad j = 1, 2$$

then

$$u_1(x) \cdot u_2(x) = T_1 * T_2[x^\alpha] \tag{58}$$

where $T_1 * T_2$ is the convolution of T_1 and T_2 regarded as a distribution on $\mathbf{R}$ (note that the convolution exists since both distributions are supported by $\overline{\mathbf{R}_+}$).

The most elementary example of a GAF which is not analytic is the following:

Example 1′. For $0 < x < 1$ we have with $T = Y$

$$\frac{1}{-\ln x} = \int_0^\infty x^\alpha \, d\alpha.$$

It can be generalized as

Example 2. If $a \in \mathbf{C} \setminus \mathbf{N}_0$ then

$$(-\ln x)^a = \frac{1}{\Gamma(-a)} \alpha_+^{-a-1}[x^\alpha] \qquad \text{for } 0 < x < 1$$

(α_+^{-a-1} denotes the homogeneous distribution of order $-a-1$ supported by $\overline{\mathbf{R}_+}$; see Subsection 3.5). For $j \in \mathbf{N}_0$ we have for $0 < x$

$$(-\ln x)^j = \delta^{(j)}[x^\alpha].$$

(1) Observe that Property A does not hold in general if T is a Fourier hyperfunction ([K]) due to the existence of Fourier hyperfunctions supported by the point $\{\infty\}$.

Proof: Let $\operatorname{Re} a < 0$. Substituting $t = \alpha y$ (for a fixed $y > 0$) in

$$\Gamma(-a) = \int_0^\infty e^{-t} t^{-a-1}\, dt$$

we get

$$y^a = \frac{1}{\Gamma(-a)} \int_0^\infty e^{-y\alpha} \alpha^{-a-1}\, d\alpha$$

which gives the desired result in variable $y = -\ln x$. For the remaining $a \in \mathbf{C} \setminus \mathbf{N}_0$ the result follows by analytic continuation. The second statement of the example is clear.

We can now show that all the special functions listed at the beginning are generalized analytic functions. Indeed in view of Example 2 and the convolution formula (58) we have

$$\operatorname{Erf}(\sqrt{-\ln x}) = \frac{1}{2\Gamma(-1/2)} \alpha_+^{-3/2} * \alpha_+^{-1/2} Y(1-\alpha)[x^\alpha] \qquad \text{for } 0 < x < 1,$$

$$\operatorname{Erfc}(\sqrt{-\ln x}) = \frac{1}{2\Gamma(-1/2)} \alpha_+^{-3/2} * \alpha^{-1/2} Y(\alpha-1)[x^\alpha] \qquad \text{for } 0 < x < 1.$$

Further

$$\gamma(a, -\ln x) = T[x^\alpha] \qquad \text{for } 0 < x < 1,$$

where

$$T = \begin{cases} \dfrac{1}{\Gamma(-a)} \alpha_+^{-a-1} * \alpha_+^{a-1} Y(1-\alpha) & \text{for } a \in \mathbf{C} \setminus \mathbf{N}_0, \\ \delta^{(j)} * \alpha_+^{j-1} Y(1-\alpha) & \text{for } j \in \mathbf{N}_0; \end{cases}$$

$$\Gamma(a, -\ln x) = T[x^\alpha] \qquad \text{for } 0 < x < 1,$$

where

$$T = \begin{cases} \dfrac{1}{\Gamma(-a)} \alpha_+^{-a-1} * \alpha^{a-1} Y(\alpha-1) & \text{for } a \in \mathbf{C} \setminus \mathbf{N}_0, \\ \delta^{(j)} * \alpha_+^{j-1} Y(\alpha-1) & \text{for } j \in \mathbf{N}_0. \end{cases}$$

Finally

$$\operatorname{Ei}(\ln x) = -\frac{1}{\alpha} Y(\alpha-1)[x^\alpha] \qquad \text{for } 0 < x < 1.$$

The idea of studying functions of the form (44) appeared already in the classical monograph of Widder [Wi]. The monograph provides us with a series of basic facts on functions (44) generalizing the well known properties of power series and holomorphic functions:

Let u be of the form (43) with T integrable on $(0,r)$ for any $r>0$. Then $u(x)$ makes sense for $0<x<\varrho$ where ϱ is given by the formula

$$-\ln\varrho = \overline{\lim_{\alpha\to\infty}} \frac{\ln|T(\alpha)|}{\alpha}. \tag{59}$$

Formula (59) extends the well known expression

$$\frac{1}{\varrho} = \overline{\lim_{j\to\infty}} \sqrt[j]{|a_j|}$$

for the convergence radius of the power series $\sum_{j=0}^{\infty} a_j x^j$. However, in distinction to the power series, one can not claim that the number ϱ is the convergence radius of the GAF u (see [Wi] for details).

Of more advanced character is the following theorem which can be viewed upon as a counterpart of the well known estimation

$$\left|\frac{d^k}{dz^k}F(z)\right| \le \frac{Mk!}{(R-|z|)^k}, \qquad |z|<R$$

for the function F holomorphic in a neighbourhood of the set $\{|z|\le R\}$:

Theorem ([Wi] p. 306). *Let $u\colon (0,1)\to \mathbf{C}$. The function u is of the form*

$$u(x) = \int_0^\infty x^\alpha T(\alpha)\,d\alpha \qquad \text{for } 0<x<1$$

with $|T(\alpha)|\le M$ for $0\le\alpha<\infty$ if and only if

$$\left|\left(x\frac{d}{dx}\right)^k u(x)\right| \le \frac{Mk!}{(-\ln x)^{k+1}} \qquad \text{for } 0<x<1,\ k=0,1,2\dots.$$

It is also worthwhile to mention the classical Abel and Tauber type theorems which relate the behaviour of a GAF u at zero and at 1 with the behaviour of T at zero and at $+\infty$ (respectively).

All the considerations conducted in this subsection carry over without essential changes if, instead of (46), we consider the expressions in which T is a Laplace distribution on a line $\zeta_0+i\mathbf{R}$ (for some $\zeta_0\in\mathbf{C}$) supported by the halfline $L=\zeta_0+i\overline{\mathbf{R}_+}$.

Before embarking on a further study of Fuchsian type ordinary differential operators in the space of GAF-s we note the following simple differential identities:

Differentiating (46) with respect to x we get

$$\frac{d}{dx}u(x) = \alpha T[x^{\alpha-1}]$$

which after multiplication by x becomes

$$\left(x\frac{d}{dx}\right)u(x) = \alpha T[x^\alpha].$$

The last formula can be generalized as

$$P\left(x\frac{d}{dx}\right)u(x) = P(\alpha)T[x^\alpha], \tag{60}$$

where P is a polynomial.

Let us come back again to the special functions considered at the beginning of this subsection. All of them are closely related to the confluent hypergeometric function traditionally denoted by ${}_1F_1(a,c;z)$ and given by the series (see [Sl])

$${}_1F_1(a,c;z) = 1 + \frac{a}{c}\frac{z}{1!} + \frac{a(a+1)}{c(c+1)}\frac{z^2}{2!} + \frac{a(a+1)(a+2)}{c(c+1)(c+2)}\frac{z^3}{3!} + \cdots$$

of convergence radius $+\infty$. The function ${}_1F_1(a,c;z)$ is the solution of the confluent hypergeometric equation also called the Kummer equation:

$$y\frac{d^2 f}{dy^2} + (c-y)\frac{df}{dy} - af = 0. \tag{61}$$

The Kummer equation has a regular singular point at zero and an irregular point at $+\infty$. The latter statement means that the equation (61) in the variable $s = \frac{1}{y}$ has a singular point at zero which is not regular. However considered in the logarithmic variable $y = -\ln x$ the Kummer equation assumes the form:

$$-\ln x\left(x\frac{d}{dx}\right)^2 u - (c+\ln x)\left(x\frac{d}{dx}\right)u - au = 0, \tag{62}$$

which after dividing both sides by $-\ln x$ and grouping the terms, becomes

$$\left(x\frac{d}{dx}\right)^2 u + \left(x\frac{d}{dx}\right)u + \frac{c}{\ln x}\left(x\frac{d}{dx}\right)u + \frac{a}{\ln x}u = 0. \tag{62'}$$

Note that the irregular singular point at ∞ for the equation (61) got transformed to a regular singular point at zero for the equation (62′) and the price we paid was the replacement of the coefficients of (61) by functions of the form $v(x) = (-\ln x)^{-1}$. Since $v(0) = 0$, the operator $\left(x\frac{d}{dx}\right)^2 + x\frac{d}{dx}$ can be regarded as a principal part of the equation (62′). Thus (62′) is a Fuchsian type equation whose coefficients are generalized analytic functions. Leaving aside the delicate general problem of solvability of such equation (see [Zie9]; also cf. [Tou]) we solve (62′) directly. We seek solutions of (62′) in the form $u(x) = T[x^\alpha]$. From Exercise 1′ and formulas (58) and (60) we get the equation

$$(63) \qquad (\alpha^2 + \alpha)T = Y * ((c\alpha + a)T)$$

for the Laplace distribution T. Differentiating (63) with respect to α we arrive at the equation (called by Malgrange the Laplace equation)

$$(63') \qquad (\alpha^2 + \alpha)\frac{dT}{d\alpha} = ((c-2)\alpha + a - 1)T.$$

Outside the points $\alpha = 0$, $\alpha = -1$ the equation (63′) is solved by the functions

$$T(\alpha) = C\,|\alpha|^{a-1}\,|\alpha+1|^{c-a-1}.$$

It is easy to check that if $a - 1 \notin -\mathbb{N}_0$, $c - a - 1 \notin -\mathbb{N}_0$ then (63′) has the following two distributional solutions supported by $[-1, 0]$ and $[0, +\infty)$, respectively:

$$T_1 = \alpha_-^{a-1}(\alpha+1)_+^{c-a-1}$$
$$T_2 = (\alpha+1)^{c-a-1}\alpha_+^{a-1}$$

($(\alpha+1)_+^b$ denotes the translate by -1 of the homogeneous distribution α_+^b). Thus equation (62′) and consequently the equivalent equation (62) possesses two solutions which are GAF-s:

$$u_1(x) = \alpha_-^{a-1}(\alpha+1)_+^{c-a-1}[x^\alpha],$$
$$u_2(x) = (\alpha+1)^{c-a-1}\alpha_+^{a-1}[x^\alpha].$$

This result is compatible with the so called Barens formulas for the function ${}_1F_1(a, c; y)$ (which satisfies the Kummer equation)

$${}_1F_1(a, c; -\ln x) = \frac{\Gamma(c)}{\Gamma(c-a)\Gamma(a)} \int_{-1}^{0} x^\alpha(-\alpha)^{a-1}(\alpha+1)^{c-a-1}\, d\alpha$$

for $a > 0$ and $c - a > 0$, and for the function $U(a, c; y)$ which is another solution of the Kummer equation

$$U(a, c; -\ln x) = \frac{1}{\Gamma(a)} \int_0^\infty x^\alpha \alpha^{a-1}(\alpha+1)^{c-a-1}\, d\alpha$$

for $a > 0$. Further study of the GAF-s u_1 and u_2 is carried in Appendix which also provides introduction to the theory of resurgent functions of J. Ecalle.

To end this section we complete the list of special functions with those expressible by the values of the functions ${}_1F_1$ and U on the imaginary axis (see [Sl]): the integral sine

$$\operatorname{Si} y = \int_0^y \frac{\sin t}{t}\, dt = \frac{\pi}{2} - \frac{1}{2} i (e^{-iy} U(1,1;iy) + e^{iy} U(1,1;-iy)),$$

the integral cosine

$$\operatorname{Ci} y = -\int_y^\infty \frac{\cos t}{t}\, dt = \frac{\pi}{2} - \frac{1}{2} i (e^{-iy} U(1,1;iy) - e^{iy} U(1,1;-iy)),$$

the Bessel functions

$$J_\nu(y) = \frac{y^\nu e^{-iy}}{2^\nu \Gamma(\nu+1)}\, {}_1F_1\left(\frac{1}{2}+\nu, 1+2\nu; 2iy\right),$$

the Fresnel integrals

$$\begin{aligned} S(y) &= \frac{1}{\sqrt{2\pi}} \int_0^y \frac{\sin t}{\sqrt{t}}\, dt \\ &= \sqrt{\frac{y}{2\pi}}\, i \left({}_1F_1\left(\frac{1}{2}, \frac{3}{2}; -ye^{\frac{1}{2}i\pi}\right) - {}_1F_1\left(\frac{1}{2}, \frac{3}{2}; -ye^{-\frac{1}{2}i\pi}\right)\right), \end{aligned}$$

$$\begin{aligned} C(y) &= \frac{1}{\sqrt{2\pi}} \int_0^y \frac{\cos t}{\sqrt{t}}\, dt \\ &= \sqrt{\frac{y}{2\pi}} \left({}_1F_1\left(\frac{1}{2}, \frac{3}{2}; -ye^{\frac{1}{2}i\pi}\right) + {}_1F_1\left(\frac{1}{2}, \frac{3}{2}; -ye^{-\frac{1}{2}i\pi}\right)\right). \end{aligned}$$

Exercises

1. Let $m \in \mathbb{N}$, $-\infty \le \alpha < \beta \le +\infty$, $A = (\alpha, \beta)$. Suppose $a_j \in C^\infty(A)$ ($j = 0, 1, \dots, m-1$), $f \in C^0(A)$. Show that every solution $u \in D'(A)$ of the equation

$$(*) \qquad \left(\frac{d}{dx}\right)^m u = \sum_{j=0}^{m-1} a_j \left(\frac{d}{dx}\right)^j u + f$$

is a function of class C^m, i.e. a classical solution. If $f \in C^s(A)$, $s \in \mathbb{N}$ ($f \in C^\infty(A)$) then every solution of $(*)$ is a C^{s+m}-function (a C^∞-function).
HINT. Let $u \in D'(A)$ satisfy $(*)$ and take any interval $I = (a, b)$ with $\alpha < a < b < \beta$. By Theorem 3.4 $u = \left(\frac{d}{dx}\right)^k g$ on I with $k \in \mathbb{N}_0$, and g continuous in a neighbourhood

of $\bar{I}$. Introduce the terminology: a distribution $w \in D'(I)$ is said to be of class $C^{-r}(I)$ ($r \in \mathbf{N}_0$) iff there exists a continuous function h such that $w = \left(\frac{d}{dx}\right)^r h$ on I. Assume that the solution u of equation $(*)$ is of class $C^{-k}(I)$ and that the right-hand side of the identity $(*)$ is of class $C^{-k-m+1}(I)$. Deduce that then $u \in C^{-k+1}(I)$. The proof follows by a repeated application of this argument.

2. Let $u(x) = \sum_{j=0}^{\infty} a_j x^j$ be an analytic function with radius of convergence ϱ at zero. Take $\chi_{\tilde{r},r}$ with $r < \varrho$ and let $\chi_{\tilde{r}} = \chi_{(0,\tilde{r}]}$, $\chi = \chi_{\tilde{r},r} - \chi_{\tilde{r}}$. Let $C = \sum |a_j| \tilde{r}^j$. Show that (cf. Example 7.2) $|\mathcal{M}(\chi_{\tilde{r}} u)(\alpha + i\beta)| \le \frac{C}{|\beta|} \tilde{r}^{-\alpha}$ and that $|\mathcal{M}(\chi u)(\alpha + i\beta)| \le C\tilde{r}^{-\alpha}$ for $\alpha \ge -1$.

3. Show that the functions (34), with the series $\sum_{l=0}^{\infty} c_{ljk} x^l$ ($j = 0, \dots, a$, $k = 0, \dots, b$) of positive convergence radius, are generalized analytic functions.

4. Show that when restricted to the set $(0, \tilde{r})$, any Mellin distribution w solving the equation $Rw = v$ with $v \in D'_{[\tilde{r},r]}((0, \mathring{r}))$, $0 < \tilde{r} < r < \mathring{r}$, coincides with the function (34).

HINT. Assume $v \in D'_{[\tilde{r},r]}((0, \mathring{r}))$ in the considerations preceding Theorem 3. Apply Corollary 9.1.

§14. ELLIPTIC FUCHSIAN TYPE PARTIAL DIFFERENTIAL EQUATIONS IN SPACES $M'^{s}_{(\omega)}$

We introduce below the concept of a singular elliptic operator of order m with regular singularities along the walls $\{x_1 = 0\}, \dots, \{x_n = 0\}$ with edge 0 (cf. [Schl], [K–K]). It is a generalization to the case of several dimensions of an ordinary Fuchsian type differential operator (Subsection 13.2). Since the generalization is straightforward we outline only the essential novelties.

Let U be an open set $0 \in U \subset \mathbf{R}^n$. Consider a linear differential operator R of order m of the form

$$R = R\Big(x_1, \dots, x_n, x_1 \frac{\partial}{\partial x_1}, \dots, x_n \frac{\partial}{\partial x_n}\Big) = R\Big(x, x\frac{\partial}{\partial x}\Big)$$

with smooth coefficients on U. Denoting $P(z) = R(0, z)$ (the principal Mellin symbol of R at zero) we can write R as

$$R = P\Big(x\frac{\partial}{\partial x}\Big) - Q\Big(x, x\frac{\partial}{\partial x}\Big), \tag{1}$$

where

$$Q\Big(x, x\frac{\partial}{\partial x}\Big) = x_1 Q^1\Big(x, x\frac{\partial}{\partial x}\Big) + \dots + x_n Q^n\Big(x, x\frac{\partial}{\partial x}\Big) \tag{2}$$

and $Q^1, \ldots, Q^n$ are differential operators of order m. The operators $x_j Q^j$ in (2) can be written more explicitly as

$$x_j Q^j\left(x, x\frac{\partial}{\partial x}\right)u = \sum_{|\nu| \le m} \left(x\frac{\partial}{\partial x}\right)^\nu (x_j b^j_\nu u) \tag{3}$$

where $b^j_\nu \in C^\infty(U)$. We assume the condition of "ellipticity" in the following form: for every $\mathring{\alpha} \in \mathbf{R}^n$ there exist $A < \infty$ and $C_0 > 0$ such that

$$|P(\mathring{\alpha} + i\beta)| > C_0 \langle \beta \rangle^m \qquad \text{for} \quad \|\beta\| \ge A. \tag{4}$$

Hence P is a differential operator of order m.

This is in fact what we mean by regular singular elliptic operator R. We also say that R is an elliptic Fuchsian type partial differential operator.

1. Existence and Regularity of Solutions on Tangent Cones Ξ_K

The subsection based on Subsections 10.2–10.4 and 11.1 contains the proof of the following theorem

Theorem 1. *Let R be a singular operator (1) of order m satisfying the ellipticity condition (4). Let Ω be an arbitrary open bounded set $\Omega \subset\subset \mathbf{R}^{n-1}_+$. Let Ξ_K, where $K = \overline{\Omega}$, be the "tangent cone" defined by (10.15). Suppose $w \in M'^s_{(\omega)}$ Ξ_K-locally, where $\omega \in \mathbf{R}^n$, $s \in \mathbf{R}$. Then for an arbitrary $\alpha <_K \omega$ there exist $v \in M'^{s-m}_{(\alpha)}$ Ξ_K-locally and an open neighbourhood W of zero in $\mathbf{R}^n$ such that*

$$Rv = w \qquad \text{on} \quad \Xi_\Omega \cap W.$$

Proof: Let $\Omega \subset\subset \mathbf{R}^{n-1}_+$, $K = \overline{\Omega}$, $\omega \in \mathbf{R}^n$, $\alpha <_K \omega$. By Exercise 10.1 there exists $\mathring{\alpha}, \alpha < \mathring{\alpha} <_K \omega$. We choose a slightly bigger bounded open set Ω_1 such that $K \subset \Omega_1 \subset\subset \mathbf{R}^{n-1}_+$ and $\mathring{\alpha} <_{K_1} \omega$ where $K_1 = \overline{\Omega}_1$.

First we shall show (in several steps) that for every $\mathring{\alpha} \in \mathbf{R}^n$, $\mathring{\alpha} <_{K_1} \omega$, there exists $u \in \mathfrak{M}'_{\mathring{\alpha}}$ such that $Ru = w$ on $\Xi_\Omega \cap W$ where W is an open neighbourhood of zero in $\mathbf{R}^n$. Then by multiplication of u by a convenient cut-off function $\tilde{\kappa}^K_r$, we obtain the distribution $v \in M'^{s-m}_{(\alpha)}$ required in Theorem 1.

Let $\tilde{\kappa}^K_r$ be a fixed cut-off function corresponding to w as in Definition 11.3 with $K = \overline{\Omega}$, $K_1 = \overline{\Omega}_1$. Then $\tilde{\kappa}^K_r \equiv 1$ close to zero on Ξ_K, $\operatorname{supp} \tilde{\kappa}^K_r \subset \Xi_{K_1} \cap \{x \colon \|x\| \le r\}$. Set $\kappa^K_r = (\tilde{\kappa}^K_r)^2$ and

$$\begin{aligned} &w_r = \kappa^K_r w, \\ Q_r = x_1 Q^1_r + \cdots + x_n Q^n_r \qquad &\text{where} \quad Q^j_r u = Q^j(\kappa^K_r u), \quad j = 1, \ldots, n. \end{aligned} \tag{5}$$

The first step of the proof is to solve the following:

PROBLEM A. Find a solution of the equation

$$Pu = Q_r u + w_r \tag{6}$$

in $\mathfrak{M}'_{\mathring{\alpha}}(\mathbf{R}^n_+)$ for $\mathring{\alpha} <_{K_1} \omega$.

Note that by Exercise 10.8 we get

$$w_r \in \mathfrak{M}'_{\mathring{\alpha}}(\mathbf{R}^n_+), \quad \mathcal{M}_{\mathring{\alpha}} w_r(\beta) = \mathcal{M} w_r(\mathring{\alpha} + i\beta) \qquad \text{for } \beta \in \mathbf{R}^n. \tag{7}$$

If $u \in \mathfrak{M}'_{\mathring{\alpha}}(\mathbf{R}^n_+)$ then by Proposition 4.3

$$Pu \in \mathfrak{M}'_{\mathring{\alpha}}, \quad \mathcal{M}_{\mathring{\alpha}} Pu(\beta) = P(\mathring{\alpha} + i\beta)\mathcal{M}_{\mathring{\alpha}} u(\beta) \qquad \text{for } \beta \in \mathbf{R}^n \tag{8}$$

and by Proposition 10.5 and formula (6.2) for every $\gamma \leq_{K_1} \mathring{\alpha}$, $x_j b^j_\nu \kappa^K_r u \in M'_{(\gamma+e^j)}$ with e^j given by (10.13) ($j = 1, \dots, n$). Since by Exercise 10.1 (ii) there exists γ such that $\gamma \leq_{K_1} \mathring{\alpha} < \gamma + e^j$ Proposition 5.5 yields $x_j b^j_\nu \kappa^K_r u \in \mathfrak{M}'_{\mathring{\alpha}}$ and by Proposition 7.1 $\mathcal{M}_{\mathring{\alpha}}(x_j b^j_\nu \kappa^K_r u)(\beta) = \mathcal{M}(x_j b^j_\nu \kappa^K_r u)(\mathring{\alpha} + i\beta)$ for $\beta \in \mathbf{R}^n$, $j = 1, \dots, n$. By (2), (3), (5) it follows that

$$Q_r u \in \mathfrak{M}'_{\mathring{\alpha}} \qquad \text{and} \qquad \mathcal{M}_{\mathring{\alpha}}(Q_r u)(\beta) = \mathcal{M}(Q_r u)(\mathring{\alpha} + i\beta) \qquad \text{for } \beta \in \mathbf{R}^n. \tag{9}$$

Hence by (7)–(9) and Proposition 4.2 the equation (6) is equivalent to

$$\mathcal{M}_{\mathring{\alpha}} Pu = \mathcal{M}_{\mathring{\alpha}}(Q_r u) + \mathcal{M}_{\mathring{\alpha}} w_r \qquad \text{in } S'(\mathbf{R}^n) \qquad (\text{for } u \in \mathfrak{M}'_{\mathring{\alpha}})$$

and problem A reduces to the following:

PROBLEM B. Find $u \in \mathfrak{M}'_{\mathring{\alpha}}$, $\mathring{\alpha} <_{K_1} \omega$ such that the equality

$$P(\mathring{\alpha} + i\,\cdot\,)\mathcal{M}_{\mathring{\alpha}} u = \mathcal{M}(Q_r u)(\mathring{\alpha} + i\,\cdot\,) + \mathcal{M} w_r(\mathring{\alpha} + i\,\cdot\,)$$

holds in $S'(\mathbf{R}^n)$.

The existence proof for the solution to the above problem will be given by a perturbation method based on the Main Estimation 1 (of the perturbation term $\mathcal{M}(Q_r u)$) whose proof given below, is preceded by Lemma 1. We start by introducing pertinent notation.

Let $\mathring{\alpha} \in \mathbf{R}^n$ and denote by $\mathcal{P}_0$ the set

$$\mathcal{P}_0 = \{\beta \in \mathbf{R}^n \colon P(\mathring{\alpha} + i\beta) \neq 0\}.$$

In what follows we shall denote by T a fixed tempered distribution on $\mathbf{R}^n$ such that (see Theorem 3.5)

$$P(\mathring{\alpha} + i\,\cdot)T = 1. \tag{10}$$

Since $T \in S'(\mathbf{R}^n)$ there exist $\kappa \in \mathbf{N}_0^n$ and a continuous function λ slowly increasing at infinity such that (see Theorem 3.4)

$$T = \Big(\frac{\partial}{\partial x}\Big)^{\kappa}\lambda. \tag{11}$$

Observe that if $V = \mu T$, $\mu \in O_M^s$ then $V \in S'(\mathbf{R}^n)$,

$$P(\mathring{\alpha} + i\,\cdot)V = \mu \qquad \text{and} \qquad V = \frac{\mu}{P(\mathring{\alpha} + i\,\cdot)} \qquad \text{on } \mathcal{P}_0 \tag{12}$$

and by Proposition 4.2 there exists a unique $v \in \mathfrak{M}'_{\mathring{\alpha}}$ such that $\mathcal{M}_{\mathring{\alpha}}v = V$.

Lemma 1. *Fix a bounded open set $\Omega_1 \subset \mathbf{R}_+^{n-1}$, $K_1 = \overline{\Omega}_1$ and a cut-off function $\tilde{\kappa}_r^K$ (see (10.17) with K_1 instead of K), $\operatorname{supp}\tilde{\kappa}_r^K \subset \Xi_{K_1} \cap \{x\colon \|x\| \le r\}$. Let $\kappa_r^K = (\tilde{\kappa}_r^K)^2$. Fix $\mathring{\alpha} \in \mathbf{R}^n$ and $\varepsilon, \eta \in \mathbf{R}^n$, $0 <_{K_1} \varepsilon$, $0 <_{K_1} \eta$. Let a be a function smooth in a neighbourhood of zero and let $V = \mu T$ with $\mu \in O_M^s$ i.e. $\left|\left(\frac{\partial}{\partial x}\right)^{\delta}\mu(x)\right| \le C_\delta \langle x\rangle^s$. Put*

$$\Lambda = \max_{\delta \le \kappa} C_\delta \qquad \textit{with } \kappa \textit{ as in (11) and } C_\delta \textit{ as above.} \tag{13}$$

Let $v \in \mathfrak{M}'_{\mathring{\alpha}}$, $\mathcal{M}_{\mathring{\alpha}}v = V$. Then for every $\tau \in \mathbf{N}_0^n$ there exists a constant $C = C(a, \eta, \tau, s, \varepsilon, \mathring{\alpha}, \alpha)$ independent of $r \in (0, \frac{1}{2})$ (small enough), and of the choice of $\mu \in O_M^s$ such that

$$\left|\Big(\frac{\partial}{\partial \beta}\Big)^{\tau}\mathcal{M}(a\kappa_r^K v)(\alpha + i\beta)\right| \le C\Lambda H(r, \alpha - \mathring{\alpha} - \varepsilon) r^{-\max l(\eta + 2\varepsilon, \varrho)}\langle\beta\rangle^{s-m} \tag{14}$$
$$\textit{for } \alpha <_{K_1} \mathring{\alpha} - \varepsilon - \eta,$$

where H is defined by (10.20), l by (10.12) and maximum in the exponent is taken over $\varrho \in K_1$.

Proof: By Corollary 10.3 the function g defined by (10.23) belongs to $S(\mathbf{R}^n)$; hence

$$\tilde{g}(\beta) = \mathcal{M}(x^{\varepsilon+\eta}(\ln x)^{\tau} a\tilde{\kappa}_r^K)(i\beta) = (i)^{|\tau|}\Big(\frac{\partial}{\partial \beta}\Big)^{\tau} g(\beta)$$

also belongs to $S(\mathbf{R}^n)$. By Proposition 10.6

$$(2\pi)^n \mathcal{M}(x^{\varepsilon+\eta}(\ln x)^{\tau} a\tilde{\kappa}_r^K v)(\mathring{\alpha} + i\beta) = (\tilde{g} * V)(\beta) \qquad \text{for } \beta \in \mathbf{R}^n. \tag{15}$$

By assumption (4) $|P(\mathring{\alpha}+i\beta)| > C > 0$ for $\|\beta\| \geq A$. Fix a function $\varphi \in C_0^\infty(\mathbf{R}^n)$, $\varphi(\beta) = 1$ for $\|\beta\| \leq A$. Then

$$\tilde{g} * V = \tilde{g} * \varphi V + \tilde{g} * (1-\varphi)V. \tag{16}$$

By (12) and from the choice of the function φ we have

$$(\tilde{g}*(1-\varphi)V)(\beta) = \int_{\|\gamma\|>A} \frac{(1-\varphi(\gamma))\mu(\gamma)\mathcal{M}(x^{\varepsilon+\eta}(\ln x)^\tau a\tilde{\kappa}_r^K)(i(\beta-\gamma))}{P(\mathring{\alpha}+i\gamma)}\, d\gamma. \tag{17}$$

Put $h_1 = \ldots = h_n = k = |s-m|+2$ in (10.22). By (10.22) and Remark 10.4(i) we get with some constant $C^{(1)}$

$$\begin{aligned}|\mathcal{M}(x^{\varepsilon+\eta}(\ln x)^\tau a\tilde{\kappa}_r^K)(i\gamma)| &\leq \frac{CH(r,-\varepsilon)}{\langle i\gamma_1-\varepsilon_1\rangle^k\cdot\ldots\cdot\langle i\gamma_n-\varepsilon_n\rangle^k}\\ &\leq \frac{C}{\langle\gamma_1\rangle^k\cdot\ldots\cdot\langle\gamma_n\rangle^k}.\end{aligned}$$

Hence by the definition of Λ and assumption (4) we deduce from (17) the following inequalities with different constants C

$$\begin{aligned}|(\tilde{g}*(1-\varphi)V)(\beta)| &\leq C\Lambda \int_{\|\gamma\|\geq A} \frac{\langle\gamma\rangle^s}{\langle\beta_1-\gamma_1\rangle^k\cdot\ldots\cdot\langle\beta_n-\gamma_n\rangle)^k\,|P(\mathring{\alpha}+i\gamma)|}\, d\gamma\\ &\leq C\Lambda\int_{\mathbf{R}^n}\frac{\langle\gamma\rangle^{s-m}}{\langle\beta_1-\gamma_1\rangle^k\cdot\ldots\cdot\langle\beta_n-\gamma_n\rangle^k}\, d\gamma.\end{aligned}$$

Hence by Exercise 11.3 we get

$$|(\tilde{g}*(1-\varphi)V)(\beta)| \leq C\Lambda\langle\beta\rangle^{s-m}. \tag{18}$$

Now we shall derive the same estimate for the first summand in (16). By the Leibniz rule we get

$$\begin{aligned}|(\tilde{g}*\varphi V)(\beta)| &= \left|\left(\tilde{g}*\varphi\mu\left(\frac{\partial}{\partial\gamma}\right)^\kappa\lambda\right)(\beta)\right| \leq \sum_{\kappa_1,\kappa_2,\kappa_3\leq\kappa} C_{\kappa_1\kappa_2\kappa_3}\times\\ &\times\left|\int_{\operatorname{supp}\varphi}\lambda(\gamma)\left(\frac{\partial}{\partial\gamma}\right)^{\kappa_1}\varphi(\gamma)\left(\frac{\partial}{\partial\gamma}\right)^{\kappa_2}\mu(\gamma)\left(\frac{\partial}{\partial\gamma}\right)^{\kappa_3}\tilde{g}(\beta-\gamma)\,d\gamma\right|\\ &\leq \Lambda\sum_{\kappa_1,\kappa_2,\kappa_3\leq\kappa} C_{\kappa_1\kappa_2\kappa_3}\left(\left|\lambda\left(\frac{\partial}{\partial\gamma}\right)^{\kappa_1}\varphi\right|\langle\gamma\rangle^s * \left|\left(\frac{\partial}{\partial\gamma}\right)^{\kappa_3}\tilde{g}\right|\right)(\beta).\end{aligned} \tag{19}$$

(1) Here and in the sequel $C = C(a,\eta,\tau,s,\varepsilon,\mathring{\alpha})$ denotes a constant independent of $r \in (0,\frac{1}{2})$ and of $\mu \in O_M^s$, which may be different in different inequalities.

Since the convolution of the function $\left|(\frac{\partial}{\partial\gamma})^{\kappa_3}\tilde{g}\right|$ in S by $\left|\lambda(\frac{\partial}{\partial\gamma})^{\kappa_1}\varphi\right|\langle\cdot\rangle^s \in E'$ is in S, (19) implies that there exists a constant $C = C(a, \eta, \tau, s, \varepsilon, \mathring{\alpha})$ such that

$$|(\tilde{g} * \varphi V)(\beta)| \le C\Lambda\langle\beta\rangle^{s-m}.$$

Hence by (15), (16), (18) and foot-note (1) p.160

$$\left|\mathcal{M}(x^{\varepsilon+\eta}(\ln x)^\tau a\tilde{\kappa}_r^K v)(\mathring{\alpha} + i\beta)\right| < C\Lambda\langle\beta\rangle^{s-m}. \tag{20}$$

To get (14) for $\alpha <_{\kappa_1} \mathring{\alpha} - \varepsilon - \eta$ we recall that $\kappa_r^K = (\tilde{\kappa}_r^K)^2$ and proceed similarly as in (11.8):

$$(2\pi)^n \mathcal{M}(x^{\varepsilon+\eta}(\ln x)^\tau a\kappa_r^K v)(\alpha + i\beta) = (2\pi)^n \mathcal{M}(u \cdot \psi)(\mathring{\alpha} + i\beta),$$

where $u = x^{\varepsilon+\eta}(\ln x)^\tau a\tilde{\kappa}_r^K v$, $\psi = \tilde{\kappa}_r^K x^{\mathring{\alpha}-\alpha}$. Since $v \in \mathfrak{M}'_{\mathring{\alpha}}$ by Proposition 10.5 $a\tilde{\kappa}_r^K v \in M'_{(\gamma)}$ for every $\gamma \le_{\kappa_1} \mathring{\alpha}$. Hence $u \in M'_{(\gamma+\varepsilon+\eta)}$ for every $\gamma \le_{\kappa_1} \mathring{\alpha}$. Since $\varepsilon + \eta >_{\kappa_1} 0$ by Exercise 10.1(i) to every $\xi \in \mathbf{R}^n$ there exists $\tilde{\gamma}$ such that $\xi < \tilde{\gamma} <_{\kappa_1} \xi + \varepsilon + \eta$. Taking $\xi = \mathring{\alpha} - \varepsilon - \eta$ we get $\tilde{\gamma} \le_{\kappa_1} \mathring{\alpha}$, $\mathring{\alpha} < \tilde{\gamma} + \varepsilon + \eta$ and we see that $u \in M'_{(\tilde{\gamma}+\varepsilon+\eta)}$. Hence by Proposition 7.1 $u \in \mathfrak{M}'_{\mathring{\alpha}}$ and $(\mathcal{M}_{\mathring{\alpha}} u)(\beta) = \mathcal{M}u(\mathring{\alpha} + i\beta)$ for $\beta \in \mathbf{R}^n$ and Proposition 10.6 yields

$$\begin{aligned} &(2\pi)^n \mathcal{M}(x^{\varepsilon+\eta}(\ln x)^\tau a\kappa_r^K v)(\alpha + i\beta) \\ &\quad = \int_{\mathbf{R}^n} \mathcal{M}(x^{\mathring{\alpha}-\alpha}\tilde{\kappa}_r^K)(i\gamma)\mathcal{M}(x^{\varepsilon+\eta}(\ln x)^\tau a\tilde{\kappa}_r^K v)(\mathring{\alpha} + i(\beta - \gamma))\, d\gamma \\ &\qquad\qquad \text{for } \alpha <_{\kappa_1} \mathring{\alpha}. \end{aligned} \tag{21}$$

Take $h_j = k = |s - m| + 2$ $(j = 1, \ldots, n)$. By Corollary 10.3 we have

$$\left|\mathcal{M}(x^{\mathring{\alpha}-\alpha}\tilde{\kappa}_r^K)(i\gamma)\right| \le \frac{CH(r, \alpha - \mathring{\alpha})}{\langle\gamma_1\rangle^k \cdot \ldots \cdot \langle\gamma_n\rangle^k} \qquad \text{for } \alpha <_{\kappa_1} \mathring{\alpha}.$$

Hence from (21), (20) and Exercise 11.1 we get for suitable constants $C = C(a, \eta, \tau, s, \varepsilon, \mathring{\alpha})$

$$\begin{aligned} &\left|\mathcal{M}(x^{\varepsilon+\eta}(\ln x)^\tau a\kappa_r^K v)(\alpha + i\beta)\right| \\ &\quad \le C\Lambda H(r, \alpha - \mathring{\alpha}) \int_{\mathbf{R}^n} \frac{\langle\beta - \gamma\rangle^{s-m}}{\langle\gamma_1\rangle^k \cdot \ldots \cdot \langle\gamma_n\rangle^k}\, d\gamma \le C\Lambda H(r, \alpha - \mathring{\alpha})\langle\beta\rangle^{s-m} \\ &\qquad\qquad \text{for } \alpha <_{\kappa_1} \mathring{\alpha}. \end{aligned}$$

Consequently by (7.9) and Remark 10.4(ii) we get

$$\begin{aligned}&\left|\mathcal{M}((\ln x)^{\tau} a\kappa_r^K v)(\alpha+i\beta)\right| \le C\Lambda H(r,\alpha+\varepsilon+\eta-\mathring{\alpha})\langle\beta\rangle^{s-m}\\ &\qquad \le C\Lambda H(r,\alpha-\mathring{\alpha}-\varepsilon)r^{-\max l(2\varepsilon+\eta,\varrho)}\langle\beta\rangle^{s-m} \qquad \text{for } \alpha <_{K_1} \mathring{\alpha}-\varepsilon-\eta\end{aligned}$$

hence the assertion (14) holds.

Main Estimation 1. *Let K_1 and the cut-off function κ_r^K be as in Lemma 1 and let $\mathring{\alpha} \in \mathbf{R}^n$. Choose $\varepsilon, \eta \in \mathbf{R}^n$, $0 <_{K_1} \varepsilon$, $0 <_{K_1} \eta$ so that*

$$c = -\max_{\varrho\in K_1} l(\eta+2\varepsilon,\varrho) + \min_{\varrho\in K_1}(1,\varrho_1,\dots,\varrho_{n-1}) > 0. \tag{22}$$

Suppose $u \in \mathfrak{M}'_{\mathring{\alpha}}$, $\mathcal{M}_{\mathring{\alpha}}u = \mu T$ with $\mu \in O_M^s$ and T fixed in (10).

Then there exists a constant C independent of $r \in (0,\frac{1}{2})$ and of u satisfying the above conditions such that

$$\left|\left(\frac{\partial}{\partial\beta}\right)^{\delta}\mathcal{M}(Q_r u)(\mathring{\alpha}+i\beta)\right| \le C\Lambda r^{c}\langle\beta\rangle^{s} \qquad \text{for } \beta\in\mathbf{R}^n,\ \delta\in\mathbf{N}_0^n, \tag{23}$$

where Q_r is given by (5). Hence in particular $\mu_r = \mathcal{M}(Q_r u)(\mathring{\alpha}+i\,\cdot) \in O_M^s$ for $0 < r < \frac{1}{2}$.

Proof: By (5)

$$\mathcal{M}(Q_r u)(\mathring{\alpha}+i\beta) = \sum_{j=1}^{n}\mathcal{M}(x_j Q^j(\kappa_r^K u))(\mathring{\alpha}+i\beta) \qquad \text{for } \beta\in\mathbf{R}^n. \tag{24}$$

Using the explicit expression (3) for x_jQ^j we compute

$$\mathcal{M}(x_jQ^j(\kappa_r^K u))(z) = \sum_{|\nu|\le m} z^{\nu}\mathcal{M}(x_j b_\nu^j \kappa_r^K u)(z).$$

We easily find by induction that

$$\begin{aligned}&\left|\left(\frac{\partial}{\partial z}\right)^{\delta}\mathcal{M}(x_jQ^j(\kappa_r^K u))(z)\right|\\ &\qquad \le C\langle z\rangle^{m}\max_{|\nu|\le m}\max_{\tau\le\delta}\left|\mathcal{M}((\ln x)^{\tau} b_\nu^j\kappa_r^K u)(z_1,\dots,z_j-1,\dots,z_n)\right|\\ &\qquad\qquad (j=1\dots,n).\end{aligned} \tag{25}$$

Put $\mathring{z} = \mathring{\alpha}+i\beta$ and observe that by (22) $-e^j+2\varepsilon+\eta <_{K_1} 0$ where e^j is defined by (1.1) ($j=1,\dots,n$). By hypothesis $\mathcal{M}_{\mathring{\alpha}}u = \mu T$, $\mu\in O_M^s$ and b_ν^j are smooth in a neighbourhood of zero. Therefore by Lemma 1, Remark 10.4(ii) and assumption (22) we get for $1 \le j \le n$ with different constants C (see footnote (1) p. 160).

$$
\begin{aligned}
(26) \qquad & \left|\mathcal{M}((\ln x)^\tau b^j_\nu \kappa^K_r u)(\mathring{z}_1, \dots, \mathring{z}_j - 1, \dots, \mathring{z}_n)\right| \\
&= \left|\left(\frac{\partial}{\partial z}\right)^\tau \mathcal{M}(b^j_\nu \kappa^K_r u)(\mathring{z}_1, \dots, \mathring{z}_j - 1, \dots, \mathring{z}_n)\right| \\
&\le C\Lambda H(r, -e^j - \varepsilon) r^{-\max l(\eta + 2\varepsilon, \varrho)} \langle\beta\rangle^{s-m} \\
&\le C\Lambda H(r, -\varepsilon) r^{-\max l(-e^j, \varrho) - \max l(\eta + 2\varepsilon, \varrho)} \langle\beta\rangle^{s-m} \\
&\le C\Lambda H(r, -\varepsilon) r^c \langle\beta\rangle^{s-m}
\end{aligned}
$$

where max in the exponent are taken over K_1.

Since $\langle\mathring{\alpha} + i\beta\rangle \le \langle\mathring{\alpha}\rangle\langle\beta\rangle$ it follows from (25) and (26) that

$$\left|\left(\frac{\partial}{\partial \beta}\right)^\delta \mathcal{M}(x_j Q^j(\kappa^K_r u))(\mathring{\alpha} + i\beta)\right| < C\Lambda H(r, -\varepsilon) r^c \langle\beta\rangle^s \qquad (j = 1, \dots, n).$$

Hence from (24) and Remark 10.4 (i) we get (23).

Solution of Problem B. Let $\mathring{\alpha} <_{K_1} \omega$ and $\mathring{\mu}_r = \mathcal{M}w_r(\mathring{\alpha} + i\,\cdot)$. By Lemma 11.1 $\left|\left(\frac{\partial}{\partial\beta}\right)^\delta \mathring{\mu}_r(\beta)\right| < C_\delta \langle\beta\rangle^s$. Put $\Lambda = \max_{\delta \le \kappa} C_\delta$ (κ as in (11)). Since $\mathring{\mu}_r \in O^s_M$ we have $\mathring{\mu}_r T \in S'$ and by Proposition 4.2 there exists a unique $u_0 \in \mathfrak{M}'_{\mathring{\alpha}}$ such that $\mathcal{M}_{\mathring{\alpha}} u_0 = \mathring{\mu}_r T$. First we check that the formal definitions

$$(27) \qquad \mathcal{M}_{\mathring{\alpha}} u_{p+1} = \mu^{p+1}_r T, \qquad p \in \mathbf{N}_0,$$

where

$$(28) \qquad \mu^{p+1}_r(\beta) = \mathcal{M}(Q_r u_p)(\mathring{\alpha} + i\beta) + \mathcal{M}w_r(\mathring{\alpha} + i\beta), \qquad \beta \in \mathbf{R}^n, \quad p \in \mathbf{N}_0$$

define a sequence $u_j \in \mathfrak{M}'_{\mathring{\alpha}}$ and that

$$(29) \qquad \left|\left(\frac{\partial}{\partial \beta}\right)^\delta \mu^p_r(\beta)\right| \le \Lambda \sum_{k=0}^{p} (Cr^c)^k \langle\beta\rangle^s, \qquad \beta \in \mathbf{R}^n, \quad p \in \mathbf{N}_0, \quad \delta \le \kappa,$$

where c, C are the same as in (23), and consequently $\mu^p_r \in O^s_M$, $p \in \mathbf{N}_0$. For this purpose observe that the Main Estimation 1 yields

$$\left|\left(\frac{\partial}{\partial \beta}\right)^\delta \mathcal{M}(Q_r u_0)(\mathring{\alpha} + i\beta)\right| \le C\Lambda r^c \langle\beta\rangle^s, \qquad \beta \in \mathbf{R}^n.$$

Thus

$$\begin{aligned}\left|\left(\frac{\partial}{\partial\beta}\right)^{\delta}\mu_r^1(\beta)\right| &\le \left|\left(\frac{\partial}{\partial\beta}\right)^{\delta}\mathcal{M}(Q_r u_0)(\mathring{\alpha}+i\beta)\right| + \left|\left(\frac{\partial}{\partial\beta}\right)^{\delta}\mathcal{M}w_r(\mathring{\alpha}+i\beta)\right| \\ &\le \Lambda(Cr^c+1)\langle\beta\rangle^s, \qquad \beta\in\mathbf{R}^n\end{aligned}$$

i.e. (29) holds for $p=1$. Moreover

$$\left|\left(\frac{\partial}{\partial\beta}\right)^{\delta}\mu_r^1(\beta) - \left(\frac{\partial}{\partial\beta}\right)^{\delta}\mu_r^0(\beta)\right| \le \Lambda Cr^c\langle\beta\rangle^s \qquad \beta\in\mathbf{R}^n.$$

Clearly $\mu_r^1\in O_M^s$, hence by Proposition 4.2 there exists a unique u_1 satisfying (27). By induction we get sequences $u_p\in\mathfrak{M}'_{\mathring{\alpha}}$, $\mu_r^p\in O_M^s$ satisfying (27)–(29) and the inequalities

$$\left|\left(\frac{\partial}{\partial\beta}\right)^{\delta}\mu_r^{p+1}(\beta) - \left(\frac{\partial}{\partial\beta}\right)^{\delta}\mu_r^p(\beta)\right| \le \Lambda(Cr^c)^{p+1}\langle\beta\rangle^s \qquad \text{for } p\in\mathbf{N}_0. \tag{30}$$

Now, we proceed in a standard way: we take r so small that

$$A=\sum_{p=0}^{\infty}(Cr^c)^p<\infty.$$

Then from (29)

$$\left|\left(\frac{\partial}{\partial\beta}\right)^{\delta}\mu_r^p(\beta)\right| \le A\Lambda\langle\beta\rangle^s, \qquad p\in\mathbf{N}_0$$

and in view of (30) the sequence $\{\mu_r^p\}_{p\in\mathbf{N}_0}$ is convergent in the topology of O_M^s. Let

$$\begin{aligned}\mu(\beta)&=\lim_{p\to\infty}\mu_r^p(\beta) \qquad \text{for } \beta\in\mathbf{R}^n,\\ \Phi&=\mu T.\end{aligned}$$

We have $\mu\in O_M^s$, thus $\Phi\in S'$ and there exists a unique $u\in\mathfrak{M}'_{\mathring{\alpha}}$ such that $\mathcal{M}_{\mathring{\alpha}}u=\Phi$. We shall show that the identity

$$\mu=\mathcal{M}(Q_r u)(\mathring{\alpha}+i\,\cdot\,)+\mathcal{M}w_r(\mathring{\alpha}+i\,\cdot\,) \tag{31}$$

holds in $S'(\mathbf{R})$. Since $\mu_r^p\to\mu$ in O_M^s, then $\mu_r^pT\to\mu T=\Phi$ in S'. Further it follows from the continuity of the inverse Mellin transformation $(\mathcal{M}_{\mathring{\alpha}})^{-1}$ that $u_p\to u$ in $\mathfrak{M}'_{\mathring{\alpha}}$.

Note that $\mathcal{M}w_r(\mathring{\alpha}+i\,\cdot\,)\in S'$ (see (7.12)) and that $\mu_r^p \underset{p\to\infty}{\longrightarrow} \mu$ in S'. Thus to derive (31) from formula (28) we are to show that

$$\mathcal{M}(Q_r u_p)(\mathring{\alpha}+i\,\cdot\,) \underset{p\to\infty}{\longrightarrow} \mathcal{M}(Q_r u)(\mathring{\alpha}+i\,\cdot\,) \qquad \text{in } S'. \tag{32}$$

To this aim observe that by Proposition 10.5 $b_\nu^j \kappa_r^K u_p \underset{p\to\infty}{\longrightarrow} b_\nu^j \kappa_r^K u$ in $M'_{(\omega)}$ for $\omega \leq_{K_1} \mathring{\alpha}$ and hence by (6.2) $x_j b_\nu^j \kappa_r^K u_p \underset{p\to\infty}{\longrightarrow} x_j b_\nu^j \kappa_r^K u$ in $M'_{(\lambda)}$ for $\lambda \leq_{K_1} \mathring{\alpha} + e^j$ $(j = 1, \ldots, n)$. Observe that there exists $b > 0$ such that $\lambda \leq_{K_1} \mathring{\alpha} + be^1$ implies $\lambda <_{K_1} \mathring{\alpha} + e^j$ $(j = 1, \ldots, n)$. Hence (cf. (3), (5))

$$Q_r u_p \underset{p\to\infty}{\longrightarrow} Q_r u \qquad \text{in } M'_{(\lambda)} \qquad \text{for } \lambda <_{K_1} \mathring{\alpha} + be^1.$$

Since $b > 0$ by Corollary 10.2 there exists $\tilde{\lambda}$ such that $\mathring{\alpha} < \tilde{\lambda} <_{K_1} \mathring{\alpha} + be^1$. Hence by Proposition 5.5 $Q_r u_p \underset{p\to\infty}{\longrightarrow} Q_r u$ in $\mathfrak{M}'_{\mathring{\alpha}}$ and (32) holds. This ends the proof of (31).

Since $u \in \mathfrak{M}'_{\mathring{\alpha}}$, $\mathcal{M}_{\mathring{\alpha}} u = \Phi = \mu T$ we get by (10)

$$\mu = P(\mathring{\alpha} + i\,\cdot)\mu T = P(\mathring{\alpha} + i\,\cdot)\mathcal{M}_{\mathring{\alpha}} u$$

and (31) yields that u is a solution of Problem B for r sufficiently small and $\mathring{\alpha} <_{K_1} \omega$.

Thus u is also a solution to Problem A for r small enough. Since $\kappa_r^K \equiv 1$ on Ξ_Ω close to zero we conclude that $Ru = w$ on $\Xi_\Omega \cap W$ where W is a neighbourhood of zero small enough. Now we shall prove Theorem 1. Let $\alpha <_{K_1} \omega$ and take $\mathring{\alpha}$ such that $\alpha < \mathring{\alpha} <_{K_1} \omega$. Let $u \in \mathfrak{M}'_{\mathring{\alpha}}(\mathbf{R}^n_+)$ be a solution to problem A and let $\mathring{\kappa}_r^K$ be a cut-off function (10.17), such that $\mathring{\kappa}_r^K \equiv 1$ on $\Xi_\Omega \cap W$. Let $v \in \mathring{\kappa}_r^K u$. Then

$$Rv = w \qquad \text{on } \Xi_\Omega \cap W$$

and by Proposition 10.5 $v \in M'_{(\gamma)}$ for $\gamma \leq_{K_1} \mathring{\alpha}$, in particular $v \in M'_{(\mathring{\alpha})}$. Choose $\varepsilon >_{K_1} 0$, $\eta >_{K_1} 0$ such that $\alpha <_{K_1} \mathring{\alpha} - \varepsilon - \eta$. By Lemma 1 and Remark 10.4 i) we get

$$\begin{aligned} |\mathcal{M}v(\gamma + i\beta)| &= |\mathcal{M}\tilde{\kappa}_r^K u(\gamma + i\beta)| \\ &\leq C\Lambda H(r, \gamma - \mathring{\alpha} - \varepsilon) r^{-\max l(\eta + 2\varepsilon, \varrho)} \langle\beta\rangle^{s-m} \\ &\leq \widetilde{C}\langle\beta\rangle^{s-m} \qquad\qquad \text{for } \gamma <_{K_1} \mathring{\alpha} - \varepsilon - \eta. \end{aligned}$$

Since $\alpha < \mathring{\alpha}$ we have $v \in M'_{(\alpha)}$ and $|\mathcal{M}v(\gamma + i\beta)| < \widetilde{C}\langle\beta\rangle^{s-m}$ for $\gamma < \alpha$. Hence by Definition 11.1 and Remark 11.1 $v \in M'^{s-m}_{(\alpha)}$. This ends the proof of Theorem 1.

Remark 1. The solution v whose existence is asserted in Theorem 1 is non unique. In fact it is well known that the equation (10) has different solutions.

Corollary 1. *Under the assumptions of Theorem 1, for every $w \in M'^{s}_{(\omega)}$ Ξ_K-locally and every $\alpha <_K \omega$ there exists v satisfying $Rv = w$ on $\Xi_\Omega \cap W$ (for suitable W) such that for every $\delta\mathring{x} \in \mathbf{R}^n_+$ we have $v \in M'^{s-m}_{(\alpha)}$ 2-locally at $\delta\mathring{x}$.*

Proof: This is clear from Theorem 1 and the fact that the space $M'^{s}_{(\omega)}$ is 2-local i.e. for any $u \in M'^{s}_{(\omega)}$ and conical cut-off function κ $\kappa u \in M'^{s}_{(\omega)}$ (see Exercise 11.6).

2. Case of a Proper Cone

The version of Theorem 1 for the "usual" cone in $\mathbf{R}^2_+$ was proved in [Sz–Ziel]. Now we present its general version.

Theorem 1*. *Let R be as in Theorem 1. Let Γ be a proper closed cone in $\mathbf{R}^n_+$. Suppose $w \in M'^{s}_{(\omega)}$ 2-locally at δx for $\delta x \in \Gamma$, $\omega \in \mathbf{R}^n$. Then for an arbitrary $\alpha \in \mathbf{R}^n$ such that $\sum_{j=1}^{n} \alpha_j < \sum_{j=1}^{n} \omega_j$ there exist v and an open neighbourhood W of zero in $\mathbf{R}^n$ such that $Rv = w$ on $\Gamma \cap W$ and $v \in M'^{s-m}_{(\alpha)}$ 2-locally for $\delta x \in \Gamma$.*

The proof proceeds along the same lines as that of Theorem 1, therefore we omit technical details which in this case are simpler.

Proof: Let κ_{r0} be a fixed conical cut-off function corresponding to w as in Definition 11.2: $\kappa_{r0} w \in M'^{s}_{(\omega)}$, $\kappa_{r0} \equiv 1$ on Γ_K close to zero, $\overline{\operatorname{supp} \kappa_{r0}} \subset (\mathbf{R}^n_+ \cup \{0\}) \cap \overline{B}(r)$. We take an $\mathring{\alpha}$ such that $\alpha < \mathring{\alpha}$, $\sum \mathring{\alpha}_j < \sum \omega_j$ and consider problem A with

$$w_r = \kappa_{r0}^2 w.$$

We reduce the problem A to problem B as in the proof of Theorem 1 with Exercise 10.8 replaced by Exercise 10.7, Proposition 10.5 replaced by Exercise 10.5 and Exercise 10.1 (ii) replaced by an evident property that in the set $\{\gamma \in \mathbf{R}^n \colon \sum \gamma_j < \sum \mathring{\alpha}_j\}$ there exists γ such that $\mathring{\alpha} < \gamma + e^j$.

Let us formulate the following analogue of Lemma 1.

Lemma 1*. *Let κ_{r1} be given by (10.2) and $\kappa_r = \kappa_{r1}^2$. Fix $\mathring{\alpha} \in \mathbf{R}^n$ and $\varepsilon, \eta \in \mathbf{R}^n$, $\sum \varepsilon_j > 0$, $\sum \eta_j > 0$. Let a be a function smooth in a neighbourhood of zero and let $V = \mu T$ with $\mu \in O^s_M$ and T satisfying (10). Let $v \in \mathfrak{M}'_{\mathring{\alpha}}$, $\mathcal{M}_{\mathring{\alpha}} v = V$. Then for every $\tau \in \mathbf{N}^n_0$ there exists a constant $C(a, \eta, \tau, s, \varepsilon, \mathring{\alpha}, \alpha)$ independent of $r \in (0, \frac{1}{2})$ and of the choice of $\mu \in O^s_M$ such that*

$$(33) \quad \left| \left(\frac{\partial}{\partial \beta} \right)^{\tau} \mathcal{M}(a \kappa_r v)(\alpha + i\beta) \right| \leq C \Lambda G(r, \alpha - \mathring{\alpha} - \varepsilon) r^{-\sum 2\varepsilon_j + \eta_j} \langle \beta \rangle^{s-m}$$
$$\text{for } \sum \alpha_j < \sum \mathring{\alpha}_j - \varepsilon_j - \eta_j,$$

where Λ is defined by (13) and G by (10.6).

Proof: By Corollary 10.1 the function g defined by (10.11) belongs to $S(\mathbf{R}^n)$, hence $\tilde{g}(\beta) = (i)^{|\tau|}\left(\frac{\partial}{\partial\beta}\right)^\tau g(\beta)$ also belongs to $S(\mathbf{R}^n)$. By Exercise 10.6

$$(34) \qquad (2\pi)^n \mathcal{M}(x^{\varepsilon+\eta}(\ln x)^\tau a\kappa_{r1}v)(\mathring{\alpha}+i\beta) = (\tilde{g} * V)(\beta) \qquad \text{for } \beta \in \mathbf{R}^n.$$

We fix a function φ as in the proof of Lemma 1 and get (16) and (17) with $\tilde{\kappa}_K$ replaced by κ_{r1}. By (10.10) with $h_1 = \cdots = h_n = k = |s-m|+2$ and Remark 10.2 we get

$$\left|\mathcal{M}(x^{\varepsilon+\eta}(\ln x)^\tau a\kappa_{r1})(i\gamma)\right| \le \frac{C}{\langle\gamma_1\rangle^k \cdot \ldots \cdot \langle\gamma_n\rangle^k}$$

and then as in the proof of Lemma 1 we obtain (18). Next (34), (16), (18) yield

$$(35) \qquad \left|(\mathcal{M}(x^{\varepsilon+\eta}(\ln x)^\tau a\kappa_{r1}v)(\mathring{\alpha}+i\beta)\right| < C\Lambda\langle\beta\rangle^{s-m}.$$

Observe that

$$(2\pi)^n \mathcal{M}(x^{\varepsilon+\eta}(\ln x)^\tau a\kappa_r v)(\alpha+i\beta) = (2\pi)^n \mathcal{M}(u\cdot\psi)(\mathring{\alpha}+i\beta),$$

where $u = x^{\varepsilon+\eta}(\ln x)^\tau a\kappa_{r1}v$, $\psi = \kappa_{r1}x^{\mathring{\alpha}-\alpha}$. Since $v \in \mathfrak{M}'_{\mathring{\alpha}}$ by Exercise 10.5 $a\kappa_{r1}v \in M'_{(\gamma)}$ for every $\sum\gamma_j < \sum\mathring{\alpha}_j$. Hence $u \in M'_{(\gamma+\varepsilon+\eta)}$ for every $\sum\gamma_j < \sum\mathring{\alpha}_j$. Since $\sum\varepsilon_j+\eta_j > 0$ taking $\tilde{\gamma}_j = \mathring{\alpha}_j - (\varepsilon_j+\eta_j) + \frac{1}{2n}\sum\varepsilon_j+\eta_j$ $(j=1,\ldots,n)$ we obtain $\sum\tilde{\gamma}_j < \sum\mathring{\alpha}_j$, $\tilde{\gamma} > \mathring{\alpha}-\varepsilon-\eta$ and we see that $u \in M'_{(\tilde{\gamma}+\varepsilon+\eta)}$. Hence by Proposition 5.5 $u \in \mathfrak{M}'_{\mathring{\alpha}}$ and $(\mathcal{M}_{\mathring{\alpha}}u)(\beta) = \mathcal{M}u(\mathring{\alpha}+i\beta)$ for $\beta \in \mathbf{R}^n$. Next we apply Exercise 10.6 (instead of Proposition 10.6) and observe that by Corollary 10.1 (with $\tau = \eta = 0$)

$$\left|\mathcal{M}(x^{\mathring{\alpha}-\alpha}\kappa_{r1})(i\gamma)\right| \le \frac{CG(r,\alpha-\mathring{\alpha})}{\langle\gamma_1\rangle^k \cdot \ldots \cdot \langle\gamma_n\rangle^k} \qquad \text{for } \textstyle\sum\alpha_j - \mathring{\alpha}_j < 0,$$

where $k = |s-m|+2$.

Then by (35) we get for suitable constants $C = C(a,\eta,\tau,s,\varepsilon,\mathring{\alpha})$

$$\left|\mathcal{M}(x^{\varepsilon+\eta}(\ln x)^\tau a\kappa_r v)(\alpha+i\beta)\right| \le C\Lambda G(r,\alpha-\mathring{\alpha})\langle\beta\rangle^{s-m} \qquad \text{for } \textstyle\sum\alpha_j < \sum\mathring{\alpha}_j.$$

Hence by (7.9) and Remark 10.2 we get

$$\begin{aligned} |\mathcal{M}((\ln x)^\tau a\kappa_r v)(\alpha+i\beta)| &\le C\Lambda G(r,\alpha+\varepsilon+\eta-\mathring{\alpha})\langle\beta\rangle^{s-m} \\ &\le C\Lambda G(r,\alpha-\mathring{\alpha}-\varepsilon)r^{-\sum 2\varepsilon_j+\eta_j}\langle\beta\rangle^{s-m} \\ &\qquad \text{for } \textstyle\sum\alpha_j < \sum\mathring{\alpha}_j - \varepsilon_j - \eta_j \end{aligned}$$

hence assertion (33) follows.

Main Estimation 1*. *Let a conical cut-off function κ_r be as in Lemma 1* and let $\mathring{\alpha} \in \mathbf{R}^n$. Choose $\varepsilon, \eta \in \mathbf{R}^n$, $0 < \sum \varepsilon_j$, $0 < \sum \eta_j$, $\sum 2\varepsilon_j + \eta_j < 1$ and put $c = 1 - \sum 2\varepsilon_j + \eta_j$.*

Then there exists a constant C independent of $r \in (0, \frac{1}{2})$ and of $u \in \mathfrak{M}'_{\mathring{\alpha}}$, $\mathcal{M}_{\mathring{\alpha}} u = \mu T$ with $\mu \in O^s_M$ and T as in (10), such that (23) holds. Hence in particular $\mu_r = \mathcal{M}(Q_r u)(\mathring{\alpha} + i\,\cdot) \in O^s_M$ for $0 < r < \frac{1}{2}$.

Proof: Proceeding as in the proof of Main Estimation 1 we get (25) with κ_r^K, replaced by κ_r and by Lemma 1*, Remark 10.2 we get with $c = 1 - \sum 2\varepsilon_j + \eta_j > 0$

$$\left|\mathcal{M}((\ln x)^r b^j_\nu \kappa_r u)(\mathring{z}_1, \ldots, \mathring{z}_j - 1, \ldots, \mathring{z}_n)\right|$$
$$\leq C\Lambda G(r, -e^j - \varepsilon) r^{-\sum 2\varepsilon_j + \eta_j} \langle\beta\rangle^{s-m} < C\Lambda r^c \langle\beta\rangle^{s-m}$$
$$(j = 1, \ldots, n).$$

Hence from (25) (with κ_r instead of κ_r^K) we find

$$\left|\left(\frac{\partial}{\partial\beta}\right)^\delta \mathcal{M}(x_j Q^j(\kappa_r u))(\mathring{\alpha} + i\beta)\right| < C\Lambda r^c \langle\beta\rangle^s \qquad (j = 1, \ldots, n),$$

which combined with (24) leads to (23).

Now we prove the existence of a solution of Problem B (and hence of Problem A) and then end the proof of Theorem 1* in the same way as we did in Subsection 1 in the case of Theorem 1. Now, we replace Lemma 1, Main Estimation 1, Lemma 11.1, Proposition 10.3 by Lemma 1*, Main Estimation 1*, Exercise 11.6, Proposition 10.1 and instead of K-inequalities $\alpha <_K \omega$ we write $\sum \alpha_j < \sum \omega_j$.

Exercise

1. Write the operator R introduced at the beginning of the Section 14 in the explicit form

$$R\left(x, x\frac{\partial}{\partial x}\right) = \sum_{|\nu| \leq m} a_\nu(x) \left(x\frac{\partial}{\partial x}\right)^\nu \qquad (a_\nu \text{ smooth on } U).$$

Show that Q^j from the right hand side of (2) have the form

$$Q^j\left(x, x\frac{\partial}{\partial x}\right) = -\sum_{|\nu| \leq m} c_{j\nu} \left(x\frac{\partial}{\partial x}\right)^\nu$$

with $c_{j\nu}$ defined by $a_\nu(x) - a_\nu(0) = \sum_{j=1}^{n} x_j c_{j\nu}(x)$.

§15. FUCHSIAN TYPE PARTIAL DIFFERENTIAL EQUATIONS IN SPACES WITH CONTINUOUS RADIAL ASYMPTOTICS

Section 15 presents a geometric theory of the radial regularity of solutions to a Fuchsian partial differential equation $R(x, x\frac{\partial}{\partial x})u = v$. This is done by assigning to the principal Mellin symbol P of R ($P(z) = R(0, z)$) the radial characteristic set $\mathrm{Char}_{\mathring{\alpha}} P$ introduced in Subsection 1. Intuitively (and rigorously, see [Zie5], [Zie6]) the set $\mathrm{Char}_{\mathring{\alpha}} P$ is the set of the (complex) exponents which enter into the generalized asymptotic expansion at zero with respect to the radial variable, of a solution $u \in \mathfrak{M}'_{\mathring{\alpha}}$ of the equation $Pu = \delta_1$. Note that in dimension more than 1 the equation $Pu = \delta_1$ may have solutions of arbitrary order of singularity, therefore it is important to assume some regularity in advance by requiring $u \in \mathfrak{M}'_{\mathring{\alpha}}$ for a fixed $\mathring{\alpha} \in \mathbf{R}^n$. In dimension 2 the set $\mathrm{Char}_{\mathring{\alpha}} P$ consists of several half lines parallel to the real axis and we derive explicit formulas for the jump of the Mellin transforms across those lines (see Appendix for the interpretation of the jumps). We start this section with an example which is representative of the phenomena which we investigate below.

Example 1. Let

$$u(x_1, x_2) = \frac{1}{4\pi} \ln\big((\ln x_1)^2 + (\ln x_2)^2\big) \qquad \text{for } x \in \mathbf{R}^2_+ - \{(1,1)\}. \tag{1}$$

The distribution u solves the equation

$$Pu = \delta_1 \otimes \delta_1$$

where $P\big(x\frac{\partial}{\partial x}\big) = \big(x_1\frac{\partial}{\partial x_1}\big)^2 + \big(x_2\frac{\partial}{\partial x_2}\big)^2$. Let κ be a conical cut-off function of Definition 10.1. Then $\mathcal{M}(\kappa u) \in \mathcal{O}(\Omega)$ where Ω is of the form (11.14) with $\Omega^1 = \mathbf{C} \setminus \overline{\mathbf{R}_+}$, and the function $H(\zeta_1, \zeta_2) = \mathcal{M}(\kappa u)(\zeta_1 - \zeta_2, \zeta_2)$ satisfies (iii) in Definition 11.6 with $m = 1$.

Proof: We shall study the Mellin transform of κu in the variables $\zeta_1 = z_1 + z_2$, $\zeta_2 = z_2$. As we know (see Theorem 8.4 and Exercise 8.6), this corresponds to the change of the variables x given by $x_1 = y_1$, $x_2 = y_1 y_2$. In view of the properties of $\operatorname{supp}\kappa$ we have

$$\mathcal{M}\big(\kappa u(y_1, y_1 y_2)\big)(\zeta_1, \zeta_2) = \frac{1}{4\pi}\int_0^{\tilde{\tau}}\int_{\omega_1}^{\omega_2} \varphi(y_1, y_1 y_2)\tilde{\kappa}(1, y_2) \times$$
$$\times \ln\big\{2\big((-\ln y_1 - \tfrac{1}{2}\ln y_2)^2 + (-\tfrac{1}{2}\ln y_2)^2\big)\big\} y_1^{-\zeta_1 - 1} y_2^{-\zeta_2 - 1}\, dy_1\, dy_2$$

for some $\tilde{\tau} > 0$, $0 < \omega_1 < \omega_2$. Since $\omega_1 > 0$ the Mellin transform with respect to y_2 gives an entire function of ζ_2. Since we are interested in the set of holomorphy of the Mellin transform over the variable y_1 we may replace $\tilde{\tau}$ by a suitable $t < 1$ such that $\varphi(y_1, y_1 y_2) = 1$ for $0 < y_1 < t$ and $y_2 \in [\omega_1, \omega_2]$. Thus it is enough to compute the one-dimensional Mellin transform

$$\mathcal{M}^1(\chi_I \ln\{2((-\ln y_1 + C)^2 + C^2)\})(\zeta_1)$$
$$= \int_0^t \ln((-\ln y_1 + C)^2 + C^2) y_1^{-\zeta_1 - 1} dy_1 + \ln 2 \cdot \int_0^t y_1^{-\zeta_1 - 1} dy_1$$

where χ_I is the characteristic function of the interval $I = (0, t]$ and $C = C(y_2) = -\frac{1}{2}\ln y_2$. Since by (7.19)

$$\int_0^t y_1^{-\zeta_1 - 1} dy_1 = \frac{t^{-\zeta_1}}{-\zeta_1}$$

we have to compute the first integral. We know that C varies in a bounded set thus for y_1 sufficiently small we have

$$\ln((-\ln y_1 + C)^2 + C^2) = 2\ln(-\ln y_1 + C) + \sum_{j=1}^{\infty} \frac{(-1)^{j+1}}{j} C^{2j}(-\ln y_1 + C)^{-2j}$$

and the series is absolutely convergent. We have by Example 7.7 and formula (7.20)

$$\mathcal{M}^1(\chi_I \ln(-\ln y_1 + C))(\zeta_1) = \frac{\gamma}{\zeta_1} t^{-\zeta_1} + \int_0^{\infty} \ln \varrho \frac{d}{d\varrho}\left(\frac{e^{-C\varrho} t^{\varrho - \zeta_1}}{\varrho - \zeta_1}\right) d\varrho,$$
$$\mathcal{M}^1(\chi_I(-\ln y_1 + C)^{-2j})(\zeta_1) = \frac{1}{\Gamma(2j)} \int_0^{\infty} \frac{\varrho^{2j-1} e^{-C\varrho} t^{\varrho - \zeta_1}}{\varrho - \zeta_1} d\varrho$$
$$\text{for } \zeta_1 \in \mathbb{C} \setminus [0, +\infty),$$

where Γ is the Euler gamma function and γ is the Euler constant. Since

$$\sum_{j=0}^{\infty} \frac{(-1)^j}{(2j)!} (C\varrho)^{2j} = \cos(C\varrho)$$

we hence get

$$\mathcal{M}^1(\chi_I \ln((-\ln y_1 + C)^2 + C^2))(\zeta_1) = t^{-\zeta_1} \frac{2\gamma}{\zeta_1} +$$
$$+ 2\int_0^{\infty} \ln \varrho \frac{d}{d\varrho}\left(\frac{e^{-C\varrho} t^{\varrho - \zeta_1}}{\varrho - \zeta_1}\right) d\varrho + 2\int_0^{\infty} \frac{1 - \cos(C\varrho)}{\varrho} \frac{e^{-C\varrho} t^{\varrho - \zeta_1}}{\varrho - \zeta_1} d\varrho.$$

We have thus found

$$(2)\quad \mathcal{M}^1(\chi_I u(y_1, y_1 y_2))(\zeta_1) = \frac{1}{4\pi}\Big(\frac{2\gamma - \ln 2}{\zeta_1} + 2\int_0^\infty \ln\varrho \frac{d}{d\varrho}\Big(\frac{y_2^{\frac{\varrho}{2}} t^\varrho}{\varrho - \zeta_1}\Big)\,d\varrho + \\ + 2\int_0^\infty \frac{1 - \cos(\frac{\varrho}{2}\ln y_2)}{\varrho}\,\frac{y^{\frac{\varrho}{2}} t^\varrho}{\varrho - \zeta_1}\,d\varrho\Big)t^{-\zeta_1},$$

which easily implies the desired properties of $\mathcal{M}(\kappa u)$.

Remark 1. Define

$$T_0 = \frac{1}{2\pi i}\big(\lim_{b_1\to 0_+} \mathcal{M}^1(\chi_I u(y_1, y_1 y_2))(\,\cdot\, + ib_1) - \\ - \lim_{b_1\to 0_-} \mathcal{M}^1(\chi_I u(y_1, y_1 y_2))(\,\cdot\, + ib_1)\big),$$

where the limit is taken in the sense of distributions. Then from (2) we get for $\varphi \in C_0^\infty(\mathbf{R})$

$$T_0[\varphi] = -\frac{2\gamma - \ln 2}{4\pi}\varphi(0) + \frac{1}{2\pi}\int_0^\infty \ln\varrho \frac{d}{d\varrho}(y_2^{\frac{\varrho}{2}}\varphi(\varrho))\,d\varrho + \\ + \frac{1}{2\pi}\int_0^\infty \frac{1 - \cos(\frac{\varrho}{2}\ln y_2)}{\varrho} y_2^{\frac{\varrho}{2}}\varphi(\varrho)\,d\varrho.$$

Hence for $a_1 > 0$

$$T_0(a_1) = -\frac{1}{2\pi} y_2^{\frac{a_1}{2}} \frac{\cos(\frac{a_1}{2}\ln y_2)}{a_1}.$$

1. The Radial Characteristic Set $\mathrm{Char}_{\mathring{\alpha}} P$

The set $\mathrm{Char}_{\mathring{\alpha}} P$ defined below for a polynomial P in n variables and $\mathring{\alpha} \in \mathbf{R}^n$ generalizes to the case of several variables the concept of the set of the characteristic roots for an ordinary Fuchsian differential operator (see Section 13). Observe that in dimension higher than 1, $\mathrm{Char}_{\mathring{\alpha}} P$ need not be discrete and is quite different from $\{z \in \mathbf{C}^n\colon P(z) = 0\}$.

Let P be a polynomial in $\mathbf{C}^n$, $n > 1$ and $\mathring{\alpha} \in \mathbf{R}^n$. Denote $\mathring{a} = \mathbf{A}\mathring{\alpha}$ and $\mathcal{P}(\zeta) = P \circ \mathbf{A}^{-1}(\zeta)$ with $\mathbf{A}$ given by (11.12), and let $(1/\mathcal{P})_{\mathring{a}}$ be a regularization to a distribution in $S'(\mathbf{R}^n)$ of the function $b \mapsto 1/\mathcal{P}(\mathring{a} + ib)$, which always exists by the division theorem (we assume that $\mathcal{P}(\mathring{a} + ib) \not\equiv 0$ for $b \in \mathbf{R}^n$).

Definition 1. Let Ω^1 be the biggest set (as in (11.14)) such that the function

(3) $\tilde{\mathcal{C}}^-(\zeta) = \frac{1}{(2\pi)^n}\left(\frac{1}{\mathcal{P}}\right)_{\mathring{\alpha}}[F(\mathring{a}+i\gamma)K(\zeta-\mathring{a}-i\gamma)]$ for $\operatorname{Re}\zeta_1 < \mathring{a}_1,\ \zeta' \in \mathbf{C}^{n-1}$

extends to a holomorphic function on $\Omega^1 \cap \widetilde{\Omega}^1 \times \mathbf{C}^{n-1}$ for any $F \in \mathcal{O}(\widetilde{\Omega}^1 \times \mathbf{C}^{n-1})$ (with $\widetilde{\Omega}^1$ as in (11.14) and K given by (12.3)) such that for any open neighbourhood W of $\mathbf{C}\backslash\widetilde{\Omega}^1$ there exist constants C and M such that

$$|F(a+ib)| \leq C\langle b\rangle^M \quad \text{for} \quad a+ib \in (\mathbf{C}\backslash W)\times\mathbf{C}^{n-1}$$

locally uniformly in $a \in \operatorname{Re}\widetilde{\Omega}^1 \times \mathbf{R}^{n-1}$. We define $\operatorname{Char}_{\mathring{\alpha}} P = \mathbf{C}^n \backslash \mathbf{A}^{-1}(\Omega^1 \times \mathbf{C}^{n-1})$ and call it the *radial characteristic set* of P with respect to $\mathring{\alpha}$.

In Theorem 1 below we compute the set $\operatorname{Char}_{\mathring{\alpha}} P$ for a class of polynomials in two complex variables. We start with notation and preliminaries.

Let $P = \sum_{|\varrho|\leq m} b_\varrho z^\varrho$ be a polynomial in $\mathbf{C}^2$ with complex coefficients. We assume that the vector $\gamma = (-1,1)$ is *non characteristic* for P i.e. $P_m(\gamma) \neq 0$ where $P_m(z) = \sum_{|\varrho|=m} b_\varrho z^\varrho$.

Define $\mathcal{P}(\zeta_1,\zeta_2) = P(\zeta_1-\zeta_2,\,\zeta_2)$ and write $\mathcal{P}(\zeta_1,\zeta_2) = a_m(\zeta_1)\zeta_2^m + \cdots + a_1(\zeta_1)\zeta_2 + a_0(\zeta_1)$. Observe that $a_m(\zeta_1)$ is a constant function

$$a_m = P_m((-1,1)) = \sum_{|\varrho|=m} b_\varrho(-1)^{\varrho_1} \neq 0.$$

Represent $\mathcal{P}$ as

$$\mathcal{P}(\zeta_1,\zeta_2) = a_m\prod_{j=1}^{m}((\zeta_2 - c_j(\zeta_1)), \tag{4}$$

where $c_1(\zeta_1),\ldots,c_m(\zeta_1)$ are the complex roots of $\mathcal{P}$ with ζ_1 regarded as a parameter. Define the *discriminant* of $\mathcal{P}$

$$\Delta = \prod_{j<k}(c_j(\zeta_1) - c_k(\zeta_1)).$$

If $\mathcal{P}$ has no multiple polynomial factors then it follows from Lemmas A.12 and A.13 in [Hö2] that Δ is a non-zero polynomial in ζ_1 and for every $\mathring{\zeta}_1$ such that $\Delta(\mathring{\zeta}_1) \neq 0$ the functions $c_j(\zeta_1), j = 1,\ldots,m$ are holomorphic in a neighbourhood of $\mathring{\zeta}_1$. Further, in a neighbourhood of every point $\mathring{\zeta}_1$ such that $\Delta(\mathring{\zeta}_1) = 0$ the $c_j(\zeta_1)$ have expansions into *Puiseux series*, i.e. series of the form

$$c_j(\zeta_1) = \sum_{k=l}^{\infty} d_k((\zeta_1 - \mathring{\zeta}_1)^{\frac{1}{p}})^k \tag{5}$$

for some $p \in \mathbf{N}$, $l \in \mathbf{Z}$ and since a_m is a constant function we can put $l = 0$.

Fix $\mathring{a} \in \mathbf{R}^2$. We shall consider the functions $c_j, j = 1, \ldots, m$ defined as follows. For $\operatorname{Re}\zeta_1 < \mathring{a}_1$ close to $\mathring{a}_1$, $c_j(\zeta_1)$ are the holomorphic functions satisfying (4) and we choose the following extension of c_j to $\{\operatorname{Re}\zeta_1 > \mathring{a}_1\}$. Let $\theta_\nu (\nu = 1, \ldots, N)$ be all points in $\mathbf{C}$ such that $\Delta(\theta_\nu) = 0$ and for some j, $1 \le j \le m$, c_j has Puiseux expansion at θ_ν with minimal $p > 1$ and $\operatorname{Re}\theta_\nu > \mathring{a}_1$.

At the points θ_ν, c_j has value $c_j(\theta_\nu)$. For $\mathring{\zeta}_1 \in \mathbf{R}_+ + \theta_\nu$, we define

$$c_j(\mathring{\zeta}_1) \stackrel{\text{def}}{=} c_j^+(\mathring{\zeta}_1) = \lim_{\substack{\zeta_1 \to \mathring{\zeta}_1 \\ \operatorname{Im}\zeta_1 > \operatorname{Im}\mathring{\zeta}_1}} c_j(\zeta_1).$$

We also define

$$c_j^-(\mathring{\zeta}_1) = \lim_{\substack{\zeta_1 \to \mathring{\zeta}_1 \\ \operatorname{Im}\zeta_1 < \operatorname{Im}\mathring{\zeta}_1}} c_j(\zeta_1).$$

Denote by $B_\mu (\mu = 1, \ldots, M)$[(1)] all points in $\mathbf{R}$ such that for some j, $1 \le j \le m$,

$$\operatorname{Re} c_j(\mathring{a}_1 + iB_\mu) = \mathring{a}_2. \tag{6}$$

For j satisfying (6) we define: $\operatorname{sgn}(j;\mu) = \operatorname{sgn}(c_j; B_\mu) = +$ if for $a_1 > \mathring{a}_1$ close to $\mathring{a}_1$, $b \mapsto \operatorname{Re} c_j(a_1 + ib_1)$ is an increasing function in a neighbourhood of B_μ. Otherwise we put $\operatorname{sgn}(j;\mu) = -$. Finally for $\zeta_1 \in \mathbf{C}$ we denote

$$\begin{aligned} I^o(B_\mu) &= \{j\colon \text{formula (6) holds}\},\\ I^+(\theta_\nu) &= \{j\colon c_j \text{ has a Puiseux expansion at } \theta_\nu \text{ with } p > 1\\ &\qquad \text{and } \operatorname{Re} c_j(\mathring{a}_1 + ib_1) > \mathring{a}_2 \text{ for } b_1 > \operatorname{Im}\theta_\nu \text{ close to } \operatorname{Im}\theta_\nu\}. \end{aligned}$$

Theorem 1. *Fix $\mathring{\alpha} \in \mathbf{R}^2$ and let $\mathring{a} = \mathbf{A}\mathring{\alpha}$. Under the notation and assumptions introduced above denote*

$$L_\mu = \mathbf{R} + iB_\mu \quad \text{for} \quad \mu = 1, \ldots, M,$$

$$\widetilde{L}_\nu = \overline{\mathbf{R}_+} + \theta_\nu \qquad \text{for} \quad \nu = 1, \ldots, N,$$

$$Z = \{\zeta_1\colon \Delta(\zeta_1) = 0\},$$

$$\mathbf{L} = \bigcup_{\mu=1}^{M} L_\mu \cup \bigcup_{\nu=1}^{N} \widetilde{L}_\nu \cup Z, \qquad \mathbf{L}_{\mathring{a}} = \mathbf{L} \cap \{\operatorname{Re}\zeta_1 \geqslant \mathring{a}_1\}.$$

Then $\operatorname{Char}_{\mathring{\alpha}} P = \mathbf{A}^{-1}(\mathbf{L}_{\mathring{a}} \times \mathbf{C})$. *Moreover, for any $F \in \mathcal{O}(\widetilde{\Omega}^1 \times \mathbf{C})$, such that the function $\mathbf{R} \ni \gamma_2 \mapsto F(\zeta_1, a_2 + i\gamma_2)$ is polynomially bounded at ∞ locally uniformly*

---

(1) We assume that for $j = 1, \ldots, m$, $\operatorname{Re} c_j(\mathring{\alpha}_1 + i\beta_1) \not\equiv \mathring{\alpha}_2$ for $\beta_1 \in \mathbf{R}$.

in ζ_1 and a_2, the differences of the boundary values of the Cauchy transform (3) are distributions $\Xi^{\mu}_{\zeta_2}F$ on the lines $L_\mu \cap \widetilde{\Omega}^1 (\mu = 1, \ldots, M)$ with support in $L_\mu \cap \{\operatorname{Re}\zeta_1 \geq \mathring{a}_1\}$, and distributions $\widetilde{\Xi}^{\nu}_{\zeta_2}$ on the lines $\{\mathbf{R} + \theta_\nu\} \cap \widetilde{\Omega}^1 (\nu = 1, \ldots, N)$ with support in $\widetilde{L}_\nu$ such that for any $l \in \mathbf{N}$

$$(7) \qquad |||\, \Xi^{\mu}_{a_2+i\cdot} F[\varphi]\, |||_{S;l} \leq C_p \sum_{|\alpha| \leq p} \sup \left| \left(\frac{\partial}{\partial b_1}\right)^{\alpha} \varphi \right| \qquad \text{for } \varphi \in C_0^\infty(L^\mu \cap \widetilde{\Omega}^1)$$

for some constants $C_p = C_p(l, a_2)$, $p = p(a_2)$ locally bounded in $a_2 \in \mathbf{R}$ and analogous estimations hold for $\widetilde{\Xi}^{\nu}$. Explicitly we have

$$(8) \qquad \Xi^{\mu}_{\zeta_2} F(\zeta_1) = \frac{-1}{a_m} \sum_{j \in I^0(B_\mu)} \operatorname{sgn}(j;\mu) \frac{K'(\zeta_2 - c_j^{\operatorname{sgn}(j;\mu)}(\zeta_1)) F(\zeta_1, c_j^{\operatorname{sgn}(j;\mu)}(\zeta_1))}{\prod\limits_{\substack{q=1 \\ q \neq j}}^{m} (c_j^{\operatorname{sgn}(j;\mu)}(\zeta_1) - c_q^{\operatorname{sgn}(j;\mu)}(\zeta_1))}$$

for $\zeta_1 \in (\mathbf{R}_+ + \mathring{a}_1 + iB_\mu) \cap \widetilde{\Omega}^1$, $\mu = 1, \ldots, M$ with $\Delta(\zeta_1) \neq 0$

$$(9) \qquad \widetilde{\Xi}^{\nu}_{\zeta_2} F(\zeta_1) = \frac{-1}{a_m} \sum_{j \in I^+(\theta_\nu)} \left(\frac{K'(\zeta_2 - c_j^+(\zeta_1)) F(\zeta_1, c_j^+(\zeta_1))}{\prod\limits_{\substack{q=1 \\ q \neq j}}^{m} (c_j^+(\zeta_1) - c_q^+(\zeta_1))} - \frac{K'(\zeta_2 - c_j^-(\zeta_1)) F(\zeta_1, c_j^-(\zeta_1))}{\prod\limits_{\substack{q=1 \\ q \neq j}}^{m} (c_j^-(\zeta_1) - c_q^-(\zeta_1))} \right)$$

for $\zeta_1 \in (\mathbf{R}_+ + \theta_\nu) \cap \widetilde{\Omega}^1$, $\nu = 1, \ldots, N$.

Proof: In view of Corollary 12.2 we are interested in the holomorphic extensions in variable ζ_1 of the function

$$\widetilde{C}'_{\zeta_1}(\zeta_2) = \frac{1}{2\pi i} \int_{\operatorname{Re}\theta = \mathring{a}_2} \frac{K'(\zeta_2 - \theta) F(\zeta_1, \theta)}{\mathcal{P}(\zeta_1, \theta)}\, d\theta$$

defined for $\zeta_1 = \mathring{a}_1 + ib_1$ with $b_1 \neq B_\mu$ for $\mu = 1, \ldots, M$. Since the function $\mathbf{C} \ni \theta \mapsto K'(\zeta_2 - \theta) F(\zeta_1, \theta)$ is rapidly decreasing along the imaginary axis locally uniformly in ζ_1 and ζ_2, it follows that the integral over the line $\operatorname{Re}\theta = \mathring{a}_2$ may be replaced by an integral over $\operatorname{Re}\theta = r$ (for big $r > 0$) if we add suitable residuum terms. To this end denote for $\zeta_1 \in \mathbf{C}$

$$I_r^+(\zeta_1) = \{j\colon r > \operatorname{Re}\, c_j(\mathring{a}_1 + ib_1) > \mathring{a}_2 \text{ for } b_1 > \operatorname{Im}\zeta_1, \text{ close to } \operatorname{Im}\zeta_1\}.$$

In view of (4) we have

$$(10)\quad \widetilde{\mathcal{C}}'_{\zeta_1}(\zeta_2)=\frac{-1}{a_m}\sum_{j\in I_r^+(\zeta_1)}\frac{K'(\zeta_2-c_j(\zeta_1))F(\zeta_1,c_j(\zeta_1))}{\prod\limits_{\substack{q=1\\q\neq j}}^{m}(c_j(\zeta_1)-c_q(\zeta_1))}+$$

$$+\frac{1}{2\pi i a_m}\int\limits_{\operatorname{Re}\theta=r}\frac{K'(\zeta_2-\theta)F(\zeta_1,\theta)}{\prod\limits_{j=1}^{m}(\theta-c_j(\zeta_1))}\,d\theta.$$

The integral in the second summand is holomorphic (as a function of ζ_1 for a fixed ζ_2) in the set $\hat{\Omega}_r=\{\zeta_1\in\mathbf{C}\colon \operatorname{Re}c_j(\zeta_1)<r \text{ for } j=1,\dots,m\}$ hence it follows that (10) gives on extension of $\widetilde{\mathcal{C}}'_{\zeta_1}(\zeta_2)$ to $\hat{\Omega}_r\cap\widetilde{\Omega}^1\cap\{\operatorname{Re}\zeta_1\geqslant\mathring{a}_1\}\backslash\mathbf{L}$. Since the functions c_j are locally bounded we observe (by pushing r to $+\infty$) that all singularities of the extension are contained in the residuum terms, and the computation of the "jumps"of $\widetilde{\mathcal{C}}'_{\zeta_1}(\zeta_2)$ is now simple. It follows from Corollary 12.2 that the holomorphic extension of $\widetilde{\mathcal{C}}^-$ is given by

$$(11)\quad \psi(\zeta)=\begin{cases}\widetilde{\mathcal{C}}^-(\zeta) & \text{for } \operatorname{Re}\zeta_1<\mathring{a}_1,\ \zeta_2\in\mathbf{C}\\ \widetilde{\mathcal{C}}^+(\zeta)+\widetilde{\mathcal{C}}'_{\zeta_1}(\zeta_2) & \text{for } \operatorname{Re}\zeta_1>\mathring{a}_1,\ \zeta_1\in(\mathbf{C}\backslash\mathbf{L}_{\mathring{a}})\cap\widetilde{\Omega}^1,\ \zeta_2\in\mathbf{C},\end{cases}$$

where

$$\widetilde{\mathcal{C}}^{\pm}(\zeta)=\frac{1}{(2\pi)^2}\Big(\frac{1}{\mathcal{P}}\Big)_{\mathring{a}}[F(\mathring{a}+i\gamma)K(\zeta-\mathring{a}-i\gamma)]\qquad\text{for}\quad \pm\operatorname{Re}\zeta_1>\pm\mathring{a}_1.$$

Thus for $\operatorname{Re}\zeta_1>\mathring{a}_1$ the jumps of $\psi(\,\cdot\,,\zeta_2)$ coincide with those of $\widetilde{\mathcal{C}}'_{\zeta_1}(\zeta_2)$ which gives formulas (8) and (9).

It remains to prove that $\Xi^{\mu}_{\zeta_2}$, $\widetilde{\Xi}^{\nu}_{\zeta_2}$ are distributions on the respective lines. To this end we shall modify the function $\psi(\zeta)$ to a function $\tilde{\psi}(\zeta)$ which has the same jumps as ψ but whose growth properties are easier to investigate. In view of (iii) in Proposition 10.7 we can write

$$K(\zeta)=K'(\zeta_2)K^1(\zeta_1)+\widetilde{K}(\zeta)$$

where K^1 is a modified Cauchy kernel in variable ζ_1 and $\widetilde{K}\in\mathcal{O}(\mathbf{C}^2)$ is such that $\widetilde{K}(a+i\,\cdot)\in\mathrm{S}(\mathbf{R}^2)$ locally uniformly in $a\in\mathbf{R}^2$.

Then we have

$$\widetilde{\mathcal{C}}^{\pm}(\zeta)=\psi_1(\zeta)+\psi_2(\zeta),$$

where

$$\psi_1(\zeta)=\frac{1}{2\pi}\widetilde{\mathcal{C}}'_{\mathring{a}_1}(\zeta_2)[K^1(\zeta_1-\mathring{a}_1-i\cdot)]\qquad\text{for}\quad\operatorname{Re}\zeta_1\neq\mathring{a}_1,$$

$$\psi_2(\zeta)=\frac{1}{(2\pi)^2}\Big(\frac{1}{\mathcal{P}}\Big)_{\mathring{a}}[F(\mathring{a}+i\gamma)\widetilde{K}(\zeta-\mathring{a}-i\gamma)]\qquad\text{for}\quad\zeta\in\mathbf{C}^2.$$

Since ψ_2 is an entire function on $\mathbf{C}^2$ we are interested in ψ_1. Let χ be a $C_0^\infty(\mathbf{R})$ function which is 1 in a neighbourhood of the points B_μ $(\mu = 1, \dots, M)$ and $\operatorname{Im}\theta_\nu$ $(\nu = 1, \dots, N)$. Write for $\operatorname{Re}\zeta_1 \neq \mathring{a}_1$

$$\psi_3(\zeta) = \frac{1}{2\pi}\chi\widetilde{\mathcal{C}}'_{\mathring{a}_1}(\zeta_2)[K^1(\zeta_1 - \mathring{a}_1 - i\,\cdot)],$$

$$\psi_4(\zeta) = \frac{1}{2\pi}(1-\chi)\widetilde{\mathcal{C}}'_{\mathring{a}_1}(\zeta_2)[K^1(\zeta_1 - \mathring{a}_1 - i\,\cdot)].$$

Again $\psi_4(\,\cdot\,, \zeta_2)$ is holomorphic in complex neighbourhoods of the points $\mathring{a}_1 + iB_\mu$ $(\mu = 1, \dots, M)$ and $\mathring{a}_1 + i\operatorname{Im}\theta_\nu$ $(\nu = 1, \dots, N)$ so we are reduced to ψ_3. Inserting

$$K^1(\zeta_1) = -\frac{1}{\zeta_1} + \overline{K}(\zeta_1)$$

where $\overline{K} \in \mathcal{O}(\mathbf{C})$, in the definition of ψ_3 we find that modulo a holomorphic factor we are led to consider the function

$$\psi_5(\zeta) = \frac{1}{2\pi}\chi\widetilde{\mathcal{C}}'_{\mathring{a}_1}(\zeta_2)\left[\frac{-1}{\zeta_1 - \mathring{a}_1 - i\gamma_1}\right] \qquad \text{for} \quad \operatorname{Re}\zeta_1 \neq \mathring{a}_1.$$

Summing up we find that about the points $\mathring{a}_1 + iB_\mu, \mathring{a}_1 + i\operatorname{Im}\theta_\nu$ the function ψ given by (11) has the same jumps as the function

$$\Phi(\zeta_1,\zeta_2) = \begin{cases} \dfrac{-1}{2\pi}\chi\widetilde{\mathcal{C}}'_{\mathring{a}_1}(\zeta_2)\left[\dfrac{1}{\zeta_1 - \mathring{a}_1 - i\gamma_1}\right] & \text{for } \operatorname{Re}\zeta_1 < \mathring{a}_1, \zeta_2 \in \mathbf{C} \\ \dfrac{-1}{2\pi}\chi\widetilde{\mathcal{C}}'_{\mathring{a}_1}(\zeta_2)\left[\dfrac{1}{\zeta_1 - \mathring{a}_1 - i\gamma_1}\right] + \widetilde{\mathcal{C}}'_{\zeta_1}(\zeta_2) & \text{for } \operatorname{Re}\zeta_1 > \mathring{a}_1, \zeta_2 \in \mathbf{C}. \end{cases}$$

Next from (10) and Corollary 12.2 we find that Φ may be replaced by

$$\tilde{\psi}(\zeta_1,\zeta_2) = \begin{cases} \dfrac{1}{2\pi}\displaystyle\int_{\mathbf{R}} \chi(\gamma_1)\frac{E(\mathring{a}_1 + i\gamma_1, \zeta_2)}{\zeta_1 - \mathring{a}_1 - i\gamma_1}\,d\gamma_1 & \text{for } \operatorname{Re}\zeta_1 < \mathring{a}_1 \\ \dfrac{1}{2\pi}\displaystyle\int_{\mathbf{R}} \chi(\gamma_1)\frac{E(\mathring{a}_1 + i\gamma_1, \zeta_2)}{\zeta_1 - \mathring{a}_1 - i\gamma_1}\,d\gamma_1 + E(\zeta_1,\zeta_2) & \text{for } \operatorname{Re}\zeta_1 > \mathring{a}_1, \end{cases}$$

where

$$E(\zeta_1,\zeta_2) = \frac{-1}{a_m}\sum_{j \in I^+(\zeta_1)} \frac{K'(\zeta_2 - c_j(\zeta_1))F(\zeta_1, c_j(\zeta_1))}{\prod\limits_{\substack{j=1 \\ q \neq j}}^{m} (c_j(\zeta_1) - c_q(\zeta_1))} \tag{12}$$

and

$$I^+(\zeta_1) = \{j\colon \operatorname{Re}c_j(\mathring{a}_1 + ib_1) > \mathring{a}_2 \text{ for } b_1 > \operatorname{Im}\zeta_1, \text{ close to } \operatorname{Im}\zeta_1\}.$$

In view of the properties of E the assertion (7) now follows from Corollary 12.3 and Theorem 3.3.

A remarkable feature of the distributions $\Xi^{\mu}_{\zeta_2} F$ and $\widetilde{\Xi}^{\nu}_{\zeta_2} F$ is the following:

Corollary 1. *The distribution valued holomorphic functions*

$$\mathbf{C} \ni \zeta_2 \mapsto \Xi^{\mu}_{\zeta_2} F, \ \mathbf{C} \ni \zeta_2 \mapsto \widetilde{\Xi}^{\nu}_{\zeta_2} F$$

are rapidly decreasing in $\operatorname{Im} \zeta_2$, *locally uniformly in* $\operatorname{Re} \zeta_2$ *even though the function* $\mathbf{C} \ni \zeta_2 \mapsto F(\zeta_1, \zeta_2)$ *may grow polynomially in* $\operatorname{Im} \zeta_2$.

2. Regularity of Solutions in Spaces $M(\Omega; \varrho)$ and $Z_d(\Omega; \varrho)$

This subsection is devoted to an extension of the results of Subsection 14.2. Namely, we establish more refined regularity results in general spaces $M(\Omega; \varrho)$ and $Z_d(\Omega; \varrho)$, where Ω is an arbitrary set (11.14) and $\varrho \equiv s \in \mathbf{R} \cup \{\pm\infty\}$ is a constant function (see Subsection 11.3). Thus we deal with a Fuchsian type operator (14.1) satisfying the ellipticity condition (14.4).

Theorem 2. *Let* $w \in M(\Omega; s)$ *2-locally at* $(0, \delta\mathring{x})$. *Let* κ *be a conical cut-off function at* $(0; \delta\mathring{x})$ *such that* $\kappa w \in M(\Omega; s)$ *and suppose* $u \in \mathfrak{M}'_{\mathring{\alpha}}(\mathbf{R}^n_+)$ *for some* $\mathring{\alpha} \in \mathbf{R}^n$ *satisfies the equation*

$$Pu = Q(\kappa u) + \kappa w \qquad \text{in } \mathbf{R}^n_+. \tag{13}$$

Then $\kappa u \in M(\Omega \setminus \bigcup_{j=0}^{\infty} \{\operatorname{Char}_{\mathring{\alpha}} P + \mathbf{j}\}; s - m)$, *where* $\mathbf{j} = (j, 0, \ldots, 0) \in \mathbf{N}^n_0$.

Proof: Computing the Mellin $\mathcal{M}_{\mathring{\alpha}}$ transform of both sides of (13) we get

$$P(\mathring{\alpha} + i\beta)\mathcal{M}_{\mathring{\alpha}} u(\beta) = \mathcal{M}(Q(\kappa u))(\mathring{\alpha} + i\beta) + \mathcal{M}(\kappa w)(\mathring{\alpha} + i\beta). \tag{14}$$

Since $u \in \mathfrak{M}'_{\mathring{\alpha}}$ it follows from Exercise 10.5 that $\kappa u \in M'_{(\gamma)}$ for any $\gamma \in \mathbf{R}^n$ such that $\sum \gamma_j \le \sum \mathring{\alpha}_j$. Thus from Theorem 9.4* we obtain that for any fixed $\mathring{\varepsilon} \in \mathbf{R}^n$, $\sum \mathring{\varepsilon}_j > 0$ there exists an $s_0 \in \mathbf{R}$ such that $\kappa u \in M(\Omega_0; s_0)$ where

$$\Omega_0 = \left\{ z \in \mathbf{C}^n \colon \sum \operatorname{Re} z_j < \sum \mathring{\alpha}_j - \mathring{\varepsilon}_j \right\}.$$

Write Q in the form (14.2) where as in (14.3)

$$x_j Q^j\left(x, x\frac{\partial}{\partial x}\right) u = \sum_{|\nu| \le m} \left(x \frac{\partial}{\partial x}\right)^{\nu} (x_j b^j_{\nu}(x) u)$$

with b^j_ν smooth on U. Then it follows from the operational rules for the Mellin transformation that

$$x_j Q^j(\kappa u) \in M(\Omega_1; s_0 + m), \quad j = 1, \dots, n,$$

where

$$\Omega_1 = \{z \in \mathbf{C}^n : \textstyle\sum \operatorname{Re} z_j < 1 + \sum \mathring{\alpha}_j - \mathring{\varepsilon}_j\}.$$

Thus $Q(\kappa u) \in M(\Omega_1; s_0 + m)$ and we see from (14) that $\mathcal{M}_{\mathring{\alpha}} u$ extends to a function $E(z)$ (i.e. $E(\mathring{\alpha} + i\beta) = \mathcal{M}_{\mathring{\alpha}} u(\beta)$) such that the function $F(z) = P(z)E(z)$ is holomorphic on $\Omega \cap \Omega_1$ and (due to the ellipticity condition (14.4)) locally uniformly for $\alpha \in \operatorname{Re}(\Omega \cap \Omega_1)$

(15) $$|E(\alpha + i\beta)| \le C\langle\beta\rangle^{s_1} \qquad \text{for large } \|\beta\|,$$

where $s_1 = \max(s_0, s - m)$. Since

$$\mathcal{M}_{\mathring{\alpha}} u(\beta) = \frac{F(\mathring{\alpha} + i\beta)}{P(\mathring{\alpha} + i\beta)}$$

and F is polynomially bounded as a function of β locally uniformly in $\alpha \in \operatorname{Re}(\Omega \cap \Omega_1)$ it follows from the definition of the set $\operatorname{Char}_{\mathring{\alpha}} P$ and from (15) (by the same reasoning as in the second part of the proof of Theorem 12.4 with $H = E \circ \mathbf{A}^{-1}$) that

$$\kappa u \in M(\Omega \cap \Omega_1 \setminus \operatorname{Char}_{\mathring{\alpha}} P;\, s_1).$$

By induction we find that

(16) $$\kappa u \in M\Big(\Omega \setminus \bigcup_{j=0}^{\infty} \{\operatorname{Char}_{\mathring{\alpha}} P + \mathbf{j}\};\, s_1\Big).$$

It remains to prove that we may take $s_1 = s - m$ if $s - m < s_0$. It follows from the results of Subsection 14.2 that $s_0 = s - m$ on the set Ω_0 for the particular solution $u = \tilde{u} \in \mathfrak{M}'_{\mathring{\alpha}}$ constructed there. Thus $s_1 = s - m$ in (16). If u is another distribution in $\mathfrak{M}'_{\mathring{\alpha}}(\mathbf{R}^n)$ satisfying (13) then $\mathcal{M}_{\mathring{\alpha}}(u - \tilde{u}) \in S'$ has support in the bounded set $\{\beta : P(\mathring{\alpha} + i\beta) = 0\}$ and in view of Corollary 12.2 and Lemma 12.1 for any $h \in \mathbf{Z}$

$$|\mathcal{M}(\kappa(u - \tilde{u}))(\alpha + i\beta)| < C\langle\beta\rangle^h \qquad \text{for large } \|\beta\|$$

locally uniformly for $\alpha \in \operatorname{Re} \Omega_0$. This ends the proof.

In dimension $n = 2$ we have the following better result:

Theorem 3. *Let $w \in Z_d(\Omega; s)$ 2-locally at $(0, \delta\mathring{x})$. Then for every $\mathring{\alpha} \in \mathbf{R}^2$ with $\{(z_1, z_2)\colon \operatorname{Re} z_1 + \operatorname{Re} z_2 < \mathring{\alpha}_1 + \mathring{\alpha}_2\} \subset \Omega$ there exists $u_{\mathring{\alpha}} \in Z_d(\Omega \setminus \bigcup_{j=0}^{\infty} \{\operatorname{Char}_{\mathring{\alpha}} P + \mathbf{j}\}; s - m)$ 2-locally at $(0, \delta\mathring{x})$ and $Ru_{\mathring{\alpha}} = w$ 2-locally at $(0, \delta\mathring{x})$ $(R = P - Q)$.*

Proof: As $u_{\mathring{\alpha}}$ we take the solution $u \in \mathfrak{M}'_{\mathring{\alpha}}$ of the equation (14) for a suitable cut-off function κ at $(0, \delta\mathring{x})$. Then we proceed as in the proof of Theorem 2: first we observe that by Theorem 2 $\kappa u_{\mathring{\alpha}} \in M(\Omega \setminus \bigcup_{j=0}^{\infty} \{\operatorname{Char}_{\mathring{\alpha}} P + \mathbf{j}\}; s - m)$ thus we only have to check that the function $H(\zeta) = \mathcal{M}(\kappa u_{\mathring{\alpha}}) \circ \mathbf{A}^{-1}(\zeta)$ fullfils condition (iii) in Definition 11.6. Denote

$$\Omega_j = \{z \in \mathbf{C}^n\colon \sum_{l=1}^{n} \operatorname{Re} z_l < \sum_{l=1}^{n} \mathring{\alpha}_l + j\} \qquad \text{for } j \in \mathbf{N}_0.$$

Clearly $M(\Omega_0; s - m) = Z_d(\Omega_0; s - m)$. Since $\kappa u_{\mathring{\alpha}} \in M(\Omega_0; s - m)$ it follows from (14) that

$$\mathcal{M}_{\mathring{\alpha}} u_{\mathring{\alpha}}(\beta) = \frac{F(\mathring{\alpha} + i\beta)}{P(\mathring{\alpha} + i\beta)}$$

where $H(\zeta) = F \circ \mathbf{A}^{-1}(\zeta)$ satisfies (i)–(iii) in Definition 11.6 on $\widetilde{\Omega}^1 = \{\zeta_1 < \sum \mathring{\alpha}_j + 1\} \cap \Omega^1$. Now $\mathcal{M}(\kappa u_{\mathring{\alpha}}) \circ \mathbf{A}^{-1}(\zeta) = \widetilde{C}^-(\zeta)$ and by Corollary 12.2 we are interested in $\widetilde{C}'_{\zeta_1}(\zeta_2)$ as in the proof of Theorem 1. Thus we have decomposition (10). The first term in (10) clearly possesses the property (iii) of Definition 11.6 on $\widetilde{\Omega}^1$ with $s = -\infty$. To get an estimation of the second term for $\zeta_1 \in \hat{\Omega}_r \cap \widetilde{\Omega}^1$ we apply Lemma 12.2. By induction we prove that (iii) and (iv) hold on any set Ω_j.

Theorem 4. *Let $\delta\mathring{x} \in \mathbf{R}^n_+$. Suppose $w \in M(\widetilde{\Omega}; \tilde{\varrho})$ and $u \in M(\Omega; \varrho)$ 2-locally at $(0; \delta\mathring{x})$ and $Ru = w$ in a local conical neighbourhood of $(0, \delta\mathring{x})$. Then*

1° *for any $\varepsilon > 0$ $u \in M(\Omega \cap \widetilde{\Omega}; \varrho(-\infty) + \varepsilon)$ 2-locally at $(0; \delta\mathring{x})$ if $\tilde{\varrho} - m \le \varrho(-\infty) \overset{\text{def}}{=} \lim_{a_1 \to -\infty} \varrho(a_1)$,*

2° *$u \in M(\Omega \cap \widetilde{\Omega}; \tilde{\varrho} - m)$ 2-locally at $(0; \delta\mathring{x})$ if $Q \equiv 0$.*

Proof: We may assume that $\varrho(-\infty) > -\infty$. Choose a conical cut-off function κ_1 at $(0; \delta\mathring{x})$ such that $\kappa_1 w \in M(\widetilde{\Omega}; \tilde{\varrho})$, $\kappa_1 u \in M(\Omega; \varrho)$ and $Ru = w$ on $\operatorname{supp} \kappa_1$. In view of (14.1) we have

$$R(\kappa_1 u) = P(\kappa_1 u) - Q(\kappa_1 u) = \kappa_1 w + \widetilde{P}_{\kappa_1} u - \widetilde{Q}_{\kappa_1} u, \tag{17}$$

where

$$\widetilde{P}_{\kappa_1}u = P(\kappa_1 u) - \kappa_1 Pu = \sum_{\substack{|\alpha+\beta|\le m\\ |\beta|\le m-1}} a_{\alpha\beta}\Big(\Big(x\frac{\partial}{\partial x}\Big)^{\alpha}\kappa_1\Big)\Big(\Big(x\frac{\partial}{\partial x}\Big)^{\beta}u\Big),$$

$$\widetilde{Q}_{\kappa_1}u = Q(\kappa_1 u) - \kappa_1 Qu = \sum_{j=1}^{n} x_j \widetilde{Q}^{j}_{\kappa_1}u$$

with

$$\widetilde{Q}^{j}_{\kappa_1} = \sum_{\substack{|\alpha+\beta|\le m\\ |\beta|\le m-1}} a^{j}_{\alpha\beta}\Big(\Big(x\frac{\partial}{\partial x}\Big)^{\alpha}\kappa_1\Big)\Big(\Big(x\frac{\partial}{\partial x}\Big)^{\beta}u\Big),$$

where $a_{\alpha,\beta}$ and $a^{j}_{\alpha,\beta}$ are smooth about zero. Computing the Mellin transform of (17) we thus get

$$P(z)\mathcal{M}(\kappa_1 u) = \mathcal{M}((\widetilde{P}_{\kappa_1} - \widetilde{Q}_{\kappa_1})u) + \mathcal{M}(Q(\kappa_1 u)) + \mathcal{M}(\kappa_1 w). \tag{18}$$

Choose another conical cut-off function κ_0 such that $\kappa_0 \equiv 1$ on $\operatorname{supp}\kappa_1$ and $\kappa_0 u \in M(\Omega;\varrho)$. Thus by Definition 11.5 we have for every neighbourhood W of $\Lambda = \mathbf{C}\setminus\Omega^1$

$$|\mathcal{M}(\kappa_0 u)(\alpha + i\beta)| \le C_0\langle\beta\rangle^{\varrho(\sum\alpha_j)} \qquad \text{for } \alpha + i\beta \in \mathbf{A}^{-1}((\mathbf{C}\setminus W)\times\mathbf{C}^{n-1}) \tag{19}$$

locally uniformly in $\alpha \in \operatorname{Re}\Omega$. We shall prove that $\widetilde{Q}^{j}_{\kappa_1}u \in M(\Omega;\varrho+m-1)$. Indeed we clearly have $\big(x\frac{\partial}{\partial x}\big)^{\beta}\kappa_0 u \in M(\Omega;\varrho+|\beta|)$ and from Theorem 12.4 applied to the conical cut-off functions $a^{j}_{\alpha\beta}\big(x\frac{\partial}{\partial x}\big)^{\alpha}\kappa_1$ we get

$$a^{j}_{\alpha\beta}\Big(\Big(x\frac{\partial}{\partial x}\Big)^{\alpha}\kappa_1\Big)\Big(x\frac{\partial}{\partial x}\Big)^{\beta}u \in M(\Omega;\varrho+|\beta|)$$

since $\kappa_0 \equiv 1$ on $\operatorname{supp}\kappa_1$. Summing over α,β we get $\widetilde{Q}^{j}_{\kappa_1}u \in M(\Omega;\varrho+m-1)$. Thus

$$\widetilde{Q}_{\kappa_1}u \in M\big(\Omega + (1,0,\dots,0);\ \varrho(\textstyle\sum\alpha_j - 1) + m - 1\big) \subset M(\Omega;\varrho+m-1)$$

because $\Omega \subset \Omega + (1,0,\dots,0)$ (see Remark 11.4) and ϱ is non-decreasing. In a similar way we have $\widetilde{P}_{\kappa_1}u \in M(\Omega;\varrho+m-1)$ and analogously

$$Q(\kappa_1 u) \in M\big(\Omega + (1,0,\dots,0);\ \varrho(\textstyle\sum\alpha_j - 1) + m\big).$$

From (18) we see that locally uniformly in $\alpha \in \operatorname{Re}\Omega \cap \tilde{\Omega}$ we have

$$|\mathcal{M}(\kappa_1 u)(\alpha + i\beta)| \le C\langle\beta\rangle^{\varrho(\sum\alpha_j)-1} + \tilde{C}\langle\beta\rangle^{\varrho(\sum\alpha_j - 1)} \tag{20}$$

for $\|\beta\|$ big enough (because by assumption $\tilde{\varrho}(a_1) - m \le \varrho(-\infty) \le \varrho(a_1 - 1)$).

Let $\delta_1 \in \mathbf{R} \cup \{\infty\}$ be such that $\varrho(a_1) - 1 \le \varrho(a_1 - 1)$ for $a_1 \le \delta_1$. Then

$$|\mathcal{M}(\kappa_1 u)(\alpha + i\beta)| \le C_1\langle\beta\rangle^{\varrho(\sum\alpha_j - 1)} \tag{21}$$

locally uniformly in α for $\sum\alpha_j < \delta_1$ and $\|\beta\|$ big. Now, choose a conical cut-off function κ_2 having the same properties as κ_1 and such that $\kappa_1 \equiv 1$ on $\operatorname{supp}\kappa_2$. Replacing (19) by (21) and repeating the above considerations for κ_2 instead of κ_1 and κ_1 instead of κ_0 we get

$$|\mathcal{M}(\kappa_2 u)(\alpha + i\beta)| \le C_2\langle\beta\rangle^{\varrho(\sum\alpha_j - 2))} \tag{22}$$

locally uniformly in α for $\|\beta\|$ big and $\sum\alpha_j < \delta_1$ (since again $\varrho(a_1-1)-1 \le \varrho(a_1-2)$ for $a_1 \le \delta_1$). By induction for any n and suitable conical cut-off function κ_n

$$|\mathcal{M}(\kappa_n u)(\alpha + i\beta)| \le C_n\langle\beta\rangle^{\varrho(\sum\alpha_j - n)}$$

locally uniformly in α for $\|\beta\|$ big and $\sum\alpha_j < \delta_1$. Since $\mathcal{M}(\kappa_n u)$ is holomorphic on $\Omega \cap \tilde{\Omega}$, the assumption that $\|\beta\|$ is big may be dropped so we have proved assertion 1° on $\Omega \cap \tilde{\Omega} \cap \{\sum \operatorname{Re} z_j < \delta_1\}$.

If $\delta_1 < \infty$, choose δ_2 such that for $\delta_1 \le a_1 \le \delta_2$

$$\varrho(a_1) - 2 \le \varrho(a_1 - 1) \le \varrho(a_1) - 1.$$

From (20) we have

$$|\mathcal{M}(\kappa_1 u)(\alpha + i\beta)| \le C_1\langle\beta\rangle^{\varrho(\sum\alpha_j)-1} \tag{23}$$

for $\delta_1 \le \sum\alpha_j \le \delta_2$, locally uniformly in α for $\|\beta\|$ big.

For a fixed $\varepsilon \geq 0$ so small that $\varrho(\delta_1) - 1 \geq \varrho(-\infty) + \varepsilon$ define

$$\tilde{\varrho}_1(a_1) = \begin{cases} \varrho(a_1) - 1 & \text{for } \delta_1 \leq a_1 \leq \delta_2, \\ \varrho(-\infty) + \varepsilon & \text{for } a_1 < \delta_1. \end{cases}$$

Then from (23) and the part already proved, we get

$$|\mathcal{M}(\tilde{\kappa}_1 u)(\alpha + i\beta)| \leq \widetilde{C}_1 \langle\beta\rangle^{\tilde{\varrho}_1(\sum \alpha_j)} \tag{24}$$

for $\sum \alpha_j \leq \delta_2$, locally uniformly in α for big $\|\beta\|$, and a suitable conical cut-off function $\tilde{\kappa}_1$. Choose a cut-off function $\tilde{\kappa}_2$ having the same properties as κ_1 and such that $\tilde{\kappa}_1 \equiv 1$ on $\operatorname{supp} \tilde{\kappa}_2$. Replacing (19) by (24) and repeating the considerations conducted for $\sum \alpha_j \leq \delta_1$ with κ_0 replaced by $\tilde{\kappa}_1$, κ_1 by $\tilde{\kappa}_2$, δ_1 by δ_2 and ϱ by $\tilde{\varrho}_1$ we get

$$\begin{aligned} |\mathcal{M}(\tilde{\kappa}_2 u)(\alpha + i\beta)| &\leq C' \langle\beta\rangle^{\tilde{\varrho}_1(\sum \alpha_j) - 1} + C'' \langle\beta\rangle^{\tilde{\varrho}_1(\sum \alpha_j - 1)} \\ &\leq \widetilde{C}_2' \langle\beta\rangle^{\varrho(\sum \alpha_j - 1)} \end{aligned}$$

locally uniformly in α for $\|\beta\|$ big for $\delta_1 \leq \sum \alpha_j \leq \delta_2$, (since $\varrho(a_1) - 2 \leq \varrho(a_1 - 1)$ for $a_1 \leq \delta_2$ and $\tilde{\varrho}_1(a_1 - 1) = \varrho(-\infty) + \varepsilon \leq \varrho(\delta_1) - 1 \leq \varrho(\delta_1 - 1) \leq \varrho(a_1 - 1)$ for $\delta_1 \leq a_1 \leq \delta_2$).

This together with (21) gives that if $\kappa_1 \equiv 1$ on $\operatorname{supp} \tilde{\kappa}_2$

$$|\mathcal{M}(\tilde{\kappa}_2 u)(\alpha + i\beta)| \leq \widetilde{C}_2 \langle\beta\rangle^{\varrho(\sum \alpha_j - 1)} \tag{25}$$

locally uniformly in α for $\|\beta\|$ big if $\sum \alpha_j \leq \delta_2$. Thus using (25) instead of (21) we prove the assertion 1° on $\Omega \cap \widetilde{\Omega} \cap \{\sum \operatorname{Re} z_j < \delta_2\}$. Thus by induction we see that 1° is valid without any additional restrictions on $\Omega \cap \widetilde{\Omega}$.

Concerning 2° observe that if $Q \equiv 0$ then locally uniformly in α for $\|\beta\|$ big

$$|\mathcal{M}(\kappa_1 u)(\alpha + i\beta)| \leq C \langle\beta\rangle^{\varrho(\sum \alpha_j) - 1} + \widetilde{C} \langle\beta\rangle^{\tilde{\varrho}(\sum \alpha_j) - m}. \tag{26}$$

Thus

$$|\mathcal{M}(\kappa_1 u)(\alpha + i\beta)| \leq C' \langle\beta\rangle^{\max(\varrho(\sum \alpha_j) - 1, \tilde{\varrho}(\sum \alpha_j) - m)}$$

and replacing ϱ by $\max(\varrho - 1, \tilde{\varrho} - m)$ in (26) we find

$$|\mathcal{M}(\kappa_1 u)(\alpha + i\beta)| \leq C_1 \langle\beta\rangle^{\max(\varrho(\sum \alpha_j) - 2, \tilde{\varrho}(\sum \alpha_j) - m)}.$$

Hence by induction for any $n \in \mathbf{N}$

$$|\mathcal{M}(\kappa_1 u)(\alpha + i\beta)| \leq C_n \langle\beta\rangle^{\max(\varrho(\sum \alpha_j) - n, \, \tilde{\varrho}(\sum \alpha_j) - m)}$$

which gives assertion 1°.

Corollary 2. *As it is seen from 1° in Theorem 1 when dealing with solutions to singular elliptic PDE-s it is enough to consider only constant functions ϱ.*

Example 1 (continued). Since

$$\left(x_1\frac{\partial}{\partial x_1}\right)^2 u + \left(x_2\frac{\partial}{\partial x_2}\right)^2 u = \delta_1 \in D'(\mathbf{R}^2),$$

where $u(x_1, x_2) = \frac{1}{4\pi}\ln((\ln x_1)^2 + (\ln x_2)^2) \in \mathfrak{M}_0'(\mathbf{R}_+^2)$, it follows from Theorem 2 (with $Q \equiv 0$) that for any conical cut-off function κ

$$\kappa u \in M(\mathbf{C}^2 \setminus \operatorname{Char}_0 P;\ -2),$$

where for $P(z_1, z_2) = z_1^2 + z_2^2$,

$$\operatorname{Char}_0 P = \mathbf{A}^{-1}(\overline{\mathbf{R}_+} \times \mathbf{C}), \qquad \text{where} \quad \mathbf{A} = \begin{pmatrix} 1 & 1 \\ 0 & 1 \end{pmatrix},$$

(see Example 2 below), and $\delta_1 \in M(\mathbf{C}^2; 0)$. Then since $\delta_1 = 0$ close to zero and hence $\delta_1 \in M(\mathbf{C}^2; -\infty)$ locally at zero we have by Theorem 4, 2°

$$\kappa u \in M\big(\mathbf{A}^{-1}((\mathbf{C} \setminus \overline{\mathbf{R}_+}) \times \mathbf{C}); -\infty\big).$$

Actually, in view of Theorem 3, we have $\kappa u \in Z_d(\mathbf{A}^{-1}((\mathbf{C} \setminus \overline{\mathbf{R}_+}) \times C); -\infty)$.

Example 2. Let $0 \neq \mathring{\alpha} \in \mathbf{R}^2$ and denote by $u_{\mathring{\alpha}} \in \mathfrak{M}_{\mathring{\alpha}}'(\mathbf{R}^2)$ (see Section 14) the solution of the equation $Pu_{\mathring{\alpha}} = \delta_1$ on $\mathbf{R}_+^2$ where $P(x\frac{\partial}{\partial x}) = (x_1\frac{\partial}{\partial x_1})^2 + (x_2\frac{\partial}{\partial x_2})^2$, such that

$$\mathcal{M}_{\mathring{\alpha}} u_{\mathring{\alpha}}(\beta) = \frac{1}{(\mathring{\alpha}_1 + i\beta_1)^2 + (\mathring{\alpha}_2 + i\beta_2)^2} \in L^1(\mathbf{R}^2).$$

We have

$$\mathcal{P}(\zeta_1, \zeta_2) = (\zeta_1 - \zeta_2)^2 + \zeta_2^2 = 2(\zeta_2 - c_1(\zeta_1))(\zeta_2 - c_2(\zeta_1))$$

where $c_1(\zeta_1) = \frac{1+i}{2}\zeta_1$, $c_2(\zeta_1) = \frac{1-i}{2}\zeta_1$ and $\Delta = i\zeta_1$. Further from (6) we find

$$B_1 = \mathring{a}_1 - 2\mathring{a}_2, \quad B_2 = 2\mathring{a}_2 - \mathring{a}_1$$

and since c_1 and c_2 are regular there are no points θ_ν.

Denote

$$Z = \{0\},$$

$$L_1 = \{\zeta_1 \in \mathbf{C}\colon \zeta_1 = a_1 + i(\mathring{a}_1 - 2\mathring{a}_2)\},$$

$$L_2 = \{\zeta_1 \in \mathbf{C}\colon \zeta_1 = a_1 + i(2\mathring{a}_2 - \mathring{a}_1)\},$$

$$\mathbf{L} = L_1 \cup L_2 \cup Z, \qquad \mathbf{L}_{\mathring{a}_1} = \mathbf{L} \cap \{\operatorname{Re}\zeta_1 \geq \mathring{a}_1\}.$$

Then it follows from Theorems 1, 4 and Theorem 3 that

$$\kappa u_{\mathring{\alpha}} \in Z_d(\Omega; -\infty),$$

where $\Omega = \mathbf{A}^{-1}((\mathbf{C}\backslash \mathbf{L}_{\mathring{a}_1}) \times \mathbf{C})$ and $\kappa = \varphi \cdot \tilde{\kappa}$ is any proper conical cut-off function. Moreover by Theorem 1 and Proposition 12.1 the distributions $\Xi^1_{\zeta_2}$, $\Xi^2_{\zeta_2}$ equal

$$\Xi^1_{\zeta_2}[\varphi] = \frac{1}{2i} \int\limits_{L_1} \frac{K'(\zeta_2 - \frac{1+i}{2}\zeta_1)}{\zeta_1} \varphi(\zeta_1)\, d\zeta_1 \qquad \text{for} \quad \varphi \in C_0^\infty(\mathbf{R} + iB_1),$$

$$\Xi^2_{\zeta_2}[\varphi] = \frac{1}{2i} \int\limits_{L_2} \frac{K'(\zeta_2 - \frac{1-i}{2}\zeta_1)}{\zeta_1} \varphi(\zeta_1)\, d\zeta_1 \qquad \text{for} \quad \varphi \in C_0^\infty(\mathbf{R} + iB_2),$$

where $K'(\zeta_2) = \mathcal{M}'(\tilde{\kappa}(1, y_2))(\zeta_2)$ and $a^\mu(\zeta_2)$, $\mu = 1, 2$ are some entire functions which are the Mellin transforms of some distributions $\widetilde{T}^\mu$ in $M(\mathbf{C}; -\infty)$ vanishing near zero. Then $\widetilde{T}^\mu$ are smooth functions $\widetilde{T}^\mu(y_2)$ vanishing for y_2 close to zero. Computing the inverse Mellin transforms of Ξ^μ with respect to ζ_2 and dividing it by $2\pi i$ we find

$$T_1[\varphi](y_2) = -\frac{1}{4\pi} \tilde{\kappa}(1, y_2) \int\limits_{L_1} \frac{y_2^{\frac{1+i}{2}\zeta_1}}{\zeta_1} \varphi(\zeta_1)\, d\zeta_1 \qquad \text{for} \quad \varphi \in C_0^\infty(\mathbf{R} + iB_1),$$

$$T_2[\varphi](y_2) = -\frac{1}{4\pi} \tilde{\kappa}(1, y_2) \int\limits_{L_2} \frac{y_2^{\frac{1-i}{2}\zeta_1}}{\zeta_1} \varphi(\zeta_1)\, d\zeta_1 \qquad \text{for} \quad \varphi \in C_0^\infty(\mathbf{R} + iB_2).$$

Remark 2. Observe that if $\mathring{\alpha} \to 0$ then the lines L_1, L_2 tend towards the half-line $\overline{\mathbf{R}_+}$ and the distribution $T_1 + T_2$ for $\alpha_1 > 0$ tends towards T_0 given by Remark 1. Indeed we have

$$\frac{1}{a_1}\left(y_2^{\frac{1+i}{2}a_1} + y_2^{\frac{1-i}{2}a_1}\right) = \frac{2}{a_1} y_2^{\frac{a_1}{2}} \cos\left(\frac{a_1}{2} \ln y_2\right).$$

Appendix

GENERALIZED SMOOTH FUNCTIONS AND THEORY OF RESURGENT FUNCTIONS OF JEAN ECALLE

Connections are established between the theory of the Mellin transformation, the theory of the Borel transformation, the theory of generalized smooth functions and the theory of resurgent functions of Jean Ecalle. Applications are given to the study of special functions, non linear Euler equations and solutions to elliptic Fuchsian operators.

1. Introduction

The purpose of the appendix (see [Zie7]) is to attract reader's attention to a class of generalized smooth functions. By the term "generalized" we mean that such functions need not expand into discrete powers x^j, $j \in \mathbf{N}_0$ as the smooth functions do, but have "continuous" expansions into powers x^α, $\alpha \in \mathbf{R}$ with the density being a generalized function. Such functions behave in much the same way as the smooth functions and it is possible to generalize to them the classical theorems of calculus in one and several variables (see [Zie1], [Zie2]). Of particular importance is a subclass of resurgent functions ([E1], [E2], [M2], [P–C–N]).

In Section 3, we present an outline of the theory of resurgent functions in a slightly changed setting which is suitable for establishing connections with the theory of generalized smooth and analytic functions.

We start with presenting some result of the latter theory. The exposition has an elementary character and for simplicity we consider only functions of one real variable (the reader interested in the general theory is referred to papers [Zie2], [Zie3]). The appendix ends with examples showing that the resurgent functions appear as special functions, as solutions to linear and non-linear ordinary differential equations with irregular singularities and in the study of radial asymptotics of elliptic Fuchsian operators as presented in Section 15 (see also [Zie5], [Zie6] for further developments).

2. Generalized Taylor Expansions

We start with the usual Taylor formula for smooth functions. Let f be a smooth function in a neighbourhood of $0 \in \mathbf{R}$. Then for every $r \in \mathbf{N}_0$ we have

$$f(x) = \sum_{i=0}^{r} \frac{f^{(i)}(0)}{i!} x^i + R(x), \tag{1}$$

where $R(x) = O(x^{r+1})$ is flat of order $r+1$ at zero. Equation (1) can be viewed upon in the following new way. Let δ_i denote Dirac's delta at the point $i = 0, 1, \dots$ Denote

$$T_r = \sum_{i=0}^{r} \frac{f^{(i)}(0)}{i!} \delta_i. \tag{2}$$

T_r is regarded as a measure on $\mathbf{R}$ or as a compactly supported distribution on $\mathbf{R}$ of order zero. In the latter case its action on a test function φ is denoted by $T_r[\varphi]$.

For $x > 0$, (1) is equivalent to

$$f(x) = T_r[x^\alpha] + R(x) = \int_0^\infty x^\alpha T_r(\alpha)\, d\alpha + R(x), \tag{3}$$

where T_r is understood as a distribution in variable α and x^α denotes the function $\mathbf{R} \ni \alpha \mapsto x^\alpha$ for a fixed $x > 0$. Equation (3) suggests that if we take as T_r more general objects (e.g. distributions, hyperfunctions) then (3) will hold for a class of functions much wider than the smooth functions.

Example 1. For $0 < x < 1$ and $j = 1, 2, \dots$ we have by Example 13.2

$$(-\ln x)^{-j} = T[x^\alpha]$$

where $T = Y(\alpha)\, \alpha^{j-1}/(j-1)!$ where Y is the Heaviside function.

Example 2. For $x > 0$ and $j = 1, 2 \dots$

$$(-\ln x)^j = \delta_0^{(j)}[x^\alpha].$$

Here $T = \delta_0^{(j)}$ is the j-th derivative of the Dirac delta δ_0.

Thus T can be a distribution of an arbitrary order. Actually, it need not be a distribution as the following example shows.

Example 2′. Let $H(z) = \sum_{j=0}^{\infty} a_j z^j$ be an entire function on $\mathbf{C}$. Then for the function

$$f(x) = H(-\ln x) = \sum_{j=0}^{\infty} a_j(-\ln x)^j \qquad \text{for } 0 < x < 1$$

we have, by Example 2,

$$f(x) = \Big(\sum_{j=0}^{\infty} a_j \delta_0^{(j)}\Big)[x^\alpha]$$

i.e. $T = \sum_{j=0}^{\infty} a_j \delta_0^{(j)}$ is a hyperfunction, $f = T[x^\alpha]$ (see [Kö], [Schm], [K]).

Below we introduce tools necessary to solve the problem: given a function $f : (0,1) \to \mathbf{C}$, find the "spectral measure" T.

2.1. Generalized Borel Transformation

Let f: $(0,1) \to \mathbf{C}$, $f = O(x^a)$ for some $a \in R$ be such that $\mathcal{M}f$ is holomorphic on $\mathbf{C} \setminus \mathbf{R}$ and satisfies the estimation (3.2). By the (generalized) Borel transform $B(f)$ of f we understand the difference of boundary values of the Mellin transform of f divided by $2\pi i$, i.e.

$$B(f) = \frac{1}{2\pi i} b(\mathcal{M}f).$$

By Theorem 3.3 $B(f)$ is a distribution with support in $[a, \infty)$. In Example 5 we explain the relation of $B(f)$ with the classical Borel transformation.

Remark. Observe that the Borel transform of a (bounded) function f: $(0,1) \to \mathbf{C}$ depends only on the behaviour of f in a neighbourhood of zero. This is clear from the fact that the Mellin transform of a function f: $(0,1) \to \mathbf{C}$, $f = O(x^a)$ which is zero on an interval $(0, \hat{t})$, $\hat{t} > 0$ is an entire function on $\mathbf{C}$.

Example 3. If f is a C^∞ function then it follows from Examples 7.5 and 3.1 that

$$B(f) = \frac{1}{2\pi i} b(\mathcal{M}f) = \sum_{j=0}^{\infty} \frac{f^{(j)}(0)}{j!} \delta_j$$

i.e. on the interval $(-\infty, r+1)$ $B(f)$ coincides with T_r given by (2).

Example 4. For the function $(-\ln x)^{-j}$, $j \in \mathbf{N}$ considered on an interval $0 < x < t$ for some $t < 1$ we have, in view of Examples 7.6 and 3.1,

$$B\Big((-\ln x)^{-j}\Big)(\alpha) = Y(\alpha)\, \frac{\alpha^{j-1}}{(j-1)!}.$$

Thus the Borel transform coincides with the "spectral measure" T of Example 1.

The following theorem now seems natural:

Theorem 1 (generalized Taylor formula). *Let $f\colon (0,1) \to \mathbf{C}$ be a function such that $x^{\varepsilon-a} f(x)$ is bounded for any $\varepsilon > 0$ and let $\mathcal{M}f \in \mathcal{O}(\mathbf{C} \setminus \mathbf{R})$ satisfy the estimation (3.2). Then for every $r > a$ there exists a function R such that*

$$f(x) = T_r[x^\alpha] + R(x), \tag{4}$$

where $T_r \in D'(\mathbf{R})$ satisfies

$$T_r = \begin{cases} B(f) & \text{on } (-\infty, r) \\ 0 & \text{on } (r, \infty) \end{cases}$$

and R is such that $x^{\varepsilon-a} R(x)$ is bounded for $\varepsilon > 0$ and R is r-flat at zero i.e. $b(\mathcal{M}R) = 0$ on $(-\infty, r)$. The decomposition in (4) is unique up to a distribution supported at $\{r\}$.

In the case of elements of the space $Z_d(\Omega; s)$ introduced in Section 11 we have a "parameter" version of Theorem 1, with respect to the radial variable and with a "good" control of the rest term R. For details see [Zie6], [Zie5].

The functions f which satisfy the assumptions of the theorem are called generalized smooth functions. The word generalized refers to the fact that the support of $B(f)$ (which according to (4) controls the asymptotic behaviour of f at 0_+) need not be discrete as it is the case of smooth functions.

The calculus for smooth functions generalizes to functions with continuous asymptotics. For instance we note the following generalized Leibniz formula: if

$$f(x) = T^1[x^\alpha] + R^1(x), \qquad g(x) = T^2[x^\alpha] + R^2(x),$$

then

$$f(x) \cdot g(x) = T^1 * T^2[x^\alpha] + R^3(x), \tag{5}$$

where $T^1 * T^2[\varphi] \stackrel{\text{def}}{=} T^1\left[T^2[\varphi(\alpha_1 + \alpha_2)]\right]$ is a convolution of compactly supported distributions. See [Zie1] for details and proofs in dimension 1 and [Zie2], [Zie3] for the case of several dimensions.

In the following section we consider a special subset of the set of functions with continuous asymptotics.

3. Algebra of Resurgent Functions of Jean Ecalle

Consider the space of (germs) of functions $f\colon (0,1) \to \mathbf{C}$, $f = O(x^a)$ whose Borel transforms $B(f)$ restricted to $\mathbf{R}_+$ extend holomorphically to functions Ψ that have logarithmic singularities at the points $-1, -2, \ldots$ This means that in a neighbourhood of every point $-j$, $j \in \mathbf{N}$

$$\Psi(z) = \Phi_{-j}(z+j)\,\mathrm{Log}(z+j) + \Psi_{-j}(z+j),$$

where Ψ_{-j} is meromorphic with at most a simple pole at 0 and Φ_{-j} is holomorphic in a neighbourhood of 0. The space, denoted by ${}^{+}\mathcal{R}(1)$ is called by Ecalle the (formal model of the) space of resurgent functions. ${}^{+}\mathcal{R}(1)$ is an algebra under usual pointwise multiplication of functions. One also considers the space ${}^{+}\boldsymbol{\mathcal{R}}(1) = \{B(f)\colon f \in {}^{+}\mathcal{R}(1)\}$ called the convolutive model of the resurgent functions. The name convolutive is motivated by the fact that ${}^{+}\boldsymbol{\mathcal{R}}(1)$ is a convolution algebra which is clear from (4) and the generalized Leibniz formula in (5). The idea of Ecalle consists in the observation that the functions in ${}^{+}\boldsymbol{\mathcal{R}}(1)$ admit derivations which are not the usual differentiations. Ecalle calls them "alien derivatives" and denotes by Δ_ω, $\omega = -1, -2, \ldots$ "Alien derivatives" satisfy (and are defined by) the Leibniz rule:

$$\Delta_\omega(\Psi_1 * \Psi_2) = (\Delta_\omega \Psi_1) * \Psi_2 + \Psi_1 * (\Delta_\omega \Psi_2) \qquad \text{for } \Psi_1, \Psi_2 \in^{+} \boldsymbol{\mathcal{R}}(1).$$

and for smooth functions are always zero.

Δ_{-1} can be explicitly defined as follows: take $\Psi \in {}^{+}\boldsymbol{\mathcal{R}}(1)$. Restrict Ψ to $\mathbf{R}_{+}$ and take its holomorphic extension $\widetilde{\Psi}$ to a function in $\mathbf{C} \setminus (\{\alpha \leq -2\} \cup \{\alpha \geq -1\})$. Consider the boundary value T of $\widetilde{\Psi}$ restricted to $(-2, \infty)$. Translate T by 1 and extend by holomorphicity. The result is the "alien derivative" $\Delta_{-1}\Psi$.

4. Applications

The theories outlined above find applications in the theory of singular differential equations. If one is concerned with the asymptotic behaviour of solutions of such equations (see Theorem 1) one may apply to them the (generalized) Borel transformation rather than the Mellin transformation. In the case of linear equations this may seem insignificant, but is very useful in the non-linear case, since the Borel transforms may converge even though the corresponding Mellin transforms do not. As an illustration we have the following

Example 5. Consider the equation

$$\frac{d}{dy}\varphi - \varphi = g(y) \tag{6}$$

where $g(y) = \sum_{j=1}^{\infty} b_j y^{-j}$ is analytic function of variable y^{-1} for large y. We are interested in solutions of (6) as $y \to \infty$. The point ∞ is an irregular singular point of (6) as is easily seen from the fact that, in variable $s = y^{-1}$, (6) becomes an Euler type equation

$$-s^2 \frac{d}{ds}\gamma(s) - \gamma(s) = \sum_{j=1}^{\infty} b_j s^j.$$

Consequently, (6) is solved in hyperfunctions or in formal power series in y^{-1}.

We shall investigate solutions of (6) by adapting the ideas of J.Ecalle. First, by the change of variables ($x = e^{-y}$), we transform (6) to the equivalent:

$$-x\frac{d}{dx}\psi(x) - \psi(x) = f(x), \tag{7}$$

where

$$f(x) = g(-\ln x) = \sum_{j=1}^{\infty} b_j(-\ln x)^{-j}.$$

We compute the Borel transform of (7). We find that

$$-(\alpha+1)B(\psi) = B(f) \quad \text{in} \quad \mathcal{D}'(\mathbf{R}) \tag{8}$$

where

$$B(f)(\alpha) = Y(\alpha)\sum_{j=1}^{\infty} b_j \frac{\alpha^{j-1}}{(j-1)!}, \tag{9}$$

according to Example 4. At this point we note that assigning to a (formal) series $\sum_{j=1}^{\infty} a_j s^j$ the series $\sum_{j=1}^{\infty} a_j \alpha^{j-1}/(j-1)!$ is the classical Borel transformation which, in view of (9), motivates the term "generalized Borel transformation" introduced in Subsection 2.1.

Denote

$$A(\alpha) = \sum_{j=1}^{\infty} b_j \frac{\alpha^{j-1}}{(j-1)!} \qquad \text{for} \quad \alpha \in \mathbf{R}$$

and observe that, since the series $\sum b_j s^j$ is convergent, the function A extends to an entire function $A(z)$ on $\mathbf{C}$.

By solving (8) in the space of distributions we get

$$B(\psi) = -Y(\alpha)\frac{A(\alpha)}{\alpha+1} + c\delta_{(-1)} \tag{10}$$

for some constant $c \in \mathbf{C}$. It is seen from (10) that ψ is a resurgent function and we shall compute the "alien derivative" $\Delta_{-1}(B(\psi))$.

Note that $B(\psi)$ restricted to $\mathbf{R}_+$ equals $-A(\alpha)/(\alpha+1)$ and extends to a meromorphic function $-A(z)/(z+1)$ whose boundary value on $\mathbf{R}$ is

$$b\left(-\frac{A(z)}{z+1}\right) = 2\pi i A(-1)\delta_{(-1)},$$

in view of Example 6. Thus

$$\Delta_{-1}B(\psi) = 2\pi i A(-1)\delta_{(0)}. \tag{11}$$

It follows from (11) and Theorem 1 that the general solution $\psi(x,u) = \psi(x) + ux^{-1}$ of (7) can now be written as

$$\psi(x,u) = \psi(x) + ux^{-1}\big(\Delta_{-1}/2\pi i A(-1)\big)B(\psi)[x^\alpha],$$

where

$$\psi(x) = -\int_0^\infty \frac{A(\alpha)}{\alpha+1}x^\alpha\, d\alpha.$$

This formula can be generalized to the case of a non-linear equation

$$-x\frac{d}{dx}\psi(x) - \psi(x) = g(-\ln x, \psi) \tag{12}$$

with g analytic in $-\ln x$ and ψ. Namely we have (see [P–C–N])

$$\psi(x,u) = \exp\big(u\,x^{-1}(\Delta_{-1}/2\pi i A(-1))\psi^0\big)[x^\alpha],$$

where ψ^0 is a particular (resurgent) solution of (12).

All non-trivial "alien derivatives" of $\psi(x,u)$ are indexed by $\omega \in -\mathbf{N}$ and we have (bridge equation)

$$x^\omega \Delta_\omega \psi(x,u) = u^\omega A_\omega u\frac{d}{du}\psi(x,u)$$

for some constants $A_\omega \in \mathbf{C}$. The important result of the theory of Jean Ecalle is that the set $\{A_\omega\}_{\omega\in-\mathbf{N}}$ forms a complete system of holomorphic invariants of (12). In other words, $\{A_\omega\}$ can be regarded as a symbol of the non-linear operator given by (12).

Let us now consider the generalized analytic functions of Subsection 13.5.

Example 6. Let for $a-1 \notin -\mathbf{N}_0$ and $c-a-1 \notin -\mathbf{N}_0$

$$u_1(x) = T_1[x^\alpha] \quad \text{with} \quad T_1 = \alpha_-^{a-1}(\alpha+1)_+^{c-a-1},$$
$$u_2(x) = T_2[x^\alpha] \quad \text{with} \quad T_2 = (\alpha+1)^{c-a-1}\alpha_+^{a-1}$$

be two solutions of the Kummer equation (13.62). It is easy to observe that th function $T_2(\alpha)$ for $\alpha > 0$ extends to a function $\Psi(z)$ having logarithmic singularitie at 0 and -1. Thus we can compute the difference of the boundary values of Ψ acros $\mathbf{R}$ (in analogy with the alien derivative Δ_{-1}, we denote it by $\Delta_0 T_2$)

$$\Delta_0 T_2 = \lim_{\beta\to 0_+} \Psi(\,\cdot\, + i\beta) - \lim_{\beta \to 0_+} \Psi(\,\cdot\, - i\beta).$$

It follows easily from the formula

$$z^{a-1} = |z|^{a-1} e^{i(a-1)\arg z}$$

that

$$\Delta_0 T_2 = (1 - e^{2\pi(a-1)})T_2.$$

A similar effect can be observed for the distribution $T_1 + T_2$. The jump at zer recreates the distribution $T_1 + T_2$ up to a constant factor. In other words th distributions T_1 and $T_1 + T_2$ resurge from the singular point 0 (the same is als true for the point -1). This justifies the name resurgent function introduced i Subsection 3. Since in our case u_1 and u_2 are also GAF-s it seems natural to cal them resurgent generalized analytic functions.

The reader interested in the "alien" analysis of the Kummer type equations i referred to [Maj].

Finally we shall reconsider Examples 15.1 and 15.2:

Example 7. The function $u(y_1, y_1y_2)$ where u is given by (15.1) is a resurgent function of the variable y_1 and from Remark 15.1 we know that for $\alpha_1 > 0$

$$B(u(y_1, y_1y_2))(\alpha_1) = -\frac{1}{2\pi} y_2^{\frac{\alpha_1}{2}} \frac{\cos\left(\frac{\alpha_1}{2}\ln y_2\right)}{\alpha_1}.$$

Thus $B(u(y_1, y_1y_2))$ restricted to $\mathbf{R}_+$ extends to a holomorphic function on $\mathbf{C}\setminus\{0\}$ and

$$\Delta_0(u(y_1, y_1y_2)) \stackrel{\text{def}}{=} b(B(u(y_1,y_1y_2))(\zeta_1)) = i\delta_0.$$

According to Theorem 1 we have by Remark 15.1 that for any $r > 0$ and fixed $0 < y_2 < 1$

$$u(y_1, y_1y_2) = -\frac{2\gamma - \ln 2}{4\pi} + \frac{1}{2\pi}\int_0^r \ln\varrho \frac{d}{d\varrho}(y_2^{\frac{\varrho}{2}} y_1^{\varrho})\,d\varrho + $$
$$+ \frac{1}{2\pi}\int_0^r \frac{1-\cos(\frac{\varrho}{2}\ln y_2)}{\varrho} y_2^{\frac{\varrho}{2}} y_1^{\varrho}\,d\varrho + R_r(y_1, y_1y_2)$$

with R_r flat of order r with respect to y_1. However in this case we may take $R_r \equiv 0$ and $r = +\infty$ which means that $u(y_1, y_1 y_2)$ is a generalized analytic function at zero with respect to the variable y_1.

Similarly for the functions $u_{\mathring{\alpha}}(y_1, y_1 y_2)$, $\mathring{\alpha} \neq 0$, considered in Example 15.2 we observe a "coupled" resurgence phenomena and the corresponding generalized Taylor formula reads:

$$u_{\mathring{\alpha}}(y_1, y_1 y_2) = -\frac{1}{4\pi}\left(\int_{L_1} \frac{y_2^{\frac{1+i}{2}\zeta_1} y_1^{\zeta_1}}{\zeta_1}\, d\zeta_1 + \int_{L_2} \frac{y_2^{\frac{1-i}{2}\zeta_1} y_1^{\zeta_1}}{\zeta_1}\, d\zeta_1\right),$$

where $T^1(y_2)$, $T^2(y_2)$ are certain smooth functions which can be explicitly computed.

Returning to the original variables $x_1 = y_1$, $x_2 = y_1 y_2$ we see that the function $u_{\mathring{\alpha}}(x_1, x_2)$ has the form

$$-\frac{1}{4\pi}\left(\int_{L_1} \frac{x_1^{\frac{1-i}{2}\zeta_1} x_2^{\frac{1+i}{2}\zeta_1}}{\zeta_1}\, d\zeta_1 + \int_{L_2} \frac{x_1^{\frac{1+i}{2}\zeta_1} x_2^{\frac{1-i}{2}\zeta_1}}{\zeta_1}\, d\zeta_1\right) \tag{13}$$

which makes sense for $0 < x_1 \cdot x_2 < 1$. Hence we see that $u_{\mathring{\alpha}}(x_1, x_2)$ for $0 < x_1 < 1$, $0 < x_2 < 1$ is a generalized analytic function with respect to both variables. Since $\left(\frac{1-i}{2}\zeta_1\right)^2 + \left(\frac{1+i}{2}\zeta_1\right)^2 = 0$ we observe that in (13) we integrate over two half lines

$$\widetilde{L}_1 = \left\{\left(\frac{1-i}{2}\zeta_1, \frac{1+i}{2}\zeta_1\right) : \zeta_1 \in L_1\right\},$$

$$\widetilde{L}_2 = \left\{\left(\frac{1+i}{2}\zeta_1, \frac{1-i}{2}\zeta_1\right) : \zeta_1 \in L_1\right\}$$

issuing from the (two) intersection points of the complex characteristic set $\{(z_1, z_2) \in \mathbf{C}^2 : z_1^2 + z_2^2 = 0\}$ with the imaginary plane $\mathring{\alpha} + i\mathbf{R}^2$ (for a fixed $\mathring{\alpha} \in \mathbf{R}^2$), and contained in the characteristic set. The integration over $\widetilde{L}_1$, $\widetilde{L}_2$ is performed with respect to the measure induced from that on the characteristic set. Further observe that it follows from Theorem 15.1 that the density $\frac{1}{\zeta_1}$ is the inverse of the difference over the point ζ_1 of the two sheets of the characteristic set $\{z_1^2 + z_2^2 = 0\}$ (in our case $c_1(\zeta_1) - c_2(\zeta_1) = -2i\zeta_1$). This geometric description of formula (13) carries over to all (second order) operators considered in Section 15. One can easily generalize it to dimension higher than 2 by conjecturing the following general geometric result:

For elliptic Fuchsian PDE $Pu = \delta_1$ in $\mathbf{R}_+^n$, the solution $u_{\mathring{\alpha}}$ obtained by the Mellin inversion of the distribution $\frac{1}{P(\mathring{\alpha}+i\,\cdot)}$ (suitably regularized) for a fixed $\mathring{\alpha} \in \mathbf{R}^n$, is a generalized analytic function of the form

$$\int_{S_{\mathring{\alpha}}} \frac{x^z}{w(z)}\, d\mu(z), \tag{14}$$

where the integration is carried over the subset $S_{\mathring{\alpha}}$, of the characteristic set $\{z \in \mathbf{C}^n\colon P(z) = 0\}$, obtained by issuing half-lines (-curves) from the set which is the intersection of $\{P(z) = 0\}$ with $\{\operatorname{Re} z = \mathring{\alpha}\}$, with respect to the measure induced from the characteristic set. The spectral density should be of the form $\frac{1}{w(z)}$, where $w(z)$ is related to the geometry of the characteristic set. For the operators of the second order, $w(z)$ is the integral over suitable $n-2$ dimensional characteristic subspaces. For instance, for the function

$$u(x) = \frac{-1}{(n-2)\,|S^{n-1}|}((\ln x_1)^2 + \cdots + (\ln x_n)^2)^{\frac{2-n}{2}}$$

where $|S^{n-1}|$ is the measure of the unit sphere S^{n-1} in $\mathbf{R}^n$ and $n > 2$, which satisfies the equation

$$\left(\left(x_1\frac{\partial}{\partial x_1}\right)^2 + \cdots + \left(x_n\frac{\partial}{\partial x_n}\right)^2\right)u = \delta_1 \qquad \text{in } \mathbf{R}^n_+,$$

we have

$$u(x) = C\int_{\mathbf{R}_+}\int_{S^{n-2}} x_1^{\frac{1-i}{2}\tau}\, x'^{\frac{1+i}{2}\omega'\tau}\, d\omega'\, \tau^{n-3}\, d\tau,$$

where C is a suitable constant, $x' = (x_2, \ldots, x_n)$, $\omega' = (\omega_2, \ldots, \omega_n) \in \mathbf{R}^{n-1}$, $S^{n-2} = \{\omega' \in \mathbf{R}^{n-1}\colon \|\omega'\| = 1\}$.

Observe that if $n \geq 2$ the integration in (14) is performed over the set $S_{\mathring{\alpha}}$ of real dimension $n-1$. Since the dimension of the characteristic set equals $2n-2$ there is much room in the characteristic set to vary the set $S_{\mathring{\alpha}}$. This possibility is reflected in the holomorphicity properties of the spectral density resembling those of the resurgent functions. Since the resurgence phenomenon observed here originates in the characteristic set we call it the characteristic resurgence.

Final remark. The resurgence effect described in Example 7 (see also Subsection 13.5 for similar effects) is due to the fact that the right-hand side of the equation (6) is resurgent i.e. its Borel transform has logarithmic singularities. This singularities are then preserved by the operator (bridge equation).

The characteristic resurgence is due to the operator only i.e. due to the occurrence of a complex characteristic set which is an algebraic set of complex dimension≥ 1. Clearly characteristic resurgence does not occur for ordinary Fuchsian operators.

Bibliography

[B–G] M. S. Baouendi, C. Goulaouic, Cauchy problems with characteristic initial hypersurface, Comm. Pure Appl. Math. 26 (1973) 455–475.

[B] J. M. Bony, Second microlocalization and propagation of singularities for semilinear hyperbolic equations, Taniguchi Symp. H.E.R.T. Kataka (1984), 11–49.

[Br] H. Bremermann, *Distributions, complex variables and Fourier transform*, Addison–Wesley Publ. Com. 1965.

[C–L] S. Colombo, J. Lavoine, *Transformations de Laplace et de Mellin*, Mémorial des Sciences Mathématiques 1969.

[E1] J. Ecalle, *Les fonctions résurgentes*, Publ. mat. Université de Paris–Sud, several volumes.

[E2] J. Ecalle, Singularitiés irrégulières et résurgence multiple, in Cinque applications des fonctions résurgentes, Preprint 84 T 62, Orsay 2–42.

[El] J. Elschner, *Singular ordinary differential operators*, Lecture Notes in Mathematics, 1128, Springer-Verlag, Berlin 1985.

[G–S] I. M. Gel'fand, G. E. Shilov, *Generalized functions*, Academic Press, New York, 1964.

[H] J. Horwath, *Topological vector spaces and distributions* I, Addison–Wesley, Reading, Massachusetts, 1966.

[Hö1] L. Hörmander, On the division of distributions by polynomials, Ark. Mat. 3, 555–568 (1958).

[Hö2] L. Hörmander, *The analysis of linear partial differential operators I*, Springer-Verlag, Berlin 1985.

[Hö3] L. Hörmander, *An introduction to complex analysis in several variables*, D. Van Nostrand Comp. Inc., Princeton, New Jersey 1967.

[I] H. Iwaniec, Spectral theory of automorphic functions, Lecture Notes Rutgers University, 1987.

[J] P. Jeanquartier, Transformation de Mellin et développements asymptotiques, Enseign. Math. 25 (1979), 285–308.

[K] A. Kaneko, *Introduction to hyperfunctions*, Mathematics and its Applications, Dordrecht 1988.

[K–K] M. Kashiwara, T. Kawai, Second microlocalization and asymptotic expansions, Lecture Notes in Physics, 126 (1980) 21–76.

[Ko] H. Komatsu, An introduction to the theory of hyperfunctions, Lecture Notes in Mathematics 287 (1973), 1–43.

[La–M] J. L. Lavoine, O. P. Misra, *Transform analysis of generalized functions*, North-Holland Mathematical Studies 119 (1986).

[Le–P] J. E. Lewis, C. Parenti, Pseudodifferential operators of Mellin type, Comm. Part. Diff. Equ. 8.5 (1983) 477–544.

[Ł1] S. Łojasiewicz, Division d'une distribution par une fonction analytique des variables réelles, C. R. Acad. Sci. Paris 246 (1958) 683–686.

[Ł2] S. Łojasiewicz, Sur le problème de la division, Diss. Math. 22 (1961).

[Ły1] G. Łysik, On the structure of Mellin distributions, Ann. Pol. Math. 51 (1990).

[Ły2] G. Łysik, Generalized Taylor expansions of singular functions, Studia Math. (in press).

[Maj] H. Majima, Resurgent equations and Stokes multipliers for the generalized hypergeometric differential equation, preprint Ochanomizu University, 1990.

[M1] B. Malgrange, Sur les points singuliérs des équations différentielles, L'Enseign. Math. 20 (1974), 147–176.

[M2] B. Malgrange, Introduction aux travaux de J. Ecalle, Prépublication de l'Institut de Fourier, Université de Grénoble, 20, (1984).

[Mel] R. Melrose, Analysis on manifolds with corners, Lecture Notes, preprint MIT (1988).

[Mel–Men] R. Melrose, G. A. Mendoza, Elliptic boundary problems in spaces with conical points, Journés Equ. Deriv. Part. St. Jean de Monts 1981.

[O] F. Oberhettinger, *Tables of Mellin transforms*, Springer-Verlag, Berlin 1974.

[P–C–N] E. Pham, B. Candelpergher, C. Nosmos, Visite aux sources, Prépublication de l'Université de Nice (1989).

[P–B–M] A. P. Prudnikov, Yu. A. Britchkov, O. I. Maritchev, Calculation of integrals and the Mellin transformation, in Itogi Matematiki i Techniki (Mathematical analysis) 27 Moscow 1989 (in Russian).

[Rei] H.M. Reimann, *Transformation de Fourier et integrales singuliéres*, Cours d'analyse harmonique 1982/83, Université de Berne.

[Rem–S] S. Rempel, B. W. Schulze, Complete Mellin symbols and the conormal asymptotics in boundary value problem, in Proc. Journées Equ. Dériv. Part. St Jean de Monts (1984).

[Ru1] W. Rudin, *Functional analysis*, Mc Graw-Hill Book Company 1973.

[Ru2] W. Rudin, Lectures on the edge of the wedge theorem, Regional Conference in Mathematics Supp. by NSF, 6 AMS.

[Schl] H. Schlichtkrull, *Hyperfunctions and harmonic analysis on symmetric spaces*, Progress in Math. 49, Birkhäuser, Boston, 1984.

[Schm] J. Schmets, Hyperfunctions et microfunctions d'une variable, Institut Mathématique, Université de Liege.

[Schw] L. Schwartz, *Théorie des distributions*, Hermann, Paris, 1966.

[Sl] L. J. Slater, *Confluent hypergeometric functions*, Cambridge University Press, 1960.

[Sm] L. G. Smyšlayeva, *Laplace transforms of functions of several variables*, Publ. Leningrad Univ., Leningrad 1981.

[Sn] J. H. Sneddon, *The use of integral transforms*, Mc Graw-Hill Book Comp., New York 1957.

[St] E. Stein, *Singular integrals and differentiability properties of functions*, Princeton Univ. Press, Princeton, 1970.

[Sz1] Z. Szmydt, *Fourier transformation and linear differential equations*, PWN–Pol. Sci. Publishers, Warszawa and D. Reidel Publ. Comp. Dordrecht—Holland, 1977.

[Sz2] Z. Szmydt, The Paley–Wiener theorem for the Mellin transformation, Ann. Polon. Math. 51 (1990), 313–324.

[Sz–Zie1] Z. Szmydt, B. Ziemian, Multidimensional Mellin transformation and partial differential operators with regular singularity, Bull. Pol. Ac. Sci.: Math. 35 (1987), 167–180.

[Sz–Zie2] Z. Szmydt, B. Ziemian, Solutions of singular elliptic equations via the Mellin transformation on sets of high order tangency to the singular lines, Bull. Pol. Ac.: Math. 36 (1988) 521–535.

[Sz–Zie3] Z. Szmydt, B. Ziemian, Local existence and regularity of solutions of singular elliptic operators on manifolds with corner singularities, Journal Diff. Equ. 83, (1990), 1–25.

[Sz–Zie4] Z. Szmydt, B. Ziemian, Characterization of Mellin distributions supported by certain non compact sets, Studia Math. (in press).

[T] H. Tahara, Fuchsian type equations and Fuchsian hyperbolic equations, Japan J. Math. 5 (1979), 245–347.

[Tou] J.C. Tougeron, *Gevrey expansions and applications*. Preprint, University of Toronto 1991.

[Tr] F. Treves, *Topological vector spaces, distributions and kernels*, Academic Press, New York 1967.

[V1] V. S. Vladimirov, *Methods of the theory of functions of several complex variables*, MIT Press, Cambridge, Mass. 1966.

[V2] V. S. Vladimirov, *Equations of Mathematical Physics*, Moscow 1963 (in Russian).

[W] W. Wasow, *Asymptotic expansions for ordinary differential equations*, New York 1965.

[Wa] A. Wawrzyńczyk, *Group representations and special functions*. PWN–Pol. Sci. Publishers, Warszawa and D. Reidel Publ. Comp Dordrecht 1984.

[Wi] D. V. Widder, *The Laplace transform*, Princeton Univ. Press, Princeton N. J., 1946.

[Z] A. H. Zemanian, *Generalized integral transformations*, Interscience Publishers, 1969.

[Zie1] B. Ziemian, Taylor formula for distributions, Dissertationes Math. 264 (1988).

[Zie2] B. Ziemian, The Mellin transformation and multidimensional generalized Taylor expansions of singular functions, J. Fac. Sci. Univ. Tokyo 36 (1989), 263–295.

[Zie3] B. Ziemian, The modified Cauchy transformation with applications to generalized Taylor expansions, Studia Math. (in press).

[Zie4] B. Ziemian, Elliptic corner operators in spaces with continuous radial asymptotics I, Journal of Diff. Equ. (in press).

[Zie5] B. Ziemian, Elliptic corner operators in spaces with continuous radial asymptotics II, Proc. Banach Semester on PDE-s, Banach Center Publications (in press).

[Zie6] B. Ziemian, Continuous radial asymptotics for solutions to elliptic Fuchsian equations in 2-dimensions, Proc. Symp. Microlocal analysis and its applications, Publications RIMS, Kyoto Kokyuroku 750, 1991.

[Zie7] B. Ziemian, Generalized Taylor expansions and theory of function of Jean Ecalle, in Generalized functions and convergence, Proceedings of a conference in Katowice, World Scientific, 1990.

[Zie8] B. Ziemian, Special functions and generalized analytic functions, Wiadomości Matematyczne (in Polish).

[Zie9] B. Ziemian, Generalized analytic functions (in preparation).

[Zie–K] B. Ziemian, H. Kołakowski, Second microlocalization and the Mellin transformation, Publications RIMS Kyoto University 26 (1990), 785–801.

List of symbols*

*§1 not included.

Subject index*

*§1 not included.

GPSR Compliance

The European Union's (EU) General Product Safety Regulation (GPSR) is a set of rules that requires consumer products to be safe and our obligations to ensure this.

If you have any concerns about our products, you can contact us on ProductSafety@springernature.com

In case Publisher is established outside the EU, the EU authorized representative is:

Springer Nature Customer Service Center GmbH
Europaplatz 3
69115 Heidelberg, Germany

Batch number: 10607320

Printed by Printforce, the Netherlands